Producing Streaming Video for Multiple Screen Delivery

By Jan Ozer

Producing Streaming Video for Multiple Screen Delivery

Jan Ozer

Doceo Publishing
412 West Stuart Drive
Galax, VA 24333

www.doceo.com
www.streaminglearningcenter.com

ISBN 978-0-9762595-4-1

Printed in the United States of America

For Mom

Acknowledgments

This book is the successor to *Video Compression for Flash, Apple Devices and HTML5*. As with that book, I couldn't have written this one without the backing of the Streaming Media team, for the seminars that they've sponsored, the contacts that they've fostered, and the various writing assignments that helped me become familiar with the products and technologies discussed herein. So to Eric Schumacher-Rasmussen, Stephen Nathans-Kelly, Dan Rayburn, Dick Kaser and Tom Hogan Jr. and Sr., thank you, thank you, thank you.

I also want to express my appreciation to the vendors who have provided hardware, software and bountiful assistance, including Adobe, Apple, Bambuser, BlackMagic Design, Digital Rapids, Google, HP, Haivision, Livestream, MainConcept, Matrox, MediaPlatform, Microsoft, NVIDIA, Rhozet, Seawell Networks, Sorenson Media, Telestream, Ustream, ViewCast, Wowza Media Systems and I'm sure some that I've forgotten.

This book is the fourth published by my company, Doceo Publishing. As always, budgets are tight, time is short, and the topics are fast moving, so I apologize in advance for any rough edges. Any polish that you see is wholly attributable to my copy editor/proofreader Lucy Sutton.

That said, I'm unhappy with the output quality of some of the figures in the print version of this book. To provide higher quality output, I've created a free PDF file with all figures and tables that you can download at **bit.ly/Ozer_multi**. I apologize for any inconvenience.

As always, thanks to Pat Tracy for technical and marketing assistance.

Contents

Contents

Chapter 1: Formulating a Multi-Screen Strategy

> ***Note:*** You can download a PDF file with all figures from this book at bit.ly/Ozer_multi. Since this chapter contains lots of images, now would be a good time to do so.

The most critical job of a streaming producer is to make his or her content available on as many platforms as possible. As recently as four or five years ago, producing a single file for delivery to desktop clients accomplished this function.

Today, that's nowhere close to sufficient. Mobile video consumption is growing at incredible rates. For example, for the London Olympics, NBC reported that 10.1 million unique users visited its mobile site, about 17% of the 57.1 million that visited its desktop site. While estimates of the growth of mobile viewing vary widely, it's clear that all video producers need a mobile strategy. Larger producers also need a strategy for distributing to over-the-top (OTT) players like the Roku, Boxee or Apple TV devices.

Once you incorporate mobile into the picture, a single video file strategy no longer cuts it. A stream that would look great on your 31-inch computer monitor is too "fat" to distribute to mobile clients watching via 3G. You need multiple files for multiple targets, a strategy called adaptive streaming. Unfortunately, mobile, desktop and OTT devices all use different technologies and environments to view and play these adaptive files.

In this chapter, you'll learn which technologies are available to distribute to these different platforms, and the features and tradeoffs offered by each technology. You'll also learn how producers of all sizes—small, medium and large—can access these technologies to make their videos available to the broadest possible viewer base as efficiently and economically as possible.

For those readers new to streaming, I'll start with a look at the fundamentals of streaming, defining terms and explaining concepts so you can follow the more technical conversations to follow. Specifically, you will learn:

- The definition of compression and codecs
- The definitions of streaming, progressive download and adaptive streaming
- The significance of the transition from Flash to HTML5, both on the desktop and on mobile devices
- The key features offered (and not offered) by technologies like Flash, HTML5 and other delivery platforms
- The key features and benefits of adaptive streaming technologies like Flash Dynamic Streaming (RTMP and HTTP), Apple's HTTP Live Streaming, and the nascent Dynamic Adaptive Streaming over HTTP (DASH) specification
- How to formulate a multiple-technology strategy for targeting the most relevant target platforms
- How small and mid-size producers can use service providers like Online Video Platforms (OVPs) and Live Streaming Service Providers (LSSPs) to achieve the same reach as larger companies who can afford to develop their own solutions from scratch
- Where to learn more about all of these concepts in this book.

Since the concept of compression is absolutely pervasive to all streaming media, we'll start with a quick look at the definition of codecs and compression. We're not going deep here, just providing a high-level view so you can understand the basics and the strengths and weaknesses of the technology alternatives we'll be considering throughout the rest of the chapter.

Compression and Codecs

As you probably know, video files are very bulky, which is why full-length HD Hollywood movies are delivered on Blu-ray discs that can store 25 to 50 GB of information. You probably also know from watching Internet video that some broadband and particularly mobile connections can be pretty slow, which is why video doesn't always play smoothly on your iPhone.

Compression is the technology that shrinks your streams down to sizes that you can deliver to desktop and mobile viewers. Compression is also the technology that can make your video ugly when you apply too much of it, as Figure 1-1, a video of my eldest daughter, shows.

That's because all video compression technologies are lossy, which means they throw away information during compression. Upon decompression, lossy technologies create only an approximation of the original frame, not an exact replica. The more you compress, the more information gets thrown away and the worse the approximation looks.

Figure 1-1. Too much compression equals ugly video.

If you've messed around with compression in the past, you've probably heard the term "codec." Simply stated, codecs are compression technologies with two components: an enCOder to compress the file in your studio or office and a DECoder to decode the file when played by the remote viewer. As the nifty capitalization in the previous sentence suggests, the term codec is a contraction of the terms "encoder" and "decoder."

There are lots of video codecs—like H.264, MPEG-4, VP8, VP6, Windows Media, MPEG-2 and MPEG-1—and lots of audio codecs—like MP3 and Advanced Audio Coding (AAC). How do you know which ones to choose? Well, this is where you focus on the decoder side.

Ensuring Playback

Specifically, when you're targeting a distribution platform, you have to make sure that it includes the ability to decode the file you're about to send it. For example, until around 2009, the VP6 codec was probably the most widely used codec on the web. However, the Apple iPhone didn't include a VP6 decoder; instead, it included a decoder for H.264. Accordingly, to distribute video to iPhones (and later Android and other devices), you had to encode the file using an H.264 codec.

Since the Flash Player also included an H.264 decoder (along with a VP6 decoder), many web producers switched to a single H.264 file playable on desktop and mobile devices rather than separate VP6 and H.264 files for the two platforms. So rule No. 1 to apply when encoding your files is to choose a codec that has a decoder on your target playback platforms. As you'll learn later in this chapter, this is particularly relevant when it comes to HTML5-compatible browsers.

Some of these—like Internet Explorer, Chrome and Safari—do include H.264 decoders, while others—like Firefox and Opera—do not.

While we're in definition mode, let's cover a few additional concepts.

Distribution Alternatives

When choosing and deploying web video technologies, it's important to recognize that they offer varying delivery options, including progressive download, streaming and adaptive streaming. Since these techniques are critical to streaming operation, let's describe how they work.

Progressive Download

Video delivered via progressive download is delivered by a regular HTTP web server rather than a streaming server. Since the video is delivered just like any other file on the website—be it a large graphic or a large PDF file—it's delivered as quickly as possible, as opposed to metering out the video as it's being watched.

This is shown in Figure 1-2. On the left is the Firefox browser, showing a video from Streaming Media East 2011 playing back from the user-generated-content (UGC) website Vimeo. On the upper right is a streaming video capture tool called Jaksta. On the bottom right is the file location where Jaksta is storing the files.

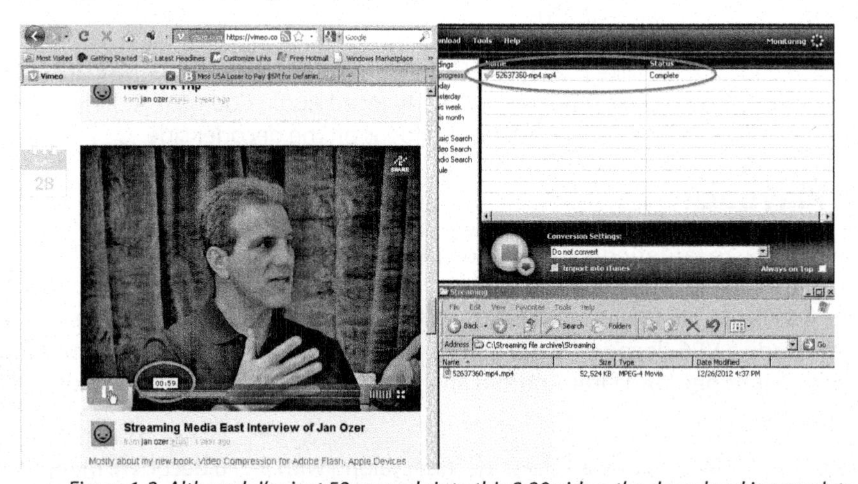

Figure 1-2. Although I'm just 59 seconds into this 8:29 video, the download is complete.

Note that all video playback technologies and platforms—whether Flash, HTML5, iOS or Android—can play files delivered via progressive download.

In the player window in the browser on the left, you can see that I'm only 59 seconds into the file, but Jaksta shows that the download is complete and the video file is stored on my disk, a total of about 52 MB. If I stopped viewing 59 seconds in, about 89% of that transfer, or 46 MB, would be wasted. This boosts Vimeo's bandwidth costs and can degrade quality of service, because while Vimeo was busy delivering the video to me as quickly as possible, it may not have had the capacity to deliver video to others.

In most instances, video delivered via progressive download is stored on the viewer's hard drive as it's received, and then it's played from the hard drive. One of the key reasons that producers use streaming servers is because once video is stored (or cached) on a hard drive, it's very easy to copy. In contrast, streaming video is usually played directly from computer memory, so it's not cached, which makes it inherently more secure.

Streaming

The technical definition of "streaming" is video delivered as it's needed. Traditionally, this was accomplished via streaming servers like the Adobe Media Server (formerly Adobe Flash Media Server), Wowza Media Server and other similar products. More recently, this has been enabled for videos distributed without a streaming server, as you'll learn about in Chapter 5 when we discuss adaptive streaming in detail.

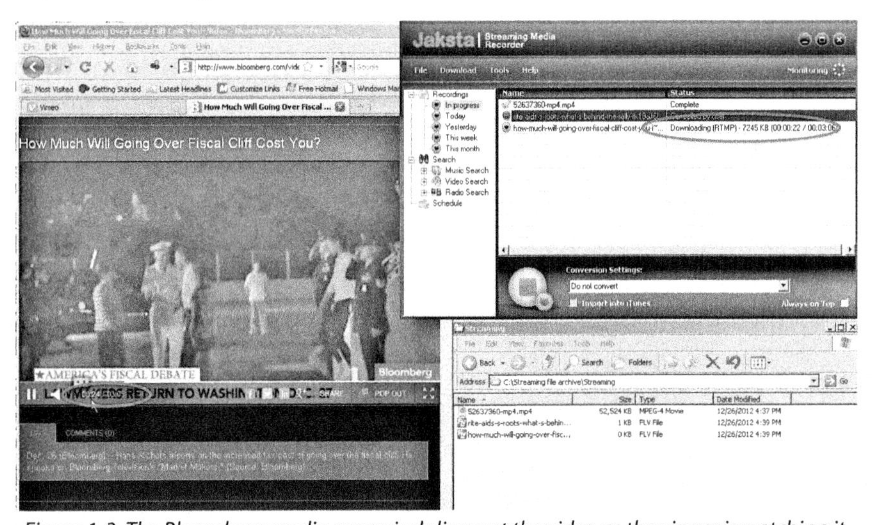

Figure 1-3. The Bloomberg media server is doling out the video as the viewer is watching it.

You can see this concept in Figure 1-3, which shows the same three windows as the previous figure, but a file downloaded from Bloomberg.com rather than Vimeo. Bloomberg delivers its videos via a streaming server that uses the RTMP delivery protocol, which you can read about in Chapter 5 but isn't really important to this discussion.

What is important is that the server doles out the video in chunks as necessary to support the viewer's continued watching. On the lower left of the figure, you see that video playback is at 24 seconds, while the upper right shows Jaksta has finished downloading a chunk starting 22 seconds in. If the viewer stopped watching at this point, very little extra bandwidth would have been wasted. In addition, this schema for distributing video as needed is much more efficient for maintaining a high quality of service over high numbers of viewers than the progressive download approach.

You can see a video tutorial showing both distribution techniques at bit.ly/prog_rtmp. This video is also the source of the two preceding figures, so it should look immediately familiar.

Adaptive Streaming

Both progressive download and streaming use a single encoded file. In contrast, adaptive streaming technologies encode a single live or on-demand file into multiple streams with varying configurations and switches them adaptively based upon changing line conditions and other variables. When the connection is good, the viewer gets a high-quality, high-data-rate stream, but if connection speed drops, the server will send a lower-data-rate file to ensure a continuous connection—albeit at lower quality. Adaptive streaming provides the best of all possible worlds: great-quality video for those with the connection speed to retrieve it (and the CPU required to play it back) and a passable-quality stream for those with Wi-Fi, mobile or other slow connections or those watching on lower-power devices.

There are multiple adaptive streaming alternatives today, including Adobe's Dynamic Streaming (either RTMP or HTTP), Apple's HTTP Live Streaming (HLS), Microsoft's Smooth Streaming, and the still nascent Dynamic Adaptive Streaming over HTTP (DASH). As you'll learn below, while Flash Dynamic Streaming enjoys the broadest support on the desktop, HLS has the broadest adoption in mobile and OTT platforms, excepting Windows Phones, which only work with Microsoft's own Smooth Streaming.

Note that supporting multiple adaptive streaming technologies is not as complex as it sounds. That's because many streaming servers, like the Wowza or Adobe Media Server, can transmux a set of adaptive files from one adaptive format to another. Specifically, these products rewrap the compressed streams into the formats used by the respective adaptive formats, and create any required metadata files.

Supporting new adaptive streaming technologies, like DASH, when they come online won't double your encoding workflow. Rather, it will be a simple matter of configuring your streaming server to produce a set of DASH-compatible files. For this reason, while convenient, the delivery of a unifying standard like DASH won't be a life changing event for most streaming producers.

Chapter 5 provides more details regarding the various adaptive streaming technologies and describes the optimal procedures for encoding for each technology. Now let's take a look at the technologies used to distribute to desktop viewers, and then mobile and OTT.

Distributing to Desktops

Where mobile platforms came into existence very rapidly, the technologies used to distribute video to computers have evolved significantly over time. A brief overview of this evolution will help you understand some critical tensions that are still relevant today—most notably Flash vs. HTML5.

The RealNetworks Era: 1996-2001

RealNetworks pioneered the streaming audio and video markets, broadcasting the first audio event over the Internet—a baseball game between the New York Yankees and Seattle Mariners—in 1995 and launching the first streaming video technology in 1997. According to some accounts, by 2000, more than 85% of streaming content on the Internet was in Real format.

Despite this success, problems arose because Real's primary business model depended upon the sale of servers, and Microsoft and Apple were giving those products away. As servers from Microsoft and Apple became more capable, Real's sales inevitably eroded. Or perhaps it was Microsoft's anticompetitive activities, for which Microsoft paid Real $761 million in a legal settlement in 2005.

In addition, consumers started balking at the intrusiveness of the free Realplayer, which installed multiple background processes, made itself the default player for all multimedia content and continuously nagged the user to upgrade. A player that once had been absolutely essential was now being uninstalled in droves, and not installed on newer computers.

The Microsoft Era: 2001-2006

Since the DOS days, Microsoft dominated the computer landscape, giving the Windows Media Player a dominant share of available desktops and notebooks. From the early 2000s to around 2007, Windows Media was the most widely used format on the Internet and in most company intranets. However, Windows Media Player had a non-customizable interface and required a third-party plug-in to play on the Mac, which was just starting to begin its resurgence.

The Flash Era: 2006-

During this same period, website design was transitioning from HTML to Flash, which offered much greater interactivity and design flexibility. Although Flash had a video component, the initial codecs offered poor video quality and sketchy audio/video synchronization. That changed when Macromedia licensed the On2 Technologies VP6 codec in 2005.

With the VP6 codec, Macromedia (and then Adobe, which acquired Macromedia in 2005), could match Microsoft's video quality in a brandable player that could be integrated with the rest of a Flash-based site, was truly cross-platform and was near ubiquitous. This proved irresistible to most broadcast and entertainment sites, where Microsoft's share in these markets dropped to single digits by 2010. Later on, when Adobe added support for H.264 decode, Flash became even more dominant.

Silverlight Stalls

In 2007, Microsoft launched Silverlight, which, very much like Flash, was both an authoring environment and a video playback environment. If you want to read up on how the two technologies compared in 2008, check out my article "Flash vs. Silverlight" (bit.ly/flashvsilverlight). The bottom line was that while Silverlight has enjoyed some impressive design wins, including both the Summer and Winter Olympics and Sunday Night Football, the Silverlight player never achieved the penetration necessary for producers with less compelling content to use it.

As I write this, in March 2013, Silverlight's penetration is up to around 70% overall, but still under 50% on platforms like the Mac (statowl.com). These penetration figures work if you have content like the Olympics that's so compelling that viewers will download the player to watch it, but probably not for your average sales or marketing video. For these reasons, Silverlight never really caught on for most public-facing websites.

There are a some scenarios where Silverlight makes a lot of sense—primarily when an organization has significant legacy content in Windows Media format, which won't play in Flash, and doesn't want to re-encode. In addition, when an organization has lots of .NET programmers around who can easily transition over to Silverlight development, that's a better option than retraining them all for Flash.

That said, both scenarios work best in an intranet or campus scenario where the organization can dictate which plug-ins the most relevant target viewers must have. In contrast, if you're delivering general-purpose content to the larger Internet community, you run a significant risk that many viewers simply won't download the plug-in to view your content. With the launch of Window 8, Microsoft is transitioning towards HTML5 and away from Silverlight, which I think will soon be totally de-emphasized by Microsoft.

The Rise of HTML5

Through early 2010, Flash was dominant on the desktop with no clear challenger in sight. Then, in April 2010, Apple shipped the iPad, which didn't support Adobe Flash. This focused lots of attention on HTML5, the technology that Apple uses for video playback on its iOS devices. Briefly, rather than using a plug-in-based player like Flash, Windows Media or Silverlight, HTML5 uses a player that's native to the browser to play back the video file.

This is a very critical distinction that deserves some highlighting. Love it or hate it, if you used Flash to distribute your videos, the Flash player functioned more or less identically on all browsers in all supported operating systems. When Adobe added H.264 support to Flash, for example, all Flash users gained H.264 decode capabilities when they next upgraded their player. If the decode didn't work, or if some other feature was broken, Adobe was obviously responsible.

With HTML5, these decode features are supplied by the browser vendor, which gets to choose which codecs are supported. Since H.264 decode capabilities cost $5 million a year, Mozilla (Firefox) and Opera Software (the Opera browser) declined to include it. As new features get added to HTML5, like adaptive streaming or digital rights management (DRM) support, the browser vendors decide if and when to include them.

Of course, until all vendors support a particular feature, it may be unusable. As an example, the only browser today that supports closed captions is Apple Safari. However, since Safari has a very low penetration rate on the Windows platform, captioning via HTML5 makes no sense, which is why all broadcast sites still use Flash. Obviously, when a feature doesn't work, or works differently on different browsers, you have to resolve the issue with the browser vendor or vendors.

About HTML5

Digging into the technology side, HTML5 is the latest specification of hypertext markup language (HTML), the language used to create websites. The specification is currently under review by the World Wide Web Consortium (W3C) and Web Hypertext Application Technology Working Group (WHATWG), and is set to be finalized in 2014. In early 2012, the W3C and WHATWG announced that the specification would be henceforth known simply as HTML. For the sake of clarity, however, we'll continue to refer to it here as HTML5 video, or HTML5.

HTML5 & Flash

Choose both

Given the Flash versus HTML5 debate, it's natural to think you need to choose one or the other. You don't. **JW Player 6 is everything you love about HTML5 and Flash rolled into one incredible player.** JW Player 6 automatically selects the best playback for your viewer's device and browser, so you deliver unparalleled video experience on iPhone, iPad, Android, and the Desktop. Click here to read our latest HTML 5 report.

Figure 1-4. The JW Player is one of many that can fall back to Flash.

For HTML5 video to function, the user must have an HTML5-compatible browser, and that browser must support the codec used to compress the file. By the start of 2013, about 65% of the installed base of browsers were HTML5-compatible. This means that an HTML5-only approach, without fallback to Flash or Silverlight, will fail to reach about 35% of potential desktop viewers.

Let's discuss this fallback concept because it's important. When you visit a website, a mini conversation takes place where the web server queries the browser about its playback capabilities. If the browser can play back HTML5, it will do so. If it can't, the web server can call the Flash Player, if it's installed, to play the video. This fallback capability is a standard feature of most pre-fab video players used by many web producers, like the JW Player shown in Figure 1-4.

So you don't have to support HTML5 to the exclusion of Flash—it's not a one-or-the-other decision. In fact, you really have to support both to reach the broadest base of potential viewers.

HTML5 and the Dual Codec Requirement

In addition, codec support within the HTML5-compatible browsers is split. Apple Safari (8% share of the HTML5-compatible market), Google Chrome (25%), and Microsoft Internet Explorer 9 and 10 (33%) all support H.264 playback, totaling about 66% of the HTML5-compatible market. Mozilla Firefox (31%) and the Opera browser (2%) do not support H.264 playback, so an H.264-only strategy won't reach about 33% of the HTML5-compatible market (with 1% lost somewhere due to rounding). Both Firefox and Opera support playback of the WebM video codec, as does Google Chrome, but IE and Safari don't, so a WebM-only strategy will reach only 58% of the HTML5-compatible market.

Note that while Mozilla has started supporting H.264 playback in some of its browser versions, this support depends upon H.264 playback capabilities delivered via other products or components. For example, the first Firefox browser to support H.264 was the Android version, and Mozilla used H.264 playback components from the Android operating systems. Mozilla could use a similar strategy for Windows 8, which includes (in the US) Internet Explorer 10 with H.264 decode.

However, previous versions of Windows—including Windows 7, Vista and XP—shipped with Internet Explorer versions that were not HTML5-compatible, so Mozilla's strategy would not be effective for Windows versions prior to Windows 8. The bottom line is that if you want to support all HTML5-compatible browsers on all relevant desktop operating systems, you have to make files available in both the WebM and H.264 codecs.

In addition, as we'll explore more fully below, HTML5 doesn't yet offer features like adaptive streaming, digital rights management, live broadcasting and many others that are proven components of Flash, Silverlight, Windows Media and QuickTime. Advanced features like peer-to-peer delivery and multicasting, launched by Adobe with Flash Media Server 4, are not yet on the HTML5 drawing board.

On the other hand, as you'll learn in the following sections on mobile video, what HTML5 combined with H.264 does offer is a simple way to send a single stream of video to iOS, Android and Windows 8 phones. For this reason, many smaller producers are using HTML5/H.264 to reach HTML5-compatible desktops and all mobile devices with a single stream of video, with fallback to Flash for desktop browsers that either aren't HTML5-compatible or don't play back H.264.

Basically, if you want to reach the desktop market in 2013 and 2014, you need to support Flash. We'll circle back to this conclusion later in the chapter. Now let's take a look at the technologies used to distribute to mobile devices.

Distributing to Mobile

The following sections will introduce you to the capabilities of the respective mobile and OTT platforms so you can see, at a high level, how to produce files for these platforms. We'll spend a lot more time focusing on specifics for producing for these platforms in Chapter 4, where you'll learn how to produce a single file for playback on these devices, and Chapter 5, where you'll learn to produce adaptive streams.

To add some perspective to our mobile analysis, as of January 2013, NetMarketShare reported that Apple had a 61% share of the mobile/tablet operating system market, with Android second at 28%, Java ME third at 7%, BlackBerry fourth at 1.5%, Symbian fifth at 1.24%, and Windows Phone sixth at 0.9% (bit.ly/mobile_share). In this chapter, I'll discuss iOS and Android for obvious reasons, and also Microsoft.

As an overview, when streaming to traditional computers, virtually all video is delivered within the browser, as opposed to using separate applications. On all three mobile platforms, you can deliver video to the browser or via apps, which are often used to enable advanced features that may not be supported by browser-based playback. If you can't support a feature via browser-based delivery to mobile platforms, you also should consider writing a custom app.

For this reason, when discussing techniques for delivering video with certain features to mobile platforms, I'll discuss the potential for delivery via a browser, and also via apps created with Adobe AIR (or Adobe Integrated Runtime). In short, AIR is a technology that enables developers to deliver Flash video on platforms that don't run the Flash Player, like iOS and Android.

AIR isn't the only technology available for creating apps for mobile platforms, and if you dig deep enough, you can likely find a tool that allows all adaptive streaming technologies to reach all mobile platforms. For the purposes of this book, however, I'll stick with formats supported natively by the mobile platform and AIR applications.

Single-File Streaming

For single-file streaming, none of the three mobile platforms support Flash in the browser, but all support HTML5 playback via the H.264 codec, making this the preferred approach.

Adaptive Streaming

Adaptive streaming is more complex. Of course, all Apple iOS devices support Apple's own adaptive streaming technology, HTTP Live Streaming (HLS). Google added HLS support to

Android version 3.1, but this and later versions make up only about 37% of total market share as of March 2013, rendering it an imperfect solution. In addition, as my *Streaming Media* colleague Dan Rayburn pointed out in his December 2012 column titled "Google Needs a Strategy for Video on Android Devices," Google's HLS implementation has multiple flaws (**bit.ly/Rayburn_android**).

> [Android's] HLS support doesn't match the specification, and buffering is common. Industry-leading HLS implementations such as those from Cisco and Akamai Technologies will not load on Android devices [so] HLS manifests must either be hand-coded or created using Android-specific tools. If the HLS video can play without buffering, you'll find that there is no way to specify the aspect ratio, so in portrait mode it looks broken. The aspect ratio problem seems to have been fixed in Android 4.1, but it will often crash if you enter video playback in landscape mode and leave in portrait. You can allow the HLS video to open and play in a separate application, but you lose the ability to communicate with the page, and exiting the video dumps users back on their home screens.

For these reasons, most serious producers attempting to reach the Android platform with adaptive streaming use an AIR app and Adobe Flash Video, though as I noted, you could create other apps with other tools to enable other technology alternatives. The AIR approach won't work with Windows 8 Phones since they only support Smooth Streaming. Note, however, that if you use Smooth Streaming and want to support all Windows 7 phones, you can't use resolution switching—all files must be encoded with the same resolution.

Distributing to Over-the-Top Devices

All OTT devices use their own unique schema for delivering to their players. All players support H.264 playback, so for single-file streaming, H.264 is only codec to consider.

In terms of adaptive streaming, Apple TV, Boxee, Google TV, and Roku all support Apple's HLS natively, without custom apps, making this the technology of choice, though Roku and Google TV both support Smooth Streaming. To adaptively distribute to the Xbox 360, you must use Microsoft's Smooth Streaming. Note that like mobile platforms, all OTT devices also enable apps that likely would allow you to deliver alternative technologies not supported in the browser, like HLS to Xbox 360.

Choosing Your Technologies

OK, now you have an overview of the relevant target platforms and the technologies you can use to reach them. Now you have to choose the best technology to reach each platform.

For simple applications like single-file streaming, almost any technology will do. But if you need features like adaptive or live streaming, or support for closed captions, you're going to have to take a closer look at each technology. In this section, we'll do just that, examining the features supported by the various technologies discussed in the preceding section. Table 1-1 provides a summary.

Single-File Streaming

All the technologies enable single-file streaming, although to reach all HTML5-compatible browsers, you'd need to produce two files: one encoded in H.264 format, the other WebM. Note that DASH doesn't break through this logjam; it is codec-agnostic. If Mozilla and Opera don't support H.264, you'll need to produce in two formats.

	Flash	Silverlight	HTML5	HLS	DASH	AIR App
Single-File Streaming	Yes	Yes	Yes	Yes	Yes	Yes
Adaptive Streaming	Dynamic Streaming	Smooth Streaming	DASH	Yes	Yes	Dynamic Streaming
Live Streaming	Yes	Yes	DASH	Yes	Yes	Yes
Digital Rights Management	Yes	Yes	DASH	Some	Yes	Yes
Closed Captions	Yes	Yes	DASH	Yes	Yes	Yes
Multicast	Yes	Yes	No	No	No	Yes
Peer-to-Peer	Yes	No	No	No	No	Yes

Table 1-1. Features and capabilities of various streaming technologies.

Adaptive Streaming

All the technologies save HTML5 enable adaptive streaming. Interestingly, although DASH is the Great White Hope for HTML5-based adaptive streaming, it most likely will be implemented through plug-ins before it's actually integrated into the various browsers. It's very ironic: One of the key benefits of HTML5 is that you don't need plug-ins, but a plug-in will almost certainly be necessary to enable adaptive streaming under HTML5 (and many other features).

Note that in November 2012, LongTail Video announced that JW Player 6 would enable desktop computers with the Flash Player installed to play back HLS streams (**bit.ly/JWP_features**). As you may know, the JW Player is a pre-fab video player that developers can license to add Flash and HTML5 playback to their websites. Since the JW Player now has HLS support, developers who use that player could create one set of adaptive files that would play back on the desktop, so long as the Flash Player is installed, and on iOS devices, and all other devices that support HLS, which includes most OTT devices. Basically, this feature delivers the key benefit of DASH—a single set of adaptive streams—without the wait.

I'm guessing that the developer of most other similar players will be adding this feature in the short term. By mid- to late-2013, if DASH isn't widely available, HLS may be the technology of choice for producers who want to use a single set of files for desktop, mobile viewers and OTT devices.

Of course, since HLS is an incomplete solution on the Android platform, perhaps the most viable approach would be to use Flash for desktop viewers and AIR-based apps for Android and iOS mobile devices. In any event, in Table 1-2, I'm including HLS as a viable alternative on the desktop—although keep in mind that this only works if the Flash player is installed, and if you use a pre-fab player, like JW Player, that delivers HLS playback to desktop and notebook computers.

Live Streaming

One of the key deficits in the HTML5 specification is the lack of live streaming, although this should be addressed by DASH when generally available. Until then, to stream live, you must use a technology like Flash, Silverlight or HLS.

Digital Rights Management (DRM)

DRM technologies enable content owners to protect their content, and common techniques include encryption and SWF Verification. Both Flash and HLS offer DRM, and DRM will be incorporated in DASH implementations once they reach the market. However, today, there are no HTML5-only DRM technologies—if you want to protect your content, you can't use HTML5.

Closed Captions

Closed captions are already required for many broadcasters, educators and government agencies, and many other organizations may use closed captions to enhance the search engine optimization of their videos or benefit their hearing-impaired visitors. Closed captions are available with Flash, HLS, Silverlight and DASH, but the only HTML5-compatible browser that currently supports closed captions is Apple Safari. Again, if you want to supply videos with closed captions to HTML5-compatible browsers, you'll have to use Flash or HLS. Note that Chapter 14 is an introduction to closed captions for streaming media.

Multicast

Multicast is the ability for multiple computers to access and play a single video stream, just like TVs used to be able to share one analog broadcast signal. Multicast is very efficient in networked environments that otherwise would have to distribute a single unicast stream to each viewer. While many technical requirements complicate large-scale multicast streaming, it's obviously highly desirable for enterprises and other large organizations with lots of desktops on a shared LAN.

Multicast streaming was made available for Flash streaming via the Flash Media Server 4.0 and later versions (see Adobe Media Server, adobe.ly/YBn9to). Multicast under HTML5 is not yet under serious discussion and is not available with HLS.

Peer-to-Peer

Peer-to-peer distribution (or P2P) allows any node on a network to serve as either a server or client respecting other nodes on the network. In video playback applications, once a single node on a network has retrieved video content, it can share that directly with other nodes in that network so the other nodes don't have to retrieve the video from the original source location.

Like multicasting, P2P can be very efficient in the enterprise space, and it is available as a feature of the Adobe Media Server (adobe.ly/YBn9to). P2P under HTML5 is not yet under serious discussion and is not a feature of HLS or DASH.

With this as background, let's turn our attention to the technology choices available for distributing to the desktop, mobile and OTT markets.

Desktop Markets

OK, so now you know the technologies and their respective features. Now it's time to apply that on a platform-by-platform basis, so you can identify the technology choices available to you for each platform. We'll start by looking at traditional computers and notebooks.

Logically, if you're serving video to desktop clients, there are three categories you need to address separately, as shown in Table 1-2. Specifically, you need to address those using older, non-HTML5-compatible browsers. You also need to address those using HTML5-compatible browsers with and without the Flash Player installed. Although I think this last group is small and will remain that way for at least two to three more years, it's instructive to consider your options for this group.

Legacy browsers, no HTML5 support

This group primarily includes those still using pre-9 versions of Internet Explorer. As of December 1, 2012, users of IE 6 through 8 included about 33% of all users as measured by NetMarketShare, with a few percentage points of other legacy browsers also in this category.

An HTML5-only approach without fallback to Flash would not reach any of these viewers. As you can see in Table 1-2, Flash does a nice job supporting these viewers, enabling both single-file and adaptive streaming and all the advanced features. If the website uses an HLS-compatible player like the JW Player, HLS will also deliver adaptive streaming, DRM and closed captions.

Overall, Flash is the best alternative for targeting these viewers.

	Legacy (Non-HTML5 Compatible)	HTML5 Compatible With Flash Player	HTML5 Compatible No Flash Player
Share	~33-35%	~65-70%	3% Windows/ 18% Mac-Safari
HTML5 Support	0%	100%	100%
Flash Player Support	Yes	Yes	No
AIR app Support	Yes	Yes	Yes
Flash Streaming	Yes	Yes	No
HTML5 - H.264	No	HTML5 (~66%)	HTML5 (~66%)
HTML5 - WebM	No	HTML5 (~59%)	HTML5 (~59%)
Adaptive Streaming	Flash, HLS	Flash, HLS	Mac only via HLS
Live Video	Flash, HLS	Flash, HLS	Mac only via HLS
Digital Rights Management	Flash/HLS	Flash/HLS	Mac only via HLS
Closed Captions	Flash/HLS	Flash/HLS	Mac only via HLS
Multicast	Flash	Flash	No
Peer-to-Peer	Flash	Flash	No

Table 1-2. Desktop markets and video-related features.

HTML5-Compatible Browsers with Flash Player

The next market includes users with HTML5-compatible browsers and the Flash player installed, which includes about 60 to 65% of all viewers. For small websites that only want to send a single video stream to their desktop and mobile viewers, this is an easy group to serve: Just deploy an HTML5 player with Flash fallback, and you can serve virtually all desktop and mobile clients. The Flash Player also enables access to all the other listed advanced features, including adaptive streaming via Flash.

If you decided to exclusively use HTML5 delivery rather than Flash, you'd have to supply two codec flavors: WebM and H.264. Plus you'd lose all the advanced features and compatibility with the legacy browsers in the first group. Definitely not a good move.

While HTML5 works well for simple videos, if you need any advanced features, Flash is still the best alternative to reach these viewers, and will be until DASH implementations become available.

HTML5-Compatible Browsers with No Flash Player installed

This group comprises users of HTML5-compatible browsers who don't have the Flash Player installed. Statistical website StatOWL shows Flash penetration on Windows to be 97%, with Mac

88% overall, although the Mac numbers are a bit skewed. Specifically, the site shows that 18% of users connecting via the Safari browser don't have Flash installed, compared with less than 1% of those connecting via Chrome and 2% of those connecting via Firefox.

On the Windows platforms, the failure to install Flash strands these users to a relatively low-level video experience: single stream only with no live video or any advanced features. Since HLS is compatible with current versions of Safari, even on desktop computers, Flash-less Mac users can access adaptive, live and all advanced features save peer-to-peer and multicast via HLS.

For simple video playback in a window, HTML5 should be fine for these users. If your goal is to deliver more advanced features, you have no alternatives on Windows and can use HLS on the Mac.

Mobile Markets

Now let's turn our attention to iOS, Android and Windows Phone devices. As you can see in Table 1-3, none of the platforms support Flash, but all support H.264 played back via HTML5, making this the preferred alternative for simple single-file delivery.

	iOS	**Android**	**Windows Phone**
HTML5 Support	100%	100%	100%
Flash Player Support	No	No	No
AIR app Support	Yes	Yes	No
Flash Streaming	AIR only	AIR only	No
H.264 Support	Yes	Yes	Yes
WebM Support	No	Yes	No
Adaptive Streaming	HTTP Live Streaming (HLS)/AIR	HLS (Android 3.0+ ~ 35%)/AIR	Smooth Streaming
Live Video	HLS, AIR	HLS, AIR	Smooth Streaming
Digital Rights Management	HLS	HLS (Android 3.0+ ~ 35%)/AIR	Smooth Streaming
Closed Captions	HLS/AIR	HLS (Android 3.0+ ~ 35%)/AIR	Smooth Streaming
Multicast	AIR	AIR	No
Peer-to-Peer	AIR	AIR	No

Table 1-3. Mobile markets and video-related features.

HTTP Live streaming provides adaptive streaming, digital rights management and closed captions for all iOS devices, as well as for Android 3.0 and later devices, although this still makes up only around 35% of Android devices. In addition, HLS support has some notable flaws on the

Android platform, as discussed above. If you're looking for a single technology to effectively reach both iOS and Android devices, an AIR application is your best bet. Unfortunately, AIR won't fly on Windows Phones, which you can only access via Smooth Streaming.

Looking at both mobile and desktop markets, for single-file streaming, supplying an H.264 file for HTML5 playback and deploying an HMTL5 player with fallback to Flash should suffice for many websites. Those publishers wishing to go beyond single-file playback will find HTML5 of little value and may have to deploy a multiple technology strategy incorporating Flash, HLS and AIR applications, and perhaps a transmux-capable server like the Wowza or Adobe Media Servers.

Again, to be clear, you can use almost any technology on most mobile platforms with a custom app. Otherwise, the recommendations above should work for most producers.

OTT Markets

To distribute video to any OTT device, you must adhere to the device's channel-like delivery schema, so there is no unifying delivery standard like HTML5. However, for encoding purposes, note that all OTT devices support H.264 playback.

	Apple TV	Boxee	Google TV	Roku	Xbox 360
H.264 Support	Yes	Yes	Yes	Yes	Yes
WebM Support	No	No	Yes	No	No
Adaptive Streaming	HLS	HLS	HLS	HLS, Smooth Streaming	Smooth Streaming
Live Video	HLS	HLS	HLS	HLS, Smooth Streaming	Smooth Streaming
Digital Rights Management	HLS	HLS	HLS, Smooth Streaming	HLS, Smooth Streaming	Smooth Streaming
Closed Captions	HLS	HLS	HLS	HLS, Smooth Streaming	Smooth Streaming

Table 1-4. OTT markets and video-related features.

For adaptive, live and advanced features, you can access four of the five OTT platforms shown in the table via HLS, although you'll have to support Smooth Streaming to access the Xbox 360 unless you want to write a custom app.

Along that vein, to be clear, you can use almost any technology on most OTT devices with a custom app. Otherwise, the recommendations above should work for most producers.

Where to Go From Here

Now you know which technologies are available, so the question turns to implementation alternatives. In the early days of the streaming media market, the mantra for virtually all companies that distributed streaming media was do it yourself (DIY). That is, companies encoded their videos, created their player, hosted their streams and maintained the streaming server that distributed the streams. This was possible because there was one de facto technology—however it evolved over time—and a relatively homogeneous target user.

As you've read, however, the streaming media market has become much more complicated, with competing technologies like HTML5 and Flash, and a more diverse universe of connection speeds and playback platforms. The ability and need to harvest effective marketing data from streaming viewers has also increased exponentially. This complexity has increased both the capex and personnel cost of effectively delivering streaming video.

On Demand Videos

As a result, many organizations are turning to online video platform (OVP) providers like Brightcove, Sorenson and Ooyala to serve as turnkey streaming providers. Operationally, you upload your videos to the OVP, which encodes them into the formats necessary to serve to your target viewers. The OVP provides customizable players for a range of target platforms—including Flash, HTML5 and iOS—and embed codes for integrating the player into your website. The OVP hosts the necessary servers for delivering the streaming media to the consumer and also provides media management and analytics packages.

Interestingly, a number of companies also use user-generated content (UGC) sites like YouTube and Vimeo for a subset of these services. For example, many of the videos presented on IBM's website are hosted by YouTube. This approach has multiple benefits, including decreased capex and personal costs associated with hosting the video. In addition, since YouTube is developing its player for millions of streaming producers, it can quickly incorporate new technologies. For example, after the launch of the Apple iPad, YouTube was one of the first services to support it, allowing IBM to deliver videos to the new device simply by updating its embed code.

IBM's videos are also exposed to the YouTube viewing community, with impressive results. For example, as of early January 2013, on the IBM Smarter Planet YouTube channel, IBM had accumulated more than 14,813 subscribers and more than 6.5 million video views. The most popular video, "IBM Centennial Film: 100 X 100," had been viewed 927,640 times. While these numbers pale next to the exposure of TV advertising, these are all opt-in viewers who chose to watch the videos, and the cost of the YouTube channel is a fraction of that of TV advertising.

Live Videos

For live videos, you have the same alternatives. That is, you can build out your own solution or use the services of a Live Streaming Service Provider (LSSP) like Livestream or Ustream. Both offer free live streaming, which may be funded by advertisements, and for-fee services that eliminate the advertising and offer some advanced features unavailable with the free service.

In this Book

With this as background, let's take a quick peek at the upcoming chapters.

Chapter 2: Technology Fundamentals. There's a good chance you know most of the concepts covered here, which include file details like resolution, frame rate and data rate. You'll also learn how aspect ratio and deinterlacing issues can degrade the quality of your video. If you're a relative newbie, you should at least scan through this chapter.

Chapter 3: H.264 Encoding Parameters. In this chapter, you'll learn all about encoding video with the H.264 video codec and audio with the AAC audio codec. The chapter starts with an overview of common H.264 encoding parameters like profiles and levels—as well as I-, B- and P-frames—and then moves into more advanced parameters controlling search depth and precision. Then it details your audio encoding options and specifics for encoding for delivery via Flash and HTML5.

Chapter 4: Configuring H.264 for Desktop, Mobile and OTT Viewers. Now that you know how to encode with H.264, we examine how to configure single files for delivery to desktop, mobile and OTT platforms.

Chapter 5: Adaptive Streaming. In this chapter, you'll learn more about the technology alternative s for adaptive streaming, factors to consider when choosing the number of streams and their configuration, and how to encode for adaptive streaming.

Chapter 6: Choosing an On-Demand Encoding Tool. Here we look at the various categories of on-demand encoder, including free tools, bundled tools, desktop tools and enterprise tools. You'll learn how to choose the best category for your needs and the best encoder within each category.

Chapter 7: Encoder-Specific Instruction. Here we'll look at the H.264 encoding interfaces from a variety of desktop, enterprise, hardware, OVP and cloud encoders so you can apply the lessons learned in previous chapters in the encoding tool of your choice.

Chapter 8: Producing for iTunes. This chapter focuses on the best practices for encoding files for distribution via iTunes.

Chapter 9: Distributing Your Video. Here we detail distribution alternatives like UGC and OVP sites, and discusses how to choose the best service provider for your videos. The final section covers encoding for uploading to an OVP or UGC site.

Chapter 10: Introduction to Live Streaming. Live streaming is a completely different animal than on-demand. This chapter introduces you to live streaming, covering how it works and technology alternatives like OVP and LSSP service providers.

Chapter 11: Distributing Your Live Video. Live streaming involves its own distribution options, whether you choose to do it yourself (DIY) or use a Live Streaming Service Provider (LSSP). This chapter discusses factors to consider when choosing a streaming server and an LSSP services.

Chapter 12: Choosing and Using a Live Encoder. This chapter discusses how to choose the best encoder for your live productions, including software, hardware, cloud and on-camera alternatives.

Chapter 13: Producing Live Events. Now that you know the technology components, this chapter details how to produce a live event, covering pre-production and game-day production tips and techniques.

Chapter 14: Introduction to Closed Captions. This chapter introduces you to closed captioning, including which organizations need to do it and how.

Chapter 15: Essential Tools. Streaming involves a range of file formats and encoding parameters. The tools discussed in this chapter analyze the files that you and others have encoded to identify these formats and parameters.

Chapter 16: Introduction to HEVC. You probably already started hearing about H.265; in this chapter, you'll learn what it is, how it works and how soon you should start thinking about using it.

Conclusion

After reading this chapter, you should have a high-level feel for the technologies that you'll use to reach your target viewers on multiple desktop, mobile and OTT targets. In the next chapter, we cover general encoding concepts and parameters you'll need to know to produce on-demand and live files to reach these targets.

Chapter 2: Technology Fundamentals

Note: You can download a PDF file with all figures from this book at bit.ly/Ozer_multi. Since this chapter contains lots of images, now would be a good time to do so.

While compressing video isn't rocket science, there are a lot of fundamentals that you need to know to create a file that looks good and plays well on your target platforms. In this chapter, we'll cover most of the fundamentals that are not specific to H.264. Specifically, you will learn:

- How to configure the optimal resolution, frame rate and data rate for streaming files
- How and when to use bit rate control techniques like constant bit rate encoding (CBR) and variable bit rate encoding (VBR)
- How to use the concept of bits per pixel to compute the best data rate for your encoded files
- How to configure audio parameters like data and sample rate, channels and sample size
- What aspect ratios are and how to prevent distorting your video during encoding
- Why, how and when to deinterlace your video.

Let's start with a bit of perspective.

Basic File Parameters

Last chapter, we learned that Apple's HTTP Live Streaming (HLS) adaptive streaming scheme dominates mobile and over-the-top (OTT) platforms, and can even be used to distribute to some desktop video players. So a reasonable starting point for all adaptive or even single-file encodes are the encoding recommendations supplied by Apple in the company's absolutely seminal Technical Note TN2224, which you can find at **bit.ly/HTTPLive**. For your convenience, I've copied the 16:9 recommendations into Table 2-1, below.

16:9 Aspect Ratio

	Dimensions	Frame Rate	Total Bit Rate	Video Bit Rate	Audio Bit Rate	Audio Sample Rate	Keyframe**
CELL	480x320	na	64	na	64	44.1	na
CELL	416x234	10 to 12	264	200	64	44.1	30 to 36
CELL	480x270	12 to 15	464	400	64	44.1	36 to 45
WIFI	640x360	29.97	664	600	64	44.1	90
WIFI	640x360	29.97	1264	1200	64	44.1	90
WIFI	960x540	29.97	1864	1800	64	44.1	90
WIFI	960x540	29.97	2564	2500	64	44.1	90
WIFI	1280x720	29.97	4564	4500	64	44.1	90
WIFI	1280x720	29.97	6564	6500	64	44.1	90
WIFI	1920x1080	29.97	8564	8500	64	44.1	90

Table 2-1. Encoding recommendations from Apple Technical Note TN2224.

As you would expect, the Technical Note describes with a great deal of specificity file parameters like dimensions (also called resolution), frame rate, bit rate (also called data rate) and key frame interval. Not surprisingly, there are all fundamentals that we'll cover in this chapter.

Figure 2-1 shows an encoding setting from Sorenson Squeeze, a popular desktop encoding program. Note that this interface is in simple view; click the Advanced button on the lower left and you'll see lots more, almost all of which relate to H.264 configuration options that we'll discuss in Chapters 3 and 7.

But if you compare the recommendations shown in Table 2-1 with the configuration options shown in Figure 2-1, you'll notice that all recommendations except the audio bit rate and sample rate are available in the encoding interface, and the audio-related options are available when you click the Audio tab on the upper left. Perhaps they use different names, like Dimensions vs. Frame Size, but they are all there.

Basically, as compressionists, this is what we do: We identify the optimal compression parameters and then plug them into an encoding program to produce our files. While all encoding programs look different, as you would expect, they all let you configure these same basic options.

As with all things, understanding what these configuration options are and how they interrelate will allow you to make better, more-informed decisions, and avoid some potholes. So let's start with the first configuration item on the Apple recommendations, which is Dimensions.

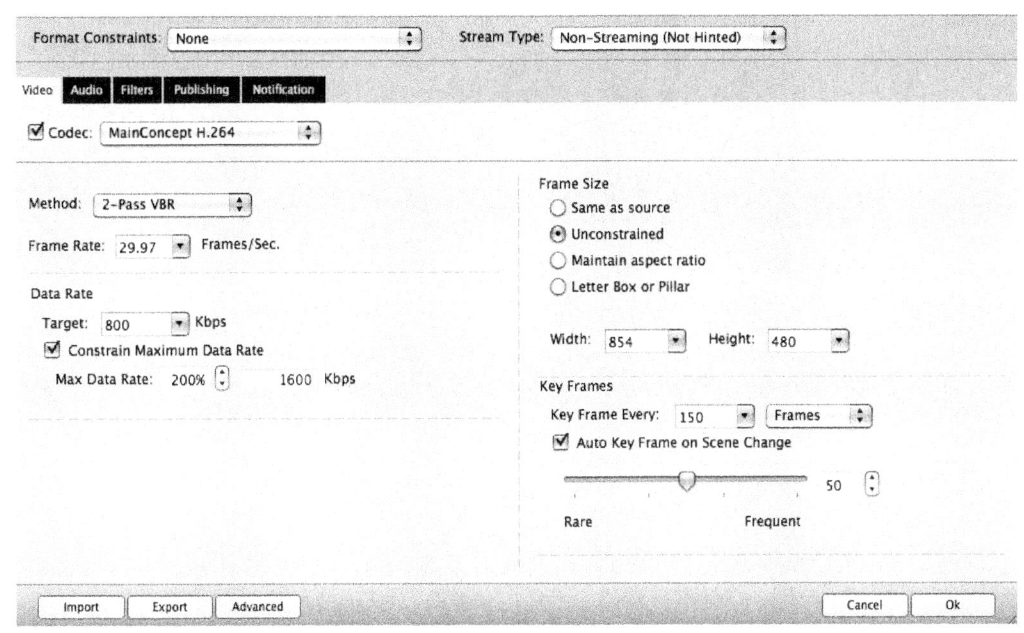

Figure 2-1. The encoding interface of Sorenson Squeeze version 8.5 in simple view.

Dimensions (or Resolution, or Frame Size)

Resolution is the height and width of the video in pixels. Most video is originally captured either at 720x480 (standard-definition) or at 1280x720 or 1920x1080 (high-definition), but often these high-resolution files get scaled to smaller resolutions for streaming. This scaling reduces the number of pixels being encoded, making the file easier to compress while retaining good quality.

For example, a 640x360 video has 230,400 pixels in each frame, while a 1280x720 video file has 921,600 pixels, or four times as many, as shown in Figure 2-2. Four times as much information means a file that's four times harder to compress. You'll notice in Table 2-1 that Apple recommends using smaller dimensions for files targeted at cellular connections; more on why in a moment.

Frame Rate

Most video starts life at 29.97 or 24 frames per second (fps), or 25 fps in Europe. Usually, producers who shoot at 24 fps deliver at that rate, while some producers who shoot at 29.97 fps deliver at 15 fps or even 10 fps when distributing to devices with a very slow Internet connection. That's because dropping the frame rate by 50% or 66% reduces the amount of pixels being encoded, just like dropping the resolution from 1280x720 to 640x360. You can see in Table 2-1 that Apple recommends dropping the frame rate on the second and third files targeted towards cellular connections.

Figure 2-2. Scaling the 1280x720 file down to 640x360 reduces the number of pixels by 75%, making the smaller file easier to compress.

Bit Rate (or Data Rate)

Data rate (or bit rate) is the amount of data per second in the encoded video file, usually expressed in kilobits (kbps) or megabits per second (Mbps). In Table 2-1, Apple recommends a total bit rate of 64 kbps for the top file, with zero allocated for video and 64 kbps for audio. That's because the lowest-quality file for HLS is often audio-only, with a JPEG image to display while the audio plays. As the files increase in resolution and frame rate, the total data rates also increase, all the way to a whopping 8,564 kbps for the highest-quality file.

By now you're seeing a relationship between data rate, resolution and, to a lesser degree, frame rate. At lower data rates, which are necessary to deliver to devices with slower connection speeds, Apple recommends lower resolutions and frame rates, which limits the number of pixels being encoded. This makes it easier for the encoder to produce a high-quality file without blockiness or other artifacts.

As an example, suppose you were producing a file using the parameters shown on the fourth line of the Apple table, 640x360 at 29.97 fps with a data rate of 600 kbps for video. Now suppose instead of compressing the file to a resolution of 640x360, you decide to compress to 1280x720. To reach the 600 kbps data rate, the encoder has to apply four times the compression, which very often noticeably degrades the image.

That's because, as discussed in Chapter 1, all streaming codecs use lossy compression, so the more you compress, the more quality you lose. For this reason, all other file characteristics (like resolution, frame rate or codec) being equal, the lower the data rate, the lower the quality of the compressed file.

Bandwidth

One concept implicit to the Apple chart is that of bandwidth. Note that the files on top, which are targeted towards cellular viewers (hence the "CELL" in the left column), have a lower total bit rate. This reflects the reality that most cellular connections are relatively slow compared with those of Wi-Fi or direct Ethernet connections used by virtually all desktops and notebook computers. Very appropriately, the table matches the total data rate to the viewer's bandwidth, which is the viewer's connection speed to the Internet.

Intuitively, you know that if the data rate of the encoded file exceeds the bandwidth capability of the Internet connection, the video will stop playing, because the video-related data can't get there fast enough to play in real time. You've seen it happen; I've seen it happen; we've all seen it happen. It's our job as compressionists to produce streams that don't exceed the bandwidth of our target viewers, and therefore play smoothly.

When you're producing and distributing only a single file, this is hard. If you select a data rate that's low enough to stream smoothly to all target viewers, even cellular viewers, you have to create a tiny, low-bit-rate file that looks bad on an iPad, awful on a decent notebook, and horrific on a 31-inch monitor or 72-inch LCD connected TV. If you produce at specs that look great on the latter two devices, it likely won't play at all on mobile devices—not only because the cellular connection can't sustain the transfer rate, but also because the file specifications exceed those playable by low-power mobile devices.

Although I'm getting ahead of myself, this dynamic is why adaptive streaming is so popular. By producing multiple files, you can match file quality with the device's connection speed and playback horsepower and produce the optimal experience on all targets.

Note that in the early days of streaming video, producers encoded video to meet the bandwidth capabilities of their target viewers. That is, back when most viewers connected via modems, you had to produce postage-stamp-sized video compressed to somewhere south of 28.8 kbps or the viewers couldn't watch it. Today, with most non-mobile viewers connecting via broadband capable of 3 to 6 Mbps or higher, most producers encode their video to meet quality and cost concerns.

For example, if you scan the websites of television networks and/or large corporations, the typical videos average about 640x360 resolution and are produced at 800 to 1,200 kbps, even though many viewers have the capacity to watch higher-bit-rate streams. That's because these producers have to pay for their bandwidth and have decided that 640x360 video at 800 to 1200 kbps provides a sufficiently high-quality experience to meet their viewers' needs. In short, back in the day, most producers encoded their files to meet the target bandwidth of their lowest-common-denominator viewer. Today, as bandwidth to the home has increased, choosing a data rate is largely a cost/quality trade-off.

Of course, just when it was safe to go back in the water, video delivery via mobile came to the fore, as did OTT devices connected to large screen LCD displays. Now it's not enough to create one file that pleases those watching on middle-of-the-road 19- to 25-inch monitors. You have to go low to satisfy mobile viewers and go high to satisfy those watching in their living room. That's why the Apple table scales from a low of 64 kbps for an audio-only stream to 8,564 kbps for a 1920x1080 movie quality stream.

Getting back to data rate for a moment, let's cover a concept not specified in the Apple table, which is that of bit rate control.

Constant vs. Variable Bit Rate Encoding

Beyond choosing a data rate for your file, you also have to choose a bit rate control technique, which is the Method list box shown on the upper left of the Figure 2-1, and also the Bitrate Encoding list box shown in Figure 2-3, which is from the Adobe Media Encoder. As you can see in the Figure, there are two options relevant to streaming: CBR, which is constant bit rate encoding, and VBR, which is variable bit rate encoding.

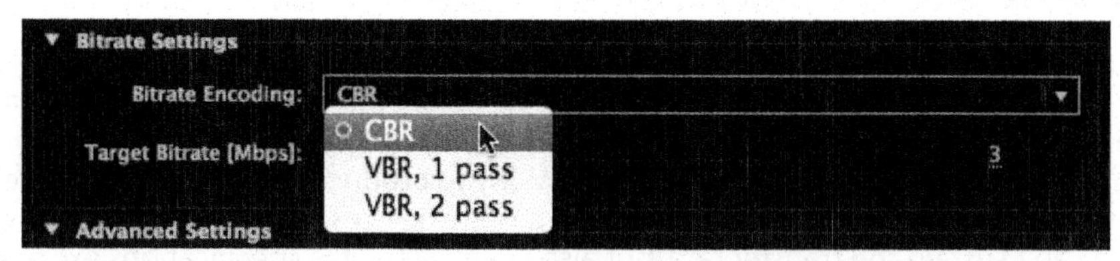

Figure 2-3. Choosing the bit rate control technique in Adobe Media Encoder.

Figure 2-4 illustrates the differences between CBR and VBR. Simply stated, encoding via CBR produces a file that has a constant bit rate throughout. In contrast, encoding via VBR varies the data rate according to the complexity of the video file, while achieving the same average data rate as that of a file produced with CBR. This is shown in Figure 2-4, which illustrates these concepts with a file with fives scenes, as follows:

- ***Low motion:*** Talking head
- ***Moderate motion:*** Woman cooking pita bread on an outdoor oven
- ***Low motion:*** An integrated-circuit chip-cutting machine in operation
- ***Moderate motion:*** A musician playing the violin
- ***High motion:*** Walking through a narrow street holding camcorder to chest while panning from side to side.

Both CBR and VBR techniques achieve the same average data rate over the duration of the file, but the CBR line, with dashes, stays constant throughout, while the solid VBR line varies with the amount of motion in the scene. Again, let me emphasize that the average bit rate and the ultimate size of the encoded file should be the same whether you use CBR or VBR. It's just that the data rate of the VBR-encoded file will vary with scene complexity, while the CBR-encoded file will have a more uniform data rate. You'll see this in action in a moment.

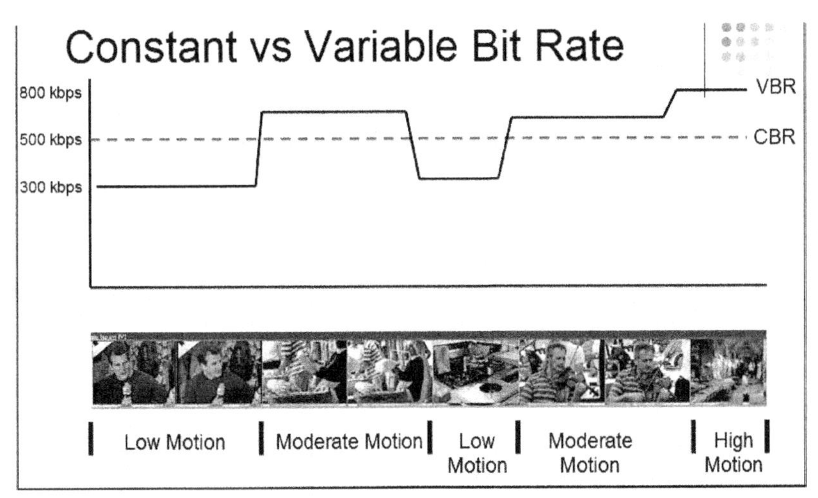

Figure 2-4. Constant and variable bit rate encoding, in theory.

In files with varying scenes with different encoding complexities, VBR encoding should produce a higher-quality file than CBR because it allocates data rate as necessary to maximize quality. The primary downside is stream variability, since the per-second bit rate can vary significantly from section to section. More on this in a moment.

Producing Optimal-Quality CBR Files

When producing files with CBR encoding, options will vary depending upon whether you're streaming live or creating on-demand files. If you're creating on-demand files, some encoders let you choose between 1-pass and 2-pass encoding, as shown in Figure 2-5, which is a screen from Microsoft Expression Encoder 4.

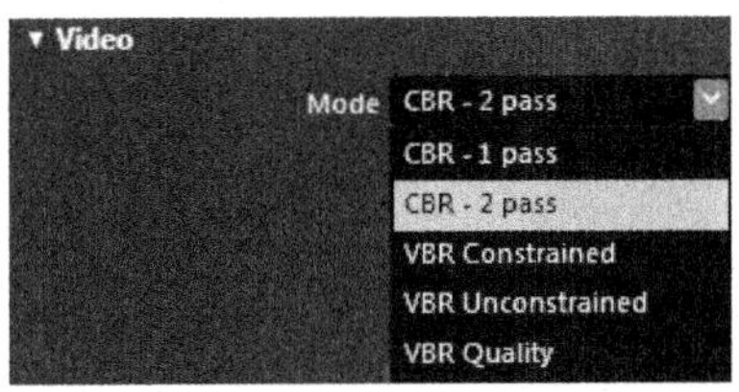

Figure 2-5. Choosing between 1-pass and 2-pass CBR in Microsoft Expression Encoder.

What happens with 2-pass CBR? Here's a blurb from Microsoft's website in an article titled "Encoding Methods" (**bit.ly/2-passcbr**).

Standard CBR uses only a single encoding pass. You provide your content as input samples, and the codec compresses the content and returns output samples. It is also possible to process input samples twice. On the first pass, the codec performs calculations to optimize encoding for your content. On the second pass, the codec uses the data gathered during the first pass to encode the content.

Two-pass CBR encoding has many advantages. It often yields significant quality gains over standard CBR encoding without changing any of the buffering requirements. This makes this encoding mode ideal for content that is streamed over a network. The only situation where two-pass CBR is not feasible is when you encode content from a live source and cannot use a second pass.

For these reasons, when 2-pass encoding is available in an encoding tool (not all tools enable this for all codecs), I always select it. As the Microsoft quote says, the only time you absolutely shouldn't elect two passes is when you're encoding a live stream.

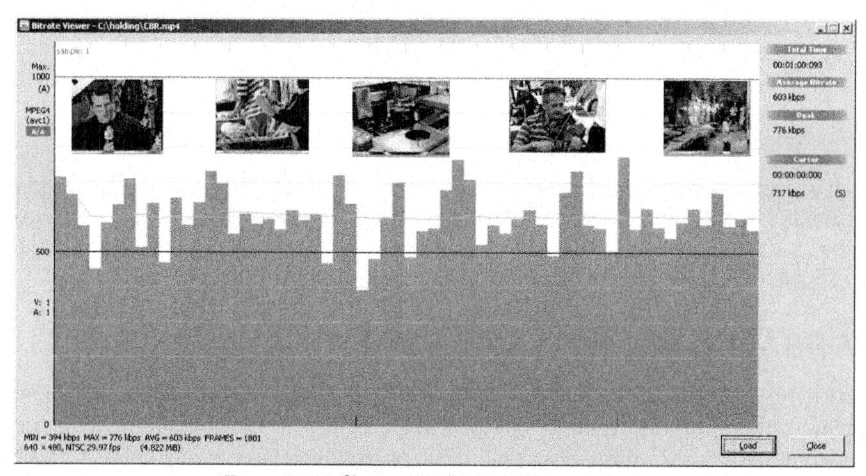

Figure 2-6. A file encoded using CBR encoding.
Note how the average data rate line remains very constant.

Now let's look at a CBR encoded file in action. Figure 2-6 shows a data rate graph of the file shown in Figure 2-4 produced using CBR encoding as displayed in a tool called Bitrate Viewer (which you can read about in Chapter 15). You can see the low-motion sequence at the start, moderate-motion next, and so on. (I inserted thumbnails from the video into the image as a reminder.)

In the figure, the spikes represent the data rate for each second of video, while the faint wavy line hovering over the black 500 kbps line represents the average data rate. While there are individual spikes in the graph, the average is pretty constant. I encoded the file to a target data rate of 600 kbps. On the upper right, you can see that the average data rate was 603 kbps, while the peak data rate was 776 kbps.

This figure shows that CBR doesn't mean a total flat line; there will be spikes in the data rate. However, when you compare this graph to Figure 2-7, which was encoded using VBR, you'll see much more variability in the VBR file.

Producing Optimal-Quality VBR Files

Figure 2-7 shows the same file encoded using VBR. Unfortunately, the scale is different for the two images, primarily because there were data rate spikes in the VBR file that extended beyond 1 Mbps. But if you ignore this and concentrate on the per-second spikes and faint average data rate line, you can see that both are low for the talking head portion of the file, then boost significantly for the moderate-motion sequence of a Druze woman tossing a pita.

Then, the per-second and average data rate settle down for the section showing the integrated-circuit cutting machine, then increase for the violinist, and peak for the high-motion walk through of a street in Jerusalem. In short, the per-second data rate and average data rate follow the underlying motion in the video file.

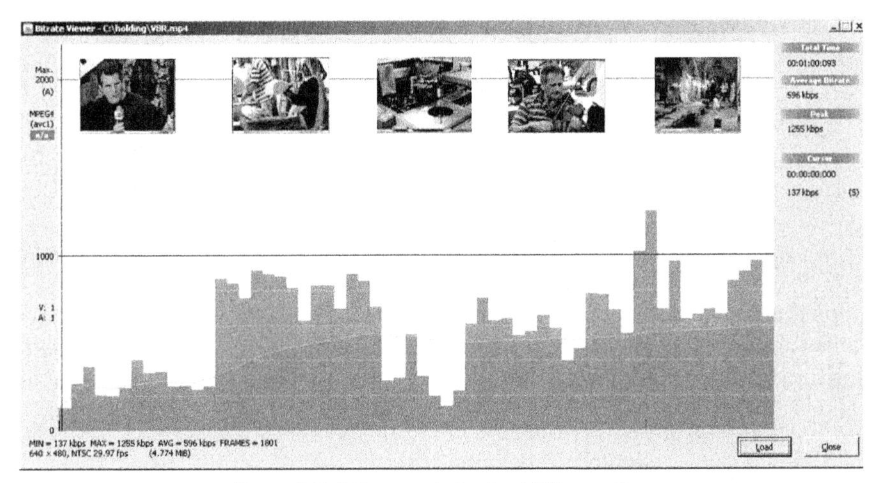

Figure 2-7. A file encoded using VBR encoding.

Again, I encoded this file to a target of 600 kbps. On the upper right of Figure 2-7, you see that the average bit rate was 596 kbps—not exactly the 603 kbps averaged with the CBR encode, but close enough to make the point that the average data rate should be very close irrespective of the technique used. However, the peak bit rate was 1255 kbps, compared with 776 kbps for CBR—quantifying, in part, the much greater stream variability shown in the VBR file.

So that's VBR. Now we'll discuss concepts relating to VBR production. First is the concept of constrained vs. unconstrained. If you look back at Figure 2-5, you'll see options for VBR Constrained and VBR Unconstrained.

The difference is this:

- ***Unconstrained VBR:*** With unconstrained VBR, you set the target data rate and the encoding tool can vary the data rate without limit; if there are exceptionally hard-to-encode scenes, the data rate can spike as high as necessary. As you would suspect, data spikes in the middle of the stream can make it difficult to deliver the stream over some connections, so when producing for streaming delivery, you never use unconstrained VBR.

- ***Constrained VBR:*** With constrained VBR, you can limit the peaks—and sometimes the valleys—in your encoded file to ease delivery over bandwidth-limited connections. You can see this in Figure 2-8, where you set the target and then the maximum and minimum percentage. For this reason, when encoding for streaming, you should always use constrained VBR.

Note that while virtually all encoding tools let you set the maximum data rate, very few let you set the minimum, and settings often vary within a tool depending upon the selected codec. So if you don't see a minimum on your screen, don't panic.

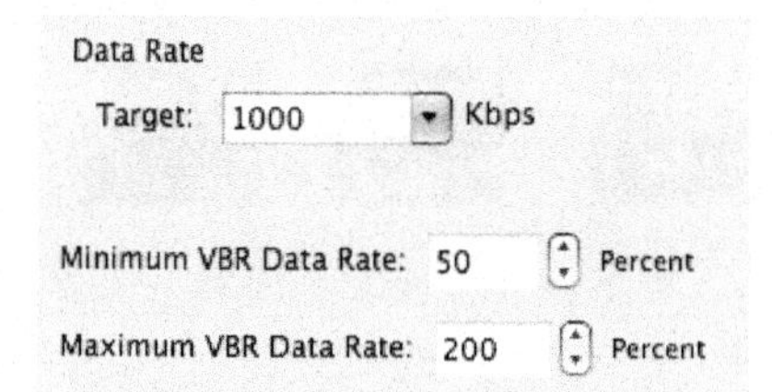

Figure 2-8. Selecting the target data rate, as well as minimum and maximum.

What options should you use for the minimum and maximum settings? Figure 2-8 shows my rule of thumb for single-file streaming, which is 50% of the target for the minimum data rate and 200% of the target for the maximum.

What about the third VBR option in Figure 2-5, VBR Quality? Here's the skinny:

- ***VBR Quality:*** VBR quality is a setting used primarily for archiving video, and is often called constant quality. When using this setting, you choose a quality level, often on an arbitrary scale like from 0 to 100, and the encoder varies the data rate to achieve the selected quality level. While this function is great for archiving, you can't use this setting for streaming production since you can't control the data rate.

Setting the Number of Passes

The next VBR-related concept is the number of passes. Some encoding tools limit you to two passes; others enable multi-pass. For example, when encoding with the MainConcept H.264 codec in Sorenson Squeeze, you can choose between 1-pass, 2-pass and multi-pass VBR. With 2-pass encoding, the encoder scans the video file in the first pass to catalog the complexity of the various sequences, and then encodes in the second pass, allocating data rate as necessary to optimize quality over the entire file.

When you encode using a single pass, the encoder doesn't have this scene-based information and can't make intelligent allocation decisions. For this reason, I almost never use 1-pass VBR.

With multi-pass VBR encoding, the encoding tool uses the additional passes to fine-tune the data rate of the file. For example, the encoder would scan the file in the first pass, then perform a test encode in the second. If a scene proved harder or easier to compress than originally predicted, the encoder would re-encode using different parameters in the third pass.

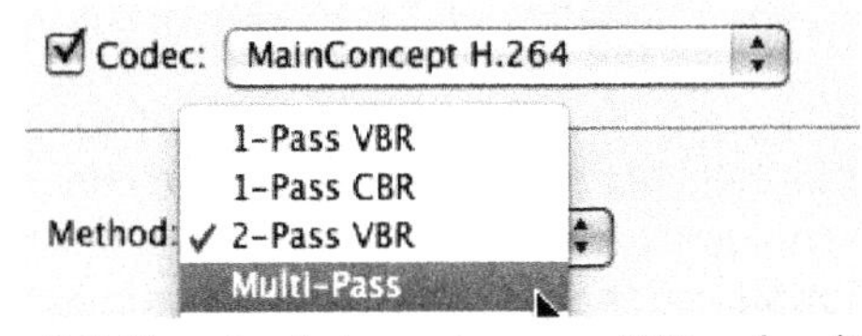

Figure 2-9. Choosing between two-pass VBR and multi-pass.

The downside of multi-pass encoding is processing time, which can get quite lengthy as some encoding tools use five passes or more. If you're in a hurry, you can run some test encodes to gauge how quality and processing time compare between multi-pass and 2-pass VBR, and see if the extra time is worth it. For example, I've run that analysis with Sorenson Squeeze and learned that the quality difference was negligible, but multi-pass encoding can take about twice as long. As a result, my default for Squeeze at this point is 2-pass, not multi-pass.

Choosing Between VBR and CBR

So, now you know the difference between VBR and CBR, when should you use the different techniques? Here are the general rules that I follow:

- Use constrained VBR (200% max/50% min) for most single-file streaming applications, whether delivered via a streaming server or via progressive download.
- Use more highly constrained VBR (150% max/75% min) when encoding for single-file delivery to mobile devices.
- Use CBR for all live streaming. More on this in Chapters 10-12.
- Use CBR or highly constrained VBR (150% max/75% min) when encoding for adaptive streaming. More on this in Chapter 5.
- Use VBR quality (or constant quality) encoding when archiving video for long-term storage.

With this behind us, let's move to our last video-related configuration option: I-, P- and B- frame settings, which is the last column in the Apple table, Table 2-1.

I-, B- and P-frames

I'll review these materials for H.264 much more extensively in Chapter 3, but I wanted to discuss these high-level concepts as early as possible. All codecs use different frame types during encoding. Most advanced codecs—like H.264, MPEG-2 and MPEG-4—use three types: I-frames (also called key frames), P-frames and B-frames.

Figure 2-10 shows all three frame types in a group of pictures (GOP), or a sequence of frames that starts with a key frame and includes all frames up to, but not including, the next key frame. Briefly, an I-frame is entirely self-contained and is compressed solely with intra-frame encoding techniques—typically a technology like JPEG, which is used for still images on the web and in many digital cameras.

P- and B-frames are different frames that reference information contained in other frames for as much content as possible. Imagine a talking head video like that shown in Figure 2-11, an animation where the only motion is in the eyes and mouth as the painter shifts the cigar.

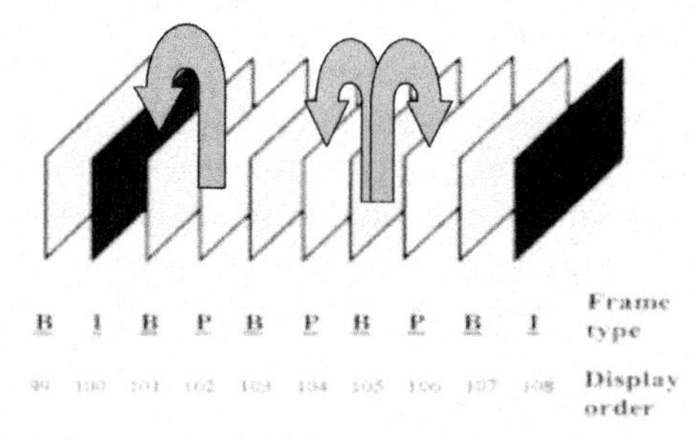

Figure 2-10. The three frame types used during encoding. P-frames look backward for redundancies, while B-frames look forward and backward.

When producing a P-frame, the encoder will look back to a previous I-frame (or P-frame) for regions in the frame that haven't changed, like the wall behind the speaker and most of his body and face. Then, it will encode only what's changed between the two frames. During playback, the player displays all pixels from the reference frame except the changed regions.

This is why talking head videos compress so efficiently: There's so much inter-frame redundancy that the P- and B-frames contain very little new information, making them easy to compress. In a fast-paced soccer game, P- and B-frames contain much more original content, which makes compressing down to the target data rate much tougher.

Back to our frame types. By definition, a P-frame looks backward to a previous P- or I-frame for redundancies, while a B-frame can look backward and forward to previous or subsequent P- or I-frames. This doubles the chance that the B-frame will find redundancies, making it the most efficient frame in the GOP.

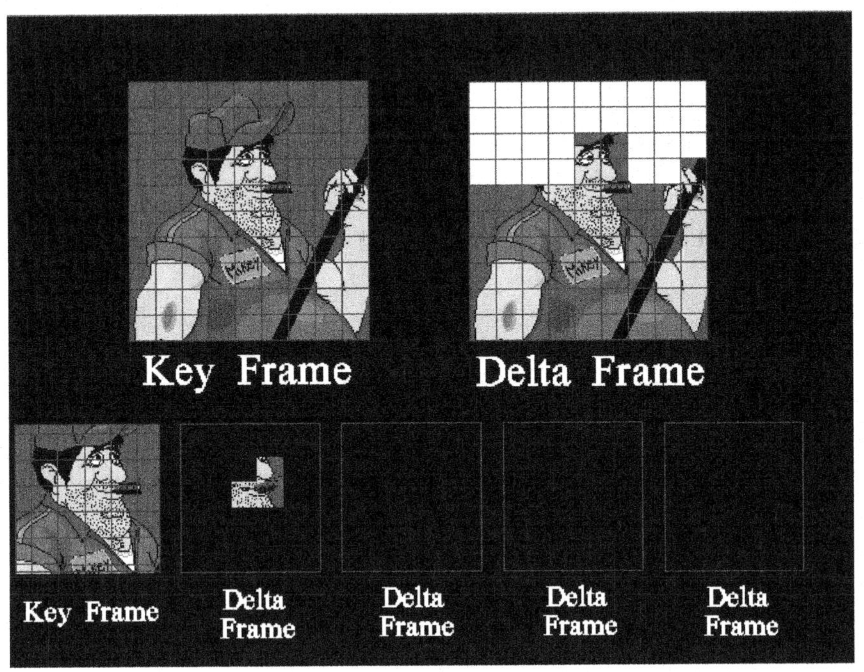

Figure 2-11. Delta frames (B- and P-frames) store the blocks that change from the reference frame (in this case a key frame).

Working with I-frames

How do you use these frame types to your advantage? With I-frames, recognize that these are the largest frames, which makes them the least efficient from a compression standpoint. Basically, you only want I-frames where they enhance either quality or interactivity.

For example, video playback must start on an I-frame since B- and P-frames don't contain sufficient content to reproduce the frame. If a viewer drags the slider on her video player to a B-frame, for example, the player must scroll back to the nearest I-frame, and then start decoding until it arrives at the B-frame. When encoding a single file, I recommend adding a key frame every 10 seconds, or every 300 frames in a 29.97 fps file, which makes the video file responsive to viewers navigating via the slider or otherwise jumping around the video file.

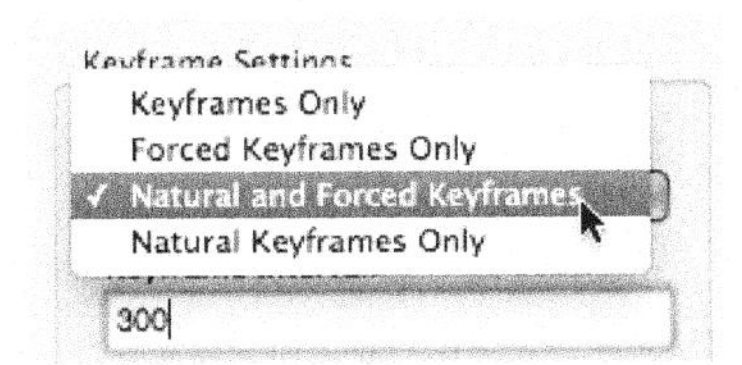

Figure 2-12. You want key frames at scene changes and about every 10 seconds.

I-frames also improve quality when inserted at a scene change, because all subsequent P- and B-frames get a high-quality frame to reference. So you also want an I-frame at scene changes. Most applications use a checkbox or similar control labeled "key frames at scene changes" or something to that effect. As you can see in Figure 2-12, Telestream Episode Pro complicates this a bit by using the term "natural key frames" to enable key frames at scene changes and "forced key frames" to enable key frames at regular intervals.

I-Frames and Adaptive Streaming

To summarize, when producing a single file for web delivery, encode using an I-frame every 10 seconds and enable the option that inserts key frames at scene changes. So why does Apple, in Table 2-1, recommend key frames at 30 to 90 frames? Because the rules change when encoding multiple files for adaptive streaming.

Specifically, as I explain in Chapter 5, the key frame interval used in adaptive streaming is generally shorter because adaptive technologies can only switch streams at key frames, and 10 seconds may be too long to be responsive when a switch is needed. Most producers prefer shorter durations—between 2 and 4 seconds—to enable more nimble switching when necessary.

In addition, when producing for adaptive streaming, you need key frames at identical locations in all files so the streams can switch seamlessly among themselves. For this reason, most producers disable the "insert key frames at scene changes" option, which could result in key frames at different locations—particularly in live events where the streams may be produced by different encoders.

> **Tip:** *For more on key frame interval, check out my article titled "What's the right key frame interval?" at bit.ly/key-frame.*

Working with B-frames

Now let's turn our attention to B-frames. As mentioned, the main benefit of B-frames is that they're very efficient from a compression perspective, so they help improve compressed quality. However, files with B-frames are harder to decode because the player has to buffer all referenced frames in memory while playing back the file, and display them in their proper order. As you'll see next chapter, for this reason, B-frames aren't enabled in the Baseline profile, which is the least complex H.264 profile used on very-low-power devices.

Otherwise, when producing for general computer and OTT playback, always use B-frames when available in the profile that you're using. Most encoding tools will let you choose the maximum number of B-frames to insert sequentially between I- and P-frames (or between P- and P-frames), and I recommended inserting three B-frames in sequence.

As you can see in Figure 2-13, which is the B-frame control from Episode Pro, you typically can also set the maximum number of reference frames from which a B-frame can search and use redundant data. My recommendation here is 5 frames.

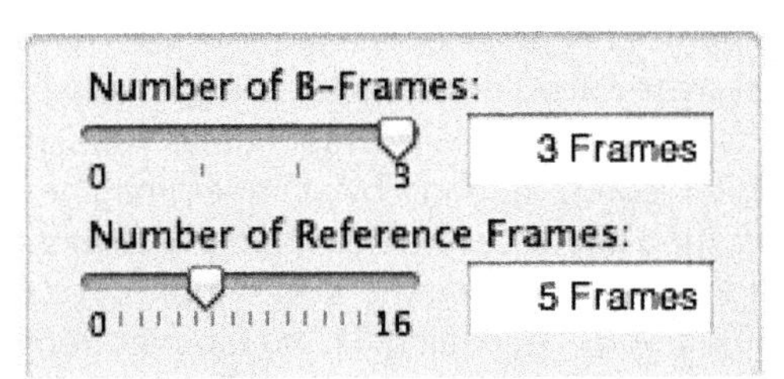

Figure 2-13. Choosing B-frame-related options in Telestream Episode Pro.

What about P-frame-related quantity options? You typically never see these because P-frames are the default. So when producing streaming files, you control your I-frame and B-frame options, and P-frames will take care of themselves.

I'll review these B-frame-related options in regard to H.264 in much more detail in Chapters 3 and 4; I just wanted to get you acquainted with them early in the process.

Configuring Your Audio

Typically, you'll compress your audio along with the video. In this section, we'll discuss the audio-related configuration parameters you'll need to set in most encoding programs.

As you'll learn in more detail in Chapter 3, when you produce H.264 video, you encode your audio in one of three flavors of Advanced Audio Coding (AAC). This is the codec selected in Figure 2-14, which shows the audio-related encoding parameters from Sorenson Squeeze.

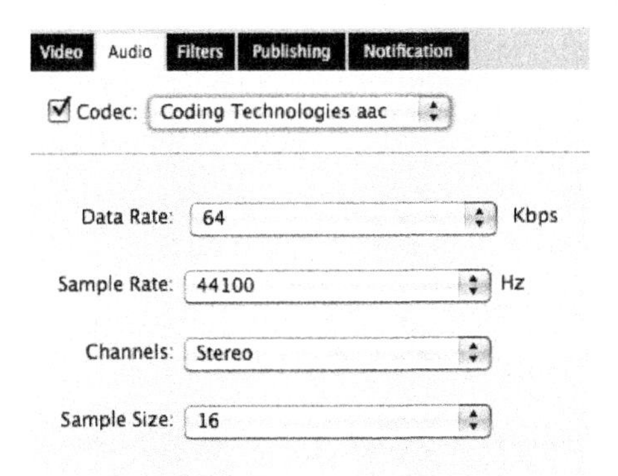

Figure 2-14. Audio-related encoding parameters in Sorenson Squeeze.

As you can see, there are four parameters:

- ***Data Rate:*** The amount of data per second of audio, usually expressed in kilobits per second (kbps). Apple uses the term "audio bit rate" to identify the data rate in Table 2-1; as with video, the terms "bit rate" and "data rate" are used interchangeably.

- ***Sample Rate:*** The number of times per second the audio is sampled. Remember that most of the audio that we're encoding starts in analog form, whether it's your boss yelling at you or a pleasant concert in the park. The more times you sample the analog source, the more accurate the digital representation. For example, imagine sampling that concert one time a second; the resulting digital recording would sound more like a series of 1-second tones than the actual song. At the recommended sample rate of 44,100 Hz, which is the same as CD-quality audio, you're encoding 44,100 samples from each second of that audio file, which is plenty for excellent sound reproduction.

- ***Channels:*** This is the stereo vs. mono decision, though some high-end users may also want to encode in surround sound. You'll note that while Apple does recommend both a data rate (64 kbps) and sample rate (44.1 kHz, or 44,100 Hz), it does not recommend stereo or mono. More on this decision in a moment.

- ***Sample Size:*** This is more often called audio bit depth, and describes the number of bits of information recorded for each sample. At the Sample Size of 16 shown in the figure, there are 16 bits of information stored for each sample. This is the recommended sample rate for web audio. If your encoding tool doesn't show this configuration option, don't sweat—it's almost certainly encoding at 16 bits per sample.

For the most part, the sample rate and sample size decisions are made for you. As I'll cover in more detail in later chapters, you'll use 44.1 kHz to ensure compatibility with your targets, and the sample size/audio bit depth should always be 16-bit. This leaves data rate and channels as the key decisions you'll have to make.

Choosing the Audio Data Rate

In terms of data rate, if the content is primarily voice-related, 32 to 64 kbps should be plenty, even if there is music in the background or at the start and finish of the video. If you're recording live music, or other presentations where audio quality is paramount, consider jumping to 96 or 128 kbps, but never higher for streams distributed over the web. As we'll explore in Chapter 8, "Producing for iTunes," you may want to use higher bit rates for files distributed via iTunes, but very few producers encode at higher than 128 kbps when distributing via the web.

Mono or Stereo? Consider the Source

Choosing the number of channels should relate to the source of your audio. One of my favorite streaming-related "what's wrong with this picture" images is shown in Figure 2-15. Specifically, although the speaker is talking into a handheld microphone, which is recording a single channel, the audio is encoded in stereo, as you can see in QuickTime's Inspector window on the upper left.

How is the stereo file created from a mono signal? By duplicating the signal into two tracks, essentially doubling the amount of information that needs to be compressed. As we discussed with video, doubling the information to be compressed forces the encoder to apply twice the compression, which could degrade the quality of the audio. Turning this around, if you publish in stereo, you need twice the data rate to produce the same audio quality as you would for a mono signal.

Concerned about producing in mono? You're not alone. Whenever I teach a seminar on streaming production, I ask the class, "What happens when you encode in mono? Who thinks the audio only plays on one of the two speakers?" Usually, about half the class raises their hands.

Figure 2-15. Encoding mono-source audio into stereo makes no sense.

Fortunately, that's not correct. When you encode a mono signal, during playback, the same signal is pushed to both the left and right speakers, which perfectly duplicates the direct listening experience. That is, the audio comes from one source, the speaker's mouth, and both ears hear the same signal.

Want proof? Surf over to an article on my website titled "Configuring your Audio - Hear it for Yourself" (bit.ly/mono_or_stereo), and you can listen for yourself. You can also learn if you can hear the difference between talking head audio encoded at 128 kbps stereo, vs. 32 and 64 kbps mono.

> **Tip:** *When you hear audio coming through a single speaker, it's not because the audio was recorded or encoded in mono. Rather, it's because mono audio was inserted into a single channel in a stereo signal, while the other is left blank. You can easily fix this in your audio or video editor by copying the mono signal into both channels and then rendering out a mono signal.*

Want more proof? How about the fact that CNN publishes its video files using 64 kbps, mono audio, as does *The New York Times*. MSNBC News and The Street are even more aggressive at 48 kbps mono. While certainly there are many sites that produce in stereo at much higher data rates, most producers of talking head videos are better served encoding in mono at 64 kbps or lower and allocating the extra bandwidth over to the video.

Putting it All Together

It's one thing to know what these configuration-related terms mean; it's quite another to use them to create a file that will stream smoothly and look good when viewed. So now that you know the basics, let's identify the optimal encoding parameters for your video files.

We'll start with a look at the streaming configurations used by media and corporate sites in the US, which is useful for several reasons. First, media sites distribute the overwhelming majority of non-user-generated content (UGC), and many have access to analytics that tell them whether their viewers can access the streams in real time. If eight three-letter networks in the US stream their episodes at a combined audio/video data rate approaching 1 Mbps, this tells us that most broadband consumers can successfully stream files encoded at that rate.

In addition, media sites help set quality and size expectations. If your target viewer keeps abreast of news by watching CNN at 640x360, then your 360x180 stream will look minuscule. If you're selling consumer electronic devices and Apple streams its iPad and iPhone advertisements at 848x480, your 320x240 stream will look quaint. Finally, if media sites are producing episodes and other content at 640x480 resolution or higher, this likely confirms that a stream using that configuration will play well on most relevant target computers.

For these reasons, every year or so I visit a few dozen prominent media and corporate sites, download their video files and analyze them. The next few tables show the results of the most recent study, which I'll pull together into a pithy, meaningful recommendation chart after discussing the individual results. Let's start with media-related websites.

Media Sites

These media sites include most three-letter networks (ABC, CBS, NBC, CNN, etc.) plus a smattering of other sites like ESPN and USA Networks. As you can see, I've divided the results into four groups according to resolution. For the record, all sites distributed at their native capture rate, which was generally 24 fps for TV episodes and 29.97 fps for news and sports.

As you can see, several media sites distribute at 1280x720, or very close thereto, including internet-only sites Bloomberg and The Street. However, the large resolution category includes multiple broadcasters that produce on the web and over the airwaves, including CNN and ABC. Clearly, these sites feel comfortable that their viewers can stream and play files encoded at higher than SD resolutions.

	Width	Height	Total Pixels
Low Resolution	456	257	118,739
SD Resolution	640	360	230,400
Large Resolution	768	432	331,776
HD Resolution	1,280	716	916,480

Table 2-2. Resolution and frame rate averages from US media sites.

Still, as mentioned above, though most viewers in the US can retrieve and play even higher-resolution/higher-data-rate files, the costs of pushing these streams dictate against their use. For CNN, 768x432@1200 kbps is sufficient quality.

Corporate Sites

In general, corporate sites show less sophistication than media sites, which you would expect, since their video is seldom mission-critical. This is reflected primarily at the low end of the spectrum, where the resolution of the videos was much smaller than that of media sties.

	Width	Height	Total Pixels
Low Resolution	397	235	100,627
SD Resolution	671	369	249,718
Large Resolution	828	442	369,691
HD Resolution	1,159	649	765,005

Table 2-3. Resolution and frame rate averages from US corporate sites.

At the high end, however, corporate sites are nearly as aggressive as media sites. For this reason, if your website is still pushing 320x240 streams out to your viewers, you'll look tiny in comparison—particularly if you're a business-to-consumer (B2C) site. In the corporate group, one site published at 15 fps, and the others at their native frame rates of either 24 or 29.97 fps.

Resolution Synthesis and Recommendations

OK, let's wrap all this data into a nice pretty bow, complete with recommendations. For media producers, the resolution sweet spot is around 640x360. If you produce at smaller sizes, you risk looking wimpy; if you go larger, you'll look bold by comparison.

US corporations should use 480x270 as a minimum, and consider producing at 640x360 or higher to make a big splash. In general, B2C sites are a bit more aggressive than business-to-business (B2B) sites, particularly at the low end, although high-profile consumer sites like Coke and Converse still publish at 480x270 or thereabouts.

What About Mod-16?

Many compressionists recommend using what's called a mod-16 resolution, where both the width and height parameters for each stream are divisible by 16. Why? Because most codecs, including H.264, encode in 16x16 blocks, and if the height and width aren't divisible by 16, the codec will create the full block anyway, adding more pixels to the file, which makes it harder to compress.

For example, a 16x16 video file would require one 16x16 block to encode, while an 18x18 video file would require four: one extra on the right and bottom to encode the extra pixels and one on the lower right to square out the video stream. Note that all of these extra blocks and pixels are automatically cropped during display, so you never see them, but the pixels must be compressed nonetheless.

Obviously, this is a worst-case scenario, and there are two schools of thought on mod-16. One treats non-mod-16 streams like "ring around the collar," evidencing a total lack of sophistication on the part of the compressionist. However, there are some very relevant arguments against the importance of mod-16. First, not all mod-16 resolutions are a perfect 16:9 aspect ratio, forcing the compressionist to either crop pixels or adapt a non-16:9 aspect ratio, which distorts the video, however slightly. Second, the extra pixels are always at the edges, so the encoder can apply less data to these blocks without a noticeable loss in quality.

In addition, the importance of mod-16 resolutions decreases as the resolution increases because extra 16x16 blocks end up making up less of the total picture. For example, according to former Microsoft employee Alex Zambelli, who helped configure the stream resolutions for NBC's Olympics and Sunday Night Football streaming offerings, 320x176 vs. 320x180 yields a 9% efficiency advantage, but 1920x1072 vs. 1920x1080 yields only a 1.5% improvement.

Finally, 640x360 is the most widely used stream resolution in existence today, and it isn't mod-16. This obviously wouldn't be the case if the lack of mod-16 compliance significantly degraded quality. When theory clashes with reality, go with reality.

Here's what Zambelli had to say regarding his Olympics and Sunday Night Football encodes:

> Most resolutions are mod-16, but in some cases, we had to settle for mod-8 or mod-4 to match a video resolution to a particular video player window size. For example, 720x404 was the Sunday Night Football player video window size, so we matched it with one of the encoded resolutions in order to ensure it played optimally without requiring any scaling.

In summary, when configuring your video, try using a mod-16 configuration, or mod-8 (where both height and width are divisible by 8) if mod-16 won't work. Note that many encoding tools prohibit encoding with odd numbers of pixels, or at mod-4 configurations and below.

Choosing the Data Rate

After choosing resolution, the next configuration decision is data rate. The most useful metric for choosing a data rate is the bits per pixel of the video file, so let's start there.

Understanding Bits per Pixel

Bits per pixel is the amount of data applied to each pixel in the video file. You compute it by dividing the per-second video data rate (e.g. 500 kbps) by the number of pixels per second (height x width x frame rate) in the video file. If you're not particularly mathematically inclined, you can use a tool called MediaInfo (shown in Figure 2-16) to do the calculation for you.

Why is bits per pixel such a valuable metric? Because it lets you assess the data rate of a file in a comparative way. In Table 2-4, for example, it's difficult to tell how Accounting 1 and Accounting 2 compare without the bits-per-pixel column because the resolutions are different. It's also difficult to compare these two sites—of two actual accounting firms whose names I withheld—with that of USA Networks, which uses another resolution entirely.

	Width	Height	Total Pixels	Data Rate (kbps)	Bits per Pixel
Accounting 1	256	144	36,864	236	0.267
Accounting 2	320	180	57,600	1,500	0.868
USA Networks	400	224	89,600	400	0.149
NY Times	504	284	143,136	650	0.152
Deloitte	720	392	282,240	1,072	0.127
Apple	848	480	407,040	1,903	0.195

Table 2-4. The bits-per-pixel value of these files lets you compare the compression applied to each video file.

However, bits per pixel tells the comparative tale. At a value of 0.868, Accounting 2 is ridiculously high, particularly compared with USA Networks at 0.149. Note that all the videos look pretty much the same, very good quality without any noticeable artifacts. However, Accounting 2 allocated close to six times the data rate per pixel as USA Networks—most of it wasted, because quality would have been identical at a much lower data rate and bits-per-pixel value. As you'll see at the end of the chapter, in addition to making these quality comparisons, you can also use bits-per-pixel values to create general rules to apply to videos across a range of resolutions.

Computing Bits per Pixel

Again, you compute bits per pixel by dividing the per-second video data rate (e.g. 500 kbps) by the number of pixels per second (height x width x fps). Or you can download MediaInfo, a free cross-platform (Windows, Mac, Linux) file analysis tool, from mediainfo.sourceforge.net. As you can see in Figure 2-16, MediaInfo also identifies the codec, multiple encoding parameters, data rate and other useful file data. I discuss where to get and how to use MediaInfo in more detail in Chapter 15, but the short answer is, if you don't have it yet, I recommend that you install it.

> **Tip:** *For a streaming tutorial on where to get and how to use MediaInfo and another video analysis tool, Bitrate Viewer, check out "Video – Two Free Video Analysis Tools: MediaInfo and Bitrate Viewer" at* bit.ly/videoanalyze.

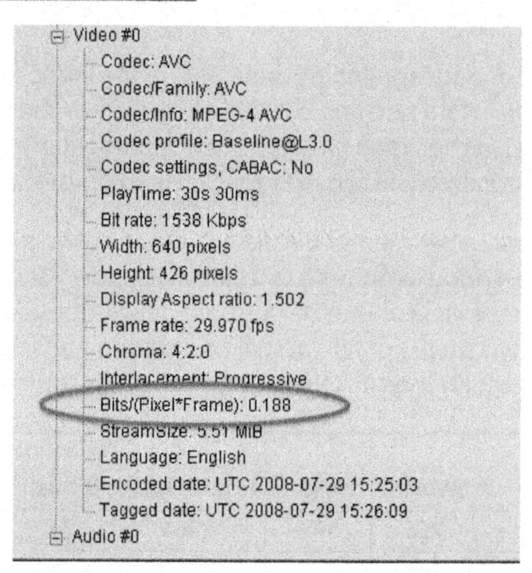

Figure 2-16. MediaInfo is the one tool on every computer in my office.

Using Bits per Pixel

Bandwidth is a major cost for media sites, yet obviously the quality of their video is very important. For this reason, media sites experiment to find a data rate that provides acceptable quality at the lowest possible data rate. By identifying the data rate used by media sites, we can create general rules to apply to media and other sites.

	Width	Height	Total Pixels	Data Rate (kbps)	Bits per Pixel Average	Bits Per Pixel Range
Low Resolution	456	257	118,739	435	0.133	0.102-0.152
SD Resolution	640	360	230,400	1,005	0.159	0.121-0.217
Large Resolution	768	432	331,776	1,283	0.134	0.082-0.201
HD Resolution	1,280	716	916,480	1,535	0.056	0.044-0.068

Table 2-5. The bits-per-pixel value of US media sites.

Table 2-5 shows the data rates and bits-per-pixel averages and ranges for the media sites we first looked at in Table 2-2. As you can see, the bits-per-pixel values used by media sites huddle around a pretty narrow range. In the low-resolution range, the highest bits-per-pixel value is

0.152. This confirms that the 0.267 and 0.868 values used by our two accounting firms was way too high.

Avoiding Wasted Bandwidth

Table 2-6 presents the bits-per-pixel values of prominent corporate sites in the US. Not surprisingly, the bits-per-pixel value for corporate sites is higher than media sites in all categories, with the greatest disparity in the low-resolution videos. Presumably, these discrepancies reflect that bandwidth is a major cost for media sites that they aggressively attempt to manage, while usually only an incidental cost to corporate sites.

	Width	Height	Total Pixels	Data Rate (kbps)	Bits per Pixel Average	Bits Per Pixel Range
Low Resolution	397	235	100,627	701	0.341	0.104-0.868
SD Resolution	671	369	249,718	1,239	0.190	0.077-0.495
Large Resolution	828	442	369,691	1,481	0.158	0.091-0.195
HD Resolution	1,159	649	765,005	1,634	0.086	0.061-0.100

Table 2-6. The bits-per-pixel value of corporate sites.

Keep in mind, however, that cost isn't the only issue with files encoded at too high a data rate. Higher-data-rate files are harder for users on sketchy connections to retrieve and play, and also makes your streaming server less efficient and less able to serve multiple viewers. As you decrease the data rate, you increase the number of viewers who can watch the videos on slower connections and the number of viewers that your streaming facility can effectively serve.

Bits per Pixel Should Decrease as Resolutions Increase

Note the in both tables, the bits-per-pixel values decrease as the resolutions increase. This is because codecs work more efficiently as video resolutions increase, so you can maintain quality at a progressively lower bits-per-pixel value. How much lower? Former Microsoft technology evangelist Ben Waggoner (now with Amazon) has quantified this as the Power of .75 rule, which he defines as follows (in a 2009 email to me):

> Using the old "Power of .75" rule, content that looks good with 500 kbps at 640x360 would need (1280x720)/(640x360)^.75*500=1414 kbps at 1280x720 to achieve roughly the same quality.

Essentially, as the resolution gets higher, a lower bits-per-pixel value will sustain the same quality level. On the off chance that you may not be up to speed on how to compute fractional exponents (I sure wasn't), I'll share tables that lay out this math after a short summary.

Applying Bits per Pixel

Taking all of this data into account, my rule of thumb is that a reasonable bits-per-pixel is around 0.1 for videos with low to average motion—your basic talking head or news video—ranging to 0.15 or higher for fast-moving videos. ESPN streams at around 0.178, which tells me that you might consider going a little higher for sports videos.

You should apply these values to video configured at 640x360 resolution (16:9) or 640x480 resolution (4:3). According to the Power of .75 rule, for lower-resolution videos, the bits-per-pixel value should increase with resolution, while it should decrease for higher-resolution videos. I used these concepts to create Table 2-7, which is computed for videos produced at 30 fps.

30 fps		Low Motion		High Motion	
Width	Height	Data Rate (kbps)	Bits per Pixel	Data Rate (kbps)	Bits per Pixel
16:9					
320	180	244,376	0.14	366,564	0.21
480	270	448,948	0.12	673,421	0.17
640	360	691,200	0.10	1,036,800	0.15
853	480	1,063,860	0.09	1,595,790	0.13
1280	720	1,955,009	0.07	2,932,513	0.11
1920	1080	3,591,581	0.06	5,387,371	0.09
4:3					
320	240	325,835	0.14	488,752	0.21
400	300	455,368	0.13	683,052	0.19
480	360	598,597	0.12	897,895	0.17
640	480	921,600	0.10	1,382,400	0.15

Table 2-7. Recommended data rates and bits-per-pixel values for video produced at 30 fps.

The top of the table deals with 16:9 videos, and normalizes the 0.1 low-motion and 0.15 high-motion rules at 640x360. I've applied the rule of .75 upward and downward to computer data rates and bits per pixels at other resolutions. The bottom four lines are for 4:3 videos.

Obviously, the data rates and bits-per-pixel values are suggestions. However, if you're producing at much higher than those values, your data rate may be excessive, unnecessarily increasing your bandwidth costs and making your videos harder to smoothly play for those on marginal connections. If you're producing at much lower and your video quality isn't up to par, perhaps you should re-evaluate. Here are tables for 24 fps and 25 fps.

24 fps		Low Motion		High Motion	
Width	Height	Data Rate (kbps)	Bits per Pixel	Data Rate (kbps)	Bits per Pixel
16:9					
320	180	195,501	0.14	293,251	0.21
480	270	359,158	0.12	538,737	0.17
640	360	552,960	0.10	829,440	0.15
853	480	851,088	0.09	1,276,632	0.13
1280	720	1,564,007	0.07	2,346,011	0.11
1920	1080	2,873,264	0.06	4,309,897	0.09
4:3					
320	240	260,668	0.14	391,002	0.21
400	300	364,294	0.13	546,442	0.19
480	360	478,877	0.12	718,316	0.17
640	480	737,280	0.10	1,105,920	0.15

Table 2-8. Recommended data rates and bits-per-pixel values for video produced at 24 fps.

25 fps		Low Motion		High Motion	
Width	Height	Data Rate (kbps)	Bits per Pixel	Data Rate (kbps)	Bits per Pixel
16:9					
320	180	203,647	0.14	305,470	0.21
480	270	374,123	0.12	561,184	0.17
640	360	576,000	0.10	864,000	0.15
853	480	886,550	0.09	1,329,825	0.13
1280	720	1,629,174	0.07	2,443,761	0.11
1920	1080	2,992,984	0.06	4,489,476	0.09
4:3					
320	240	271,529	0.14	407,294	0.21
400	300	379,473	0.13	569,210	0.19
480	360	498,831	0.12	748,246	0.17
640	480	768,000	0.10	1,152,000	0.15

Table 2-9. Recommended data rates and bits-per-pixel values for video produced at 25 fps.

Tip: *For more on the importance of bits per pixel, check out my article "The Essential Key to Producing High Quality Streaming Video" at* **bit.ly/bits-per-pixel**.

Other Production Issues

Beyond configuring your videos, there are several other production-related issues you need to be aware of, starting with those of aspect ratio.

Aspect Ratio Issues

Aspect ratio issues pop up in the most surprising places. They are evidenced by a mismatch between the appearance of a streaming video file and a digital photo of the same subject. For example, check out CNN's Anderson Cooper in Figure 2-17. On the left is a video frame grab from a CNN streaming video file; on the right, a digital picture from the CNN site. Obviously, the aspect ratio is off on the left. To be fair to CNN, the company fixed the problem since I took this screen grab in 2011, but it's too good an example to let go to waste.

Figure 2-17. A clear aspect ratio mismatch between the streaming video and a digital picture.

This also happens frequently with corporate streaming videos, as shown in Figure 2-18, which is a frame from an Accenture video, with the inset a digital shot of the same executive.

The obvious first step in resolving these issues is to recognize that you have a problem. The path toward resolving the problem depends on whether your source footage is standard-definition (SD) or high-definition (HD). Let's examine each in turn.

Figure 2-18. Same issue here in an Accenture video.

Resolving SD Aspect Ratio Issues

When working with SD footage—whether 4:3 or 16:9, NTSC or PAL—understand that the video was designed to be viewed on a television set and has a fundamentally different aspect ratio when viewed on a computer. You can see this in Figure 2-19, which shows a DV frame grab displayed on a computer on the left, and the same frame displayed on the television on the right. Essentially, the frame is horizontally squeezed about 10% when displayed on a TV set.

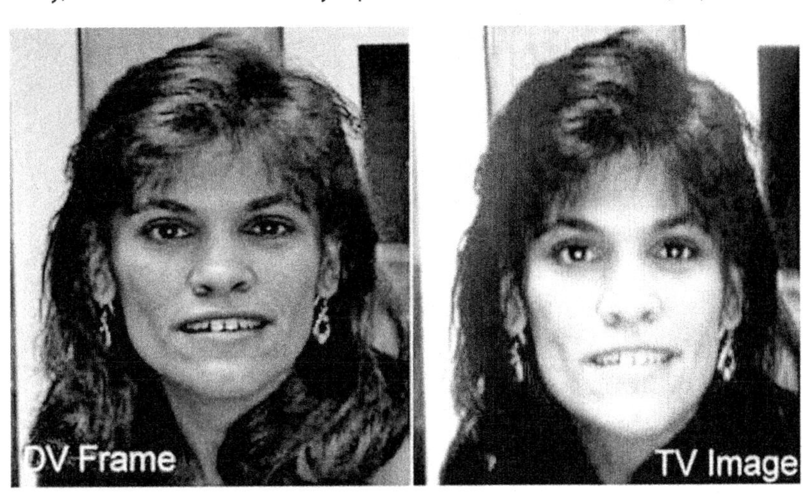

Figure 2-19. SD video displays differently on computers and TV sets.

If you analyze a standard-definition DV file in Adobe Premiere Pro, you'll see that it has an aspect ratio of 0.9091, which is the circled number in Figure 2-20. In essence, this means that each horizontal pixel in the video must be squeezed to 0.9091 of its original size to appear normal. That's a squeeze of about 10%.

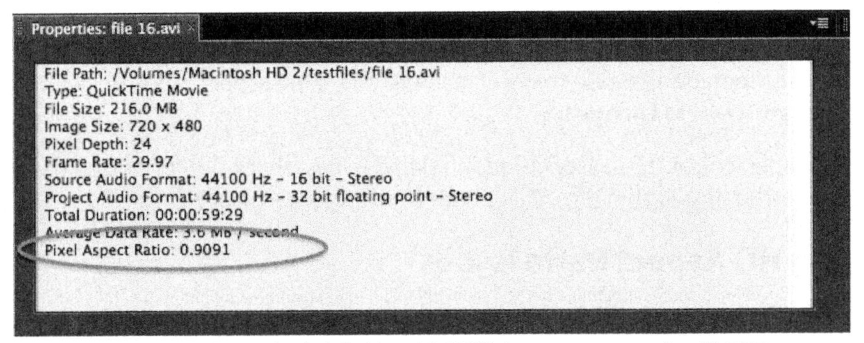

Figure 2-20. A standard-definition 4:3 DV file has an aspect ratio of 0.9091.

Television sets do this automatically, which is why the frame on the right in Figure 2-19 looks skinnier than the frame on the left. However, when producing streaming video, you have to squeeze the video by the same 10% (or, in the case of 16:9 video, expand it by about 20%). Fortunately, you don't have to do this math in your head. Although the procedure will vary by encoding tool, the general rules are the same.

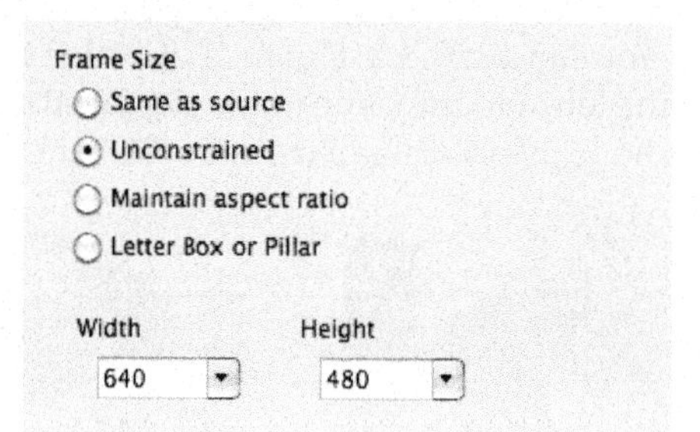

Figure 2-21. Changing the aspect ratio of SD source video in Sorenson Squeeze.

Specifically, here's the procedure:

1. Insert the desired frame size in pixels. With 4:3 source footage, the frame size should conform to 4:3, which means 640x480, 480x360, 400x300, 320x240 and so on. With 16:9 source footage, the frame size should conform to 16:9, which means 640x360, 480x270, 320x180 and so on.

2. Here's the hard part. With whatever tool you're using, make sure that the aspect ratio is changed from the source to the output. With Squeeze, shown in Figure 2-21, you select "Unconstrained," since "Maintain Aspect Ratio" would obviously maintain the source aspect ratio, which we don't want, while "Letter Box or Pillar" will insert black bars into the video, which we also don't want. With Telestream Episode Pro, you tell the program to "Distort the aspect ratio," which sounds funky but is necessary to change from 0.9091 to 1.0.

3. With Apple Compressor, shown in Figure 2-22, in the Geometry Pane, you input the desired frame size and then choose Square (or 1.0000) for the Pixel Aspect ratio. You don't want to choose NTSC CCIR 601/DV for 4:3 footage, or NTSC CCIR 601/DV (16:9) for widescreen footage, because that will maintain the same aspect ratio, and with SD source footage, you want to change it.

For a much deeper look at these aspect-ration-related issues, check out "Choosing the Optimal Video Resolution" at bit.ly/optimalres. That's it for SD; now let's look at HD.

Resolving HD Aspect Ratio Issues

With HD source footage, the solution is much simpler because there is no aspect ratio mismatch—the video should look exactly the same on a computer and television set. This means that you don't have to change the aspect ratio as part of the encoding process. Typically, if you use the "Square" output shown in Figure 2-22, or the equivalent in your encoding program, you should avoid any aspect ratio mismatches.

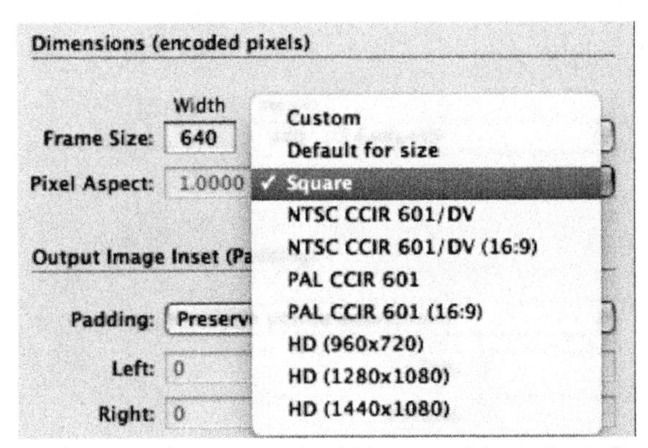

Figure 2-22. Changing the aspect ratio of SD source video in Apple Compressor.

Tip: *Lots of HD formats are shot at smaller resolutions and then expanded for display and/or editing. For example, HDV is shot at 1440x1080 pixels and then expanded to 1920x1080 during display, or when you input it into your video editor. Once expanded (which your video editing program or software encoder knows to do automatically), all HD formats are displayed at square pixel resolution. Therefore, outputting using square pixels is always the right answer, and you should never have to change the pixel aspect ratio. This is different from SD formats, which are either 0.9091 (4:3) or 1.2 (16:9). For more on aspect ratios of HD formats, see "Know Your Digital Video Formats" at* bit.ly/HDformats.

Again, fixing aspect ratio issues starts by recognizing that you have a problem. Whenever a video frame doesn't look the same as a digital picture of the same subject, you definitely have a problem. Streaming video shouldn't make a person look fatter or skinnier than a digital photo—he or she should look exactly the same.

Deinterlacing

Have you ever seen a streaming video that looked like Figure 2-23? I call those slices Venetian blind artifacts, and they are typically caused by video shot in interlaced mode with either inadequate or no deinterlacing. I got this frame from a recent ESPN story about University of Georgia linebacker Jarvis Jones, who decided to forego his final year at UGA to go pro. I'm a UGA fan from way back, and wasn't surprised at the news, but I was surprised when I zoomed the video to full screen and noticed that ESPN shot in interlaced mode and forgot to deinterlace. Sometimes, even the most experienced pros make mistakes.

You can see a subtler deinterlacing artifact in Figure 2-24, which is from a Walmart online video advertisement. Specifically, the jaggy lines on the table and back of the computer indicate that the video was shot in interlaced mode and a poor-quality deinterlacing filter was used. Note how the rest of the video is sharp and clear—it's just the diagonal lines that are affected.

Figure 2-23. This frame from an ESPN video clearly wasn't deinterlaced.

Where do these problems come from? Here's the *CliffsNotes* version:

- Most video was originally shot in interlaced format. This means that each frame consists of two fields, each shot 1/60 of a second apart (that's in NTSC-land; in Europe it's 1/50 of a second apart).

- All streaming files are progressive. This means that the two fields must be combined into a single frame.

Figure 2-24. This video frame was either inadequately deinterlaced, or wasn't deinterlaced at all.

- When combining the two fields, the most classic problem occurs in high-motion shots when there's a noticeable difference in the position of the subject in the two fields. That's what you see in Figure 2-25. During the swing, the club head moved about a foot in the 1/60 of a second between when the first and second fields were shot—hence the double image when you combine the fields without deinterlacing.

Figure 2-25. Two fields combined without deinterlacing.

Deinterlacing filters combine the two fields and apply algorithms to minimize the resulting (usually jaggy) artifacts. You can see this on the left in Figure 2-26. On the right, you can see the same frame, shot with a camera in progressive mode that was mounted alongside the camera shooting in interlaced mode.

These figures lead to several deinterlacing-related conclusions. First, the quality of deinterlaced footage will never be as good as that of progressive footage of the same source. When shooting for streaming, always shoot progressive when it's available.

Second, when working with interlaced source footage, always deinterlace your source footage before or during the encoding process. If you don't, you'll see artifacts like those in Figure 2-23 in higher-motion sequences.

Figure 2-26. On the left, two fields with deinterlacing; on the right, a progressive frame that looks cleaner.

Finally, if you see artifacts like those in Figure 2-24, deinterlacing quality isn't optimized. Try different deinterlacing settings in your encoding tool until you find the one that works best with your footage. For example, Figure 2-27 shows the options available in Telestream Episode.

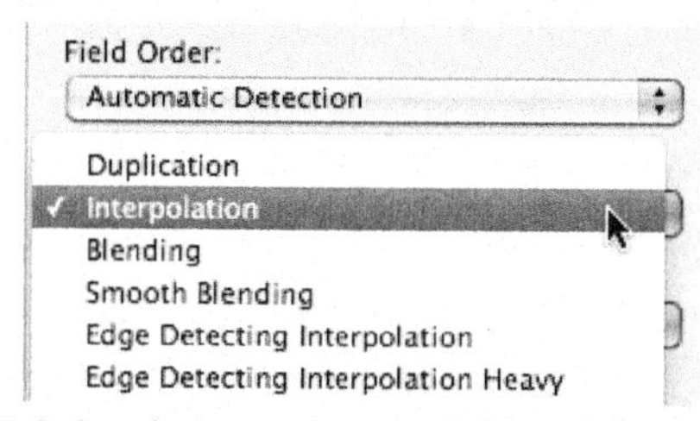

Figure 2-27. Deinterlacing options available in Telestream Episode.

Overall, when working with interlaced source footage, be on the lookout for interlacing artifacts. If you see them, try different deinterlacing settings in your encoder, and don't be afraid to change deinterlacing tools to produce the optimal result.

> **Tip:** *For more on shooting for streaming, check out my article "Shooting for Streaming – Five Key Tips" at* **bit.ly/shoot_stream**. *Or, you can find a range of articles on shooting for streaming on my website at* **bit.ly/shoot_stream_all**.

Conclusion

OK, a monster chapter in both content and concept. You're now ready to start working with the H.264 codec, which you'll do next chapter.

Chapter 3: H.264 Encoding Parameters

Note: You can download a PDF file with all figures from this book at bit.ly/Ozer_multi. Since this chapter contains lots of images, now would be a good time to do so.

If you're new to H.264 encoding, this chapter is for you. We'll start with a high-level look at H.264, covering issues like container formats and potential royalty obligations. Then you'll learn how to configure the most commonly presented H.264 encoding options.

Next, you'll see how the quality produced by the different H.264 codecs can vary, and also learn how to work with Advanced Audio Coding (AAC) audio compression. You'll walk away familiar with the most common H.264 encoding parameters and ready to tackle your own encoding. Note that Chapter 7 contains encoding-tool-specific information that supplements this chapter, so check the Table of Contents to see if I've covered the tool you're using.

Either way, you have to learn the basics first, which is what we'll address in this chapter. Specifically, you will learn:

- Which container formats you can use with the H.264 codec
- Which activities result in an obligation to pay royalties
- What profiles and levels are and how they affect quality and playback compatibility
- What CABAC and CAVLC are and when to choose one over the other
- How to configure I-, B- and P-frame-related settings
- How and why quality differs among the various H.264 codecs, and which deliver the best (and worst) quality
- Which audio codecs you can use with H.264 and how to configure them.

What Is H.264?

H.264 is a video compression technology, or codec, that was jointly developed by the International Telecommunication Union (as H.264) and International Organization for Standardization/International Electrotechnical Commission Moving Picture Experts Group (as MPEG-4 Part 10, Advanced Video Coding, or AVC). Thus, the terms H.264 and AVC mean the same thing and are interchangeable.

Architecturally, H.264 is a codec defined in the MPEG-4 specification. Of the specification's 28 parts, four are most relevant to streaming. These are:

- ***Part 2—MPEG video (the MPEG-4 codec).*** This was the first video codec incorporated into the MPEG-4 specification. Although it was adopted by Apple, the MPEG-4 codec boasted a quality below that offered by RealNetworks (RealVideo) and Microsoft (Windows Media Video), so the MPEG-4 codec never achieved much mindshare or market share in the streaming media marketplace.

- ***Part 3—MPEG-4 audio (AAC, etc).*** These are the audio compression technologies that can be included in the MP4 container format defined in Part 14. More on these below.

- ***Part 10—Advanced Video Coding (AVC/H.264).*** This is the H.264 video codec. Note that while the H.264 codec is a part of the MPEG-4 specification, the codec can be used in container formats other than that defined in Part 14. For this reason, it's common to see the H.264 codec included in files using the Flash container format (.FLV or .F4V) or the QuickTime container format (.MOV); in MPEG-2 transport streams (.TS); or in the proprietary container formats used by adaptive streaming formats like Adobe's HTTP-based Dynamic Streaming (.F4F) or Microsoft's Smooth Streaming (.ISMV).

- ***Part 14—Container Format (MP4).*** This is the container format defined by the MPEG-4 spec. I'll discuss container formats in a minute; for now, just recognize that whenever you see the file extension .MP4, the file is produced using the MPEG-4 container format.

When launched, MPEG-4 was billed as the unifying standard for broadcasters with advanced features like object orientation and two multiplex layers enabling compatibility with broadcast and streaming architectures. For more on this, check out "Will MPEG-4 Fly?" at bit.ly/MPEG-4_fly. The MPEG-4 codec itself was so bad that in 2004, I wrote a column proclaiming "MPEG-4 is Dead" (bit.ly/MPEG4_dead).

In the streaming media space, the MPEG-4 specification gained very little traction until implementations of the H.264 codec started to produce better quality than competitive codecs like VP6 and Windows Media Video, and the average computer became powerful enough to play H.264 video. Of course, it didn't hurt when Apple built H.264 playback into its little device called the iPod that morphed into the iPhone and iPad and created several huge new markets along the way. Then Adobe integrated H.264 into Flash, and Microsoft did the same with Silverlight, and the strength of the H.264 codec pulled MPEG-4 into the forefront of streaming.

Outside the streaming media marketplace, H.264 was widely used by all the devices that typically use the technology recommended by their standards body, which is the ITU for TV, radio and mobile phones, and the ISO for photography, consumer electronics and computers. This widespread adoption and usage gives H.264 significant momentum going forward. For example, its inclusion in multiple devices like mobile phones and iPods dramatically reduces the cost of chips and other components that enable H.264 playback, creating a natural barrier to entry for competitive formats.

Year	**ITU** International Telecommunications Union (TV, Radio, Phone)	**ISO** International Standards Organization (Photography, Consumer Electronics, Computers)	**Streaming Video Codecs**
1984	H.120		
1993	H.261 - video conferencing	MPEG-1	
1994		MPEG-2	
1995	H.263 - better video conferencing		
1999		MPEG-4	QuickTime 4 with MP3
2002	AVC (H.264)	AVC (MPEG-4 Part 10)	QuickTime 6 - MPEG-4
2005			QuickTime 7 - H.264 / First iPod with H.264 playback
2007			Flash - H.264 support
2008			Silverlight - H.264 Support

Table 3-1. The evolution of H.264 development and adoption.

Overall, although Google's WebM announcement created lots of noise in the computer space, it's tough to imagine any codec making significant inroads into H.264's market share over the next two to five years. Other than H.265, of course, which is the standards-based successor to H.264 discussed in Chapter 16.

H.264's audio sidekick is Advanced Audio Coding (AAC), which is designated MPEG-4 Part 3. Both H.264 and AAC are technically MPEG-4 codecs—although it's more accurate to call them by their specific names—and compatible bit streams should conform to the requirements of Part 14 of the MPEG-4 spec.

Container Formats

Every file you encode must be inserted into a container format—which, according to Wikipedia, is "a meta-file format whose specification describes how data and metadata are stored" (**bit.ly/containerformat**). You can tell which container format is used for a particular file by the extension; as mentioned above, the extension .FLV or .F4V means files using the Flash container format, .MOV means the QuickTime container format, and .MP4 means the MPEG-4 container format. But you can't tell which codec is being used by the container formats, since you can use the VP6 codec in .FLV files, any of a number of codecs in .MOV files and multiple codecs in .MP4 files.

When encoding into the H.264 codec, it's important that you choose the correct container format for your target player; otherwise, the file won't play. For example, even though QuickTime can play .MOV files encoded with the H.264 format, it may not be able to play .F4V files encoded with the H.264 format. I'll cover the container formats you should use for each target player in Chapter 4.

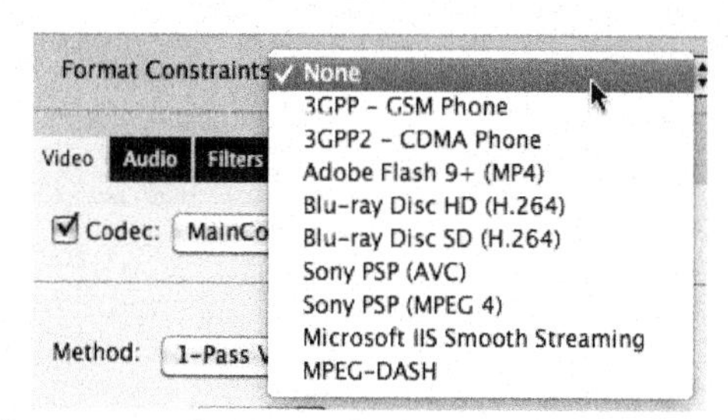

Figure 3-1. You can encode H.264 video into a number of container formats, as shown in this screen shot from Sorenson Squeeze.

If you had to choose one container format to produce a single file that plays almost everywhere, use the MPEG-4 container format to produce an .MP4 file. That's because all players—whether Flash, Silverlight, iPad, iPod, Android, or Windows Phone—play .MP4 files. As covered in more detail in Chapter 5, you can't say the same about most adaptive streaming formats because they all use their own unique formats. HTTP Live Streaming (HLS) uses MPEG-2 Transport Streams (.TS), Adobe's HTTP-based Dynamic Streaming (HDS) uses the .F4F container, while Microsoft's Smooth Streaming uses the .ISMV container. The only exception is Adobe's RTMP-based Dynamic Streaming, which uses Plain Jane H.264 files.

Understand that the H.264-specific encoding options you'll learn about in this chapter apply to all H.264-encoded files, irrespective of the intended wrapper. So although you'll have to make sure that you produce your files in the format compatible with your intended player (which I'll cover in Chapter 4), the H.264-specific encoding parameters themselves remain the same.

Transport Streams and Program Streams

Let's define transport streams and program streams before we move on. Transport streams are container formats that contain packetized elementary streams, or separate streams for audio, video, closed captions and the like. Because transport streams are also used in broadcast applications where data loss is likely, transport streams also incorporate error detection and correction techniques. The only transport stream we'll discuss is the MP2 transport stream used in Apple's HLS, which is designated with a .TS extension.

Program streams are container formats that contain elementary streams without error detection and correction, since they're used in applications with little risk of data loss, like DVDs or Blu-ray Discs. One example of a program stream is the .VOB files used on DVDs. There are no program streams used in any streaming applications discussed in this book.

Other H.264 Details

Like most video coding standards, H.264 actually standardizes only the "central decoder ... such that every decoder conforming to the standard will produce similar output when given an encoded bit stream that conforms to the constraints of the standard," according to the "Overview of the H.264/AVC video coding standard" published in *IEEE Transactions on Circuits and Systems for Video Technology*. Basically, this means that there's no standardized H.264 encoder. In fact, H.264 encoding vendors can utilize a range of different techniques to optimize video quality, so long as the file plays on the target player. This is one of the key reasons that H.264 encoding interfaces vary so significantly among the various tools.

It's also one of the key reasons that the quality of H.264-encoded files varies so significantly among the different codec vendors. Unlike VP6 and Windows Media, which come from a single vendor and are relatively uniform in terms of quality irrespective of encoding tool, there are multiple H.264 codec developers, and quality fluctuates markedly among them, as you'll see at the end of this chapter.

H.264 Royalties

H.264 was developed by a consortium of companies that (gasp!) want to get paid for their efforts. To promote this goal, they patented many of the underlying technologies and contracted with a company called MPEG LA to set up and administrate licensing and royalty collection. Typical customers of MPEG LA include consumer equipment manufacturers (Blu-ray Disc players and recorders); software developers (encoding programs, streaming players); and content developers.

To explain, companies like Adobe, Apple and Microsoft pay MPEG LA by the unit to include H.264 decoders in their respective players. There's a maximum annual fee of US$5 million, which is steep but makes including H.264 playback in hundreds of millions of players feasible. Encoding vendors like Sorenson, Telestream and Rhozet are also subject to license fees if their volumes exceed certain limits. Under certain circumstances, content producers who encode with H.264 are also subject to royalties.

MPEG LA has a licensing FAQ at **mpegla.com/main/programs/AVC/Pages/FAQ.aspx**. There's also a PDF file that you can download from that page titled "Summary of AVC/H.264 License Terms." In this section, I'll provide a brief overview of those terms, primarily focused on content producers.

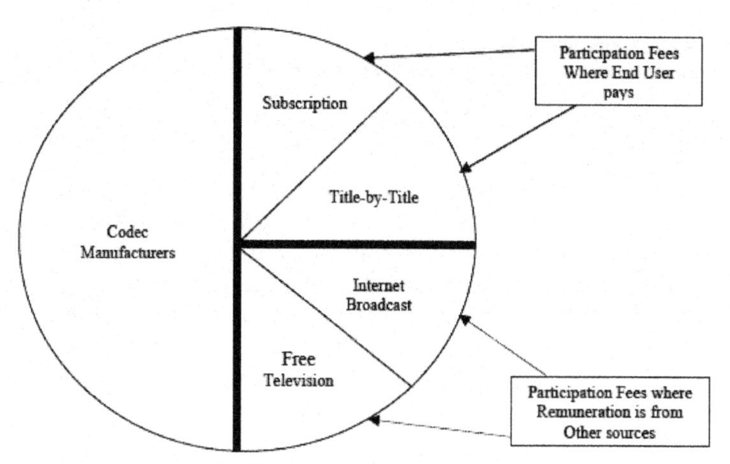

Figure 3-2. H.264 royalty categories per MPEG LA's license summary.

Figure 3-2 presents a chart from the license summary. All those on the left of the wheel are subject to royalties after shipping more than 100,000 units. On the top right, if you're selling H.264-encoded content—on either a subscription or pay-per-view basis—you owe the following, according to the license summary:

> ***Subscription:*** Royalties start at 100,001 subscribers; for 100,000 and under, there is no fee. For 100,000 to 250,000 subscribers, the annual royalty is $25,000, increasing to $50,000 for 250,001 to 500,000 subscribers. Between 500,001 and 1 million subscribers, the annual royalty is $75,000, while services with more than 1 million subscribers owe $100,000 annually.
>
> ***Title-by-Title (pay-per-view).*** There is no royalty for videos that are 12 minutes or less. For videos longer than 12 minutes, the price is "the lower of 2% of the price paid to the licensee (on first arms length sale of the video) or $0.02 per title (categories of licensees include legal entities that are [i] replicators of physical media, and [ii] service/content providers [e.g., cable, satellite, video DSL, Internet and mobile) of VOD, PPV and electronic downloads to end users]."

The Internet broadcast category is the one that's interesting to most producers. The short answer here is that it's free in perpetuity; so long as you don't charge for your videos, you can use H.264 for free (**mpegla.com/main/pages/media.aspx**).

How it got to be free is an interesting story. When the H.264 licensing policies were first announced, free Internet video had no royalty until at least January 1, 2011. Then, in February 2010, MPEG LA announced that royalties would be delayed until December 31, 2015. Then, in August 2010, MPEG LA announced that its "AVC Patent Portfolio License will continue not to charge royalties for Internet video that is free to end users (known as 'Internet Broadcast AVC Video') during the entire life of this license."

Of course, between February and August, Google launched WebM, the first real competitor to H.264. Though MPEG LA didn't specify why it waived the royalty, the decision probably related to WebM, at least in part. At a high level, though, any royalty on free Internet video could really muck up the business models for all video-intensive sites, so perhaps this was coming anyway. Either way, the bottom line is that if you're not charging for your Internet video, there is no royalty for encoding with H.264, and there won't ever be.

Comparing H.264 with Other Codecs

In the previous edition of this book, titled *Video Compression for Flash, Apple Devices and HTML5*, I spent about six pages comparing the quality and playback requirements of H.264 vs. other codecs like VP6, VC-1/WMV and WebM. For this book, I figured I would save some trees and exclude these comparisons because they're largely irrelevant; H.264 is far and away the most widely used codec. In fact, it's the only codec compatible with most applications discussed in this book. It's also the highest-quality codec with playback requirements similar to those of the other listed codecs.

For those who are interested in comparing H.264 to VC-1 and VP6, check out "Which Codec is Hardest to Play Back; VC-1, H.264 or VP6?" at **bit.ly/H264_VC-1_VP6**. For those interested in how WebM compares to H.264, check out "First Look: H.264 and VP8 Compared" at **bit.ly/h264_webm_1**, and a deeper analysis with "WebM vs. H.264: A Closer Look" at **bit.ly/h264_webm_2**.

With this behind us, let's jump into the basics of H.264 encoding.

Basic H.264 Encoding Parameters

Let's start with the basics, profiles and levels. Just for perspective, note that not all H.264 encoding tools provide access to all the controls we're going to discuss. For encoder-specific questions, check the discussion in Chapter 7.

Profiles and Levels

Profiles and levels are the most basic H.264 encoding parameters, and are available in one form or another in most H.264 encoding tools. According to Wikipedia (**en.wikipedia.org/wiki/H264**), a profile "defines a set of coding tools or algorithms that can be used in generating a conforming bit stream," whereas a level "places constraints on certain key parameters of the bit stream." In other words, a profile defines specific encoding techniques that you can or can't use when encoding a file (such as B-frames), while the level defines details such as the maximum resolutions and data rates within each profile.

	Baseline	Main	High
I and P Slices	Yes	Yes	Yes
B Slices	No	Yes	Yes
Multiple Reference Frames	Yes	Yes	Yes
In-Loop Deblocking Filter	Yes	Yes	Yes
CAVLC Entropy Coding	Yes	Yes	Yes
CABAC Entropy Coding	No	Yes	Yes
Interlaced Coding (PicAFF, MBAFF)	No	Yes	Yes
8x8 vs. 4x4 Transform Adaptivity	No	No	Yes
Quantization Scaling Matrices	No	No	Yes
Separate Cb and Cr QP control	No	No	Yes
Separate Color Plane Coding	No	No	No
Predictive Lossless Coding	No	No	No

Table 3-2. Encoding techniques enabled by profile, from Wikipedia.

Take a look at Table 3-2, which is a screen grab from Wikipedia's description of H.264. On top are H.264 profiles, including the Baseline, Main and High, which are the profiles most frequently supported in computer- and device-oriented players. On the left are the different encoding techniques available, with the table detailing which are supported by the respective profiles.

In theory, as you apply more advanced encoding algorithms, your quality should improve. What's the trade-off? Either increased encoding time, increased decoding complexity, or both. Why do profiles exist? To define different grades of H.264 that can be used by different devices depending upon the power of their CPU. For example, the original iPods and iPhones only had sufficient horsepower to play back H.264 video encoded using the Baseline profile. That's why video encoded using the Main or High Profile won't play on these devices. In contrast, most computers and all over-the-top (OTT) devices like Apple TV and Roku boxes can play video encoded using the High profile, as well as video encoded using the Baseline or Main profile.

Let's take a step back for some perspective. Whenever you produce H.264 files, you have to choose a profile, typically using a control similar to that shown in Figure 3-3. When encoding for some older or low-end mobile devices, you must use the Baseline profile; otherwise, the video won't play. However, computers and OTT devices, and even more recent mobile devices, can decode H.264 video encoded using the High profile, as well as the Baseline and Main.

Does this mean that you should produce separate files for each target playback device, each using the highest profile that will play on that device? Specifically, should you produce one file using the Baseline profile for low-end mobile devices; another using the Main profile for mid-range mobile devices; and a third using the High profile for desktops, OTT and high-end mobile devices?

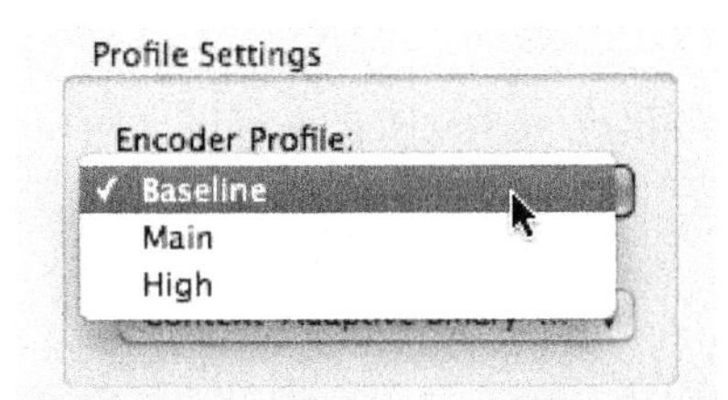

Figure 3-3. Choosing your Profile in Telestream Episode.

Well, that would depend upon the quality difference produced by the different profiles, right? If the High profile produces significantly better quality than the Baseline profile, you'd want to reward viewers watching on high-end platforms with that extra quality, even though you're doubling or even tripling your encoding, storage and related administrative costs. If there's little difference between the quality produced by the Baseline, Main and High profiles, there's no reason to create multiple files since the higher-end devices can play files encoded using the Baseline profile.

As I'm sure you're getting by now, one of the most fundamental decisions facing producers encoding for multiple screens is this exact question: Should you serve all users with Baseline-encoded video, or customize encoding for each target? And it's precisely the issue I analyzed in my article "H.264 in a Mobile World: Adios to the Main and High Profiles?" which you can read at bit.ly/h264profiles. Here's the relevant background and findings:

> Recently, I worked with a client encoding SD video for web delivery. His application was continuing education, and in addition to adaptive streaming he was offering a single file for download and local playback on computers, tablets, mobile phones and other devices. He asked whether he should encode this file using the Baseline profile, which would maximize compatibility, or the High profile, which would optimize quality.
>
> While the client could certainly produce two files using both profiles, this would cost him both encoding cycles and storage space. So the obvious question was, did the High profile produce superior quality to the Baseline profile using his encoding parameters, which were 640x480x29.97@1 Mbps? ... To answer the client's question, I produced two files to these targets, one using the Baseline profile, the other the High profile. ... The quality was pretty much identical.
>
> So, I ran several tests using three encoding configurations parameters and three software encoders. If you want to skip the commentary and look at the files themselves, I've posted them all here. The bottom line was that I found a noticeable difference in only one sequence in four sets of tests, and that was at the most aggressive compression settings, 640x360x29.97 @ 240 kbps.

Figure 3-4 shows a screen from the 640x480@500 kbps comparison. Note that at these parameters, the bits-per-pixel value is 0.054, or about 45% of that used by CNN. As the bits-per-pixel value increases, the quality of both files should improve, and you should see less difference between the files. In other words, unless you're really pushing the envelope to get the lowest data rate possible, you probably will see minimal difference between files encoded using the Baseline profile and those encoded using the High profile.

Figure 3-4. Comparing files encoded using the Baseline and the High profile. Some noticeable differences, but nothing major.

In addition, this is the highest-motion content in my SD test file, and there are no noticeable differences in the lower-motion sequences. So if you're encoding low-motion video at relatively normal bits-per-pixel values, you should expect to see little difference.

If you visit the article on my website (**bit.ly/h264profiles**), you'll see that there was a good bit of disagreement among my peers who read the article and looked at the files. Still, to my eye, there wasn't enough difference to encode separate files for the respective files, and my client agreed.

What's the takeaway? First, as I concluded in the article, recognize that your mileage will vary based upon source footage, encoding configuration and encoding tool. However, before assuming that you need different files encoded using different profiles to optimize quality for different target platforms, run your own tests and see if there's a noticeable difference.

If you encode all videos using the Baseline profile, you only have to encode one file for all targets, or one set of files for all targets if you're streaming adaptively. If you decide to customize the profile for all targets, you obviously need to know the relevant profile for all targets, which I specify in Chapter 4, and encode separately for each one.

H.264 Levels

What about H.264 levels? Again, as mentioned above, levels provide bit rate, frame rate and resolution constraints within the different profiles, which is shown in Figure 3-5.

Level number	Max video bit rate (VCL) for Baseline, Extended and Main Profiles	Max video bit rate (VCL) for High Profile	Examples for high resolution @ frame rate (max stored frames) in Level
1	64 kbit/s	80 kbit/s	128x96@30.9 (8) 176x144@15.0 (4)
1b	128 kbit/s	160 kbit/s	128x96@30.9 (8) 176x144@15.0 (4)
1.1	192 kbit/s	240 kbit/s	176x144@30.3 (9) 320x240@10.0 (3) 352x288@7.5 (2)
1.2	384 kbit/s	480 kbit/s	320x240@20.0 (7) 352x288@15.2 (6)

Figure 3-5. Levels constrain bit rate, frame rate and resolution for the different profiles. From Wikipedia.

In this role, levels enable primarily device vendors to further specify the types of streams that will play on their devices. For example, the Apple iPad will play video encoded with the Main profile, but only up to level 3.1. As with profiles, if you encode to parameters beyond the specified level, either the file won't load on the device (iTunes typically won't load non-conforming files on any iDevice) or it will load but won't play.

Accordingly, when you're producing for devices, you need to ensure that your encoding parameters don't exceed the specified level, which again should be designated by the device manufacturer. In contrast, levels are irrelevant when encoding for computer playback because all three streaming players—QuickTime, Flash and Silverlight—can play any level of the Baseline, Main or High profiles. So you can't choose the wrong level—the player will always attempt to play the file anyway.

Many encoding tools don't let you specify a level. When you're producing for devices with such encoders, you'll need to use the supplied device templates to ensure that the resolution, frame rate and data rate don't exceed those supported in the specified level. Other encoding tools do specify the level, and present an error message if you encode using parameters that exceed the selected level.

This is shown in Figure 3-6, a screen shot of Adobe Media Encoder. If you were producing for a device with this encoder, compatibility with the selected level is key, so you'd have to dial back your encoding parameters to the selected level. If you were producing for computer playback, you would simply choose a higher level that lets you produce the file at the selected parameters, and restart encoding.

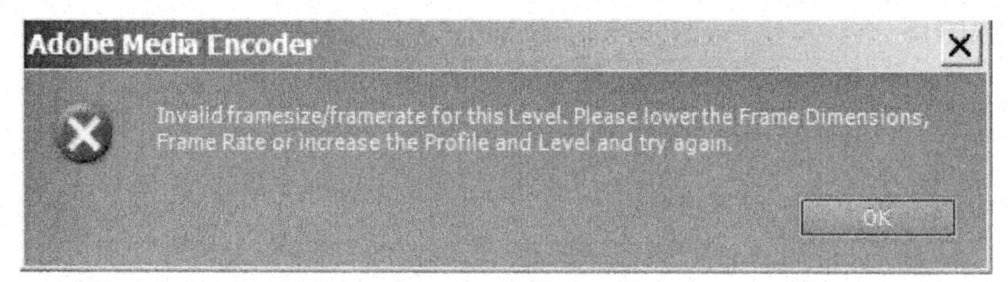

Figure 3-6. Oops, the selected frame size exceeds the parameters allowed for the selected level.

Again, however, just because the Flash Player won't refuse to play a file encoded using the High profile doesn't mean that it will play that file smoothly. While profiles and levels are critical for devices, choosing the right resolution and data rate for your streaming video file is much more important when targeting computer playback. I touch on this in Chapter 4. In summary, here's the skinny on levels:

- Critical when encoding for devices.
- Not so important when encoding for computers, where the resolution and data rate of your video are the most important considerations.

Entropy Coding

When you select the Main or High profiles, some encoding tools provide two options for Entropy Coding Mode: CAVLC, which stands for context-based adaptive variable length coding, and CABAC, which stands for context-based adaptive binary arithmetic coding. Of the two, CAVLC is the lower-quality, easier-to-decode option, while CABAC is the higher-quality, harder-to-decode option. This is shown in Figure 3-7 from Rhozet Carbon Coder.

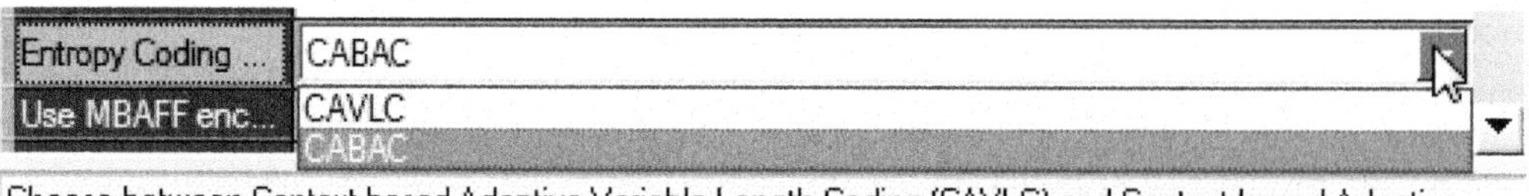

Choose between Context-based Adaptive Variable Length Coding (CAVLC) and Context-based Adaptive Binary Arithmetic Coding (CABAC). CABAC is slower for both encoding and decoding, but generally produces better output quality. CABAC is not available when encoding to the Baseline profile.

Figure 3-7. Your Entropy Coding choices: CAVLC and CABAC.

CABAC is generally regarded as being between 5 and 15% more efficient than CAVLC. This means that CABAC should deliver equivalent quality at a data rate of between 5 and 15% lower, or better quality at the same data rate.

Of course, when I ran the Baseline vs. High profile comparisons above, I enabled CABAC for the High profile and used CAVLC—my only option—with the Baseline encoded file. So the 15% differential didn't pan out for me. On the other hand, I saw very little difference in playback CPU requirements for the CABAC file: less than 1% on a five-year-old laptop, and 2% on an old

PowerPC-based dual-processor Power Mac G5 workstation.

So, little upside, but very little downside. My practice is to use CABAC whenever it's available, which is when encoding using the Main and High profile.

What Would YouTube Do?

As you likely know, YouTube streams about 70% of the video seen on the Internet, as most of us with teenagers can likely attest. I sure can. So when it comes to producing H.264 video, it's instructive to see how YouTube re-encodes videos that are uploaded.

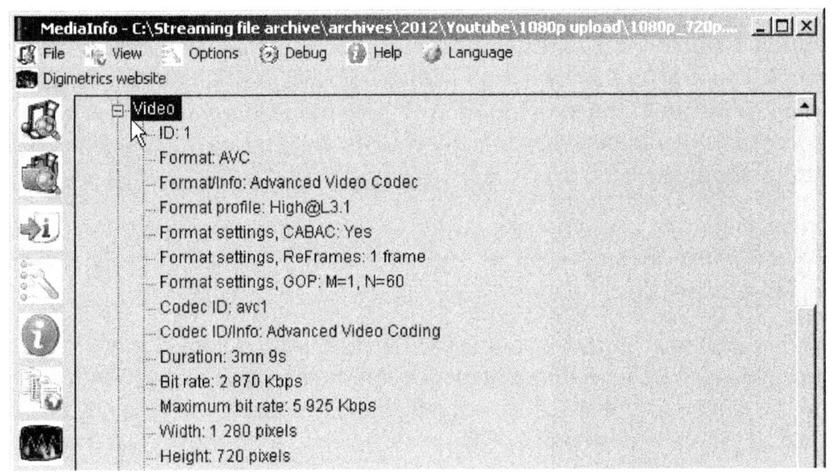

Figure 3-8. A MediaInfo analysis of a 720p YouTube file. M=1 means no B-frames

To do this yourself, download a 720p YouTube file; I use the free Firefox plug-in Download Helper (**downloadhelper.net**) for the job. Then download free file analysis tool MediaInfo (**mediainfo.sourceforge.net**), install it and analyze the downloaded YouTube file. You should see something that approximates the screen shot in Figure 3-8, showing a file encoded with the High profile with CABAC enabled, confirming my recommendations.

Take note of the line Format settings, GOP: M=1, N=60. According to Wikipedia, "The GOP structure is often referred by two numbers, for example M=3, N=12. The first one tells the distance between two anchor frames (I or P): it is the GOP size. The second one tells the distance between two full images (I-frames): it is the GOP length. For the example M=3 N=12, the GOP structure is IBBPBBPBBPBBI. Instead of the M parameter, one can use the maximal count of B-frames between two consecutive anchor frames."

In other words, YouTube is using no B-frames, and a key frame interval of 60 frames, or two seconds. YouTube is also encoding with only one reference frame, which I also found interesting. This helps set up the discussion for our next topic.

I-, B- and P-frame Controls

We looked at these in Chapter 2, but let's quickly review. I-frames, also known as key frames, are completely self-referential and don't use information from any other frames. I-frames are larger than B-frames or P-frames, and are the highest-quality of the three, but the least efficient from a compression perspective. We learned that when encoding for single-file streaming, you should use a key frame interval of 10 seconds, and insert key frames at scene changes. When encoding for adaptive streaming, as discussed in more detail in Chapter 5, you should use a shorter setting to enable more nimble stream switching.

That's why YouTube uses a key frame interval of two rather than a higher value. Although YouTube doesn't use automatic adaptive streaming, you can manually choose stream quality using the familiar control on the lower right. When you switch stream quality, playback continues more or less from the same point. With a keyframe interval of 10 seconds, that relative seamlessness would be tough to achieve; with a key frame interval of 2, the stream switch is more responsive.

P-frames are "predicted" frames. When producing a P-frame, the encoder can look backward to previous I- or P-frames for redundant picture information. P-frames are more efficient than I-frames, but less efficient than B-frames.

B-frames are "bi-directional predicted" frames. When producing B-frames, the encoder can look both forward and backward for redundant picture information. This makes B-frames the most efficient frame of the three, but also the most difficult to decode, so they are not available when producing using H.264's Baseline profile. Our default interval for B-frames was 3, which would produce a stream configured like this:

IBBBPBBBPBBBPBBBPBBBPBBB ...

and so on until the next I-frame. Those are the basics; let's look at some H.264-specific I- and B-frame-related controls.

Instantaneous Decode Refresh (IDR) Frames

Figure 3-9 shows the key frame controls from Sorenson Squeeze. Again, unless you're producing for adaptive streaming, use an interval of 10 seconds and enable key frames at scene changes. I leave Squeeze's Rare/Frequent slider at the default 50 value, which works well.

One other I-frame configuration option that you'll see in Telestream Episode and some higher-end encoders is the IDR frame (Figure 3-10). Briefly, the H.264-specification enables two types of I-frames: normal I-frames and IDR frames. If an I-frame is an IDR frame, no frame after it can refer back to any frame before it. In contrast, with regular I-frames, B- and P-frames located after the I-frame can refer back to reference frames located before it.

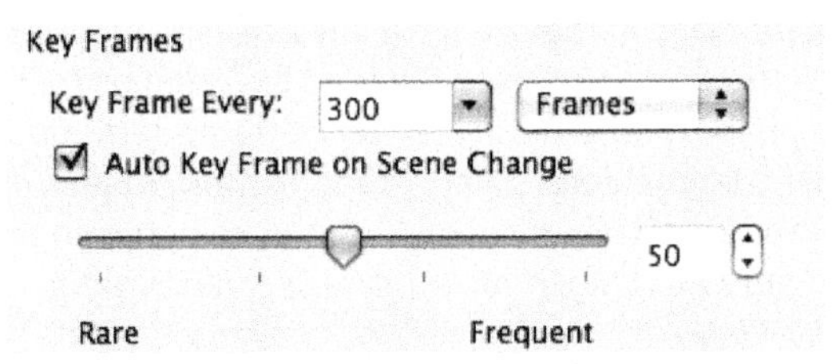

Figure 3-9. Sorenson Squeeze's key frame settings.

In terms of random access within the video stream, playback can always start on an IDR frame because no frame refers to any frames behind it. However, playback cannot always start on a non-IDR I-frame because subsequent frames may reference previous frames. Here's a blurb from the Telestream Episode help file:

> An IDR frame is an I-frame whose preceding frames cannot be used by predictive frames. More distant IDR frames may allow more efficient compression but limits the ability of a player to move to arbitrary points in the video. In particular, QuickTime Player may show image artifacts when you scrub the timeline unless every I-frame is an IDR frame.

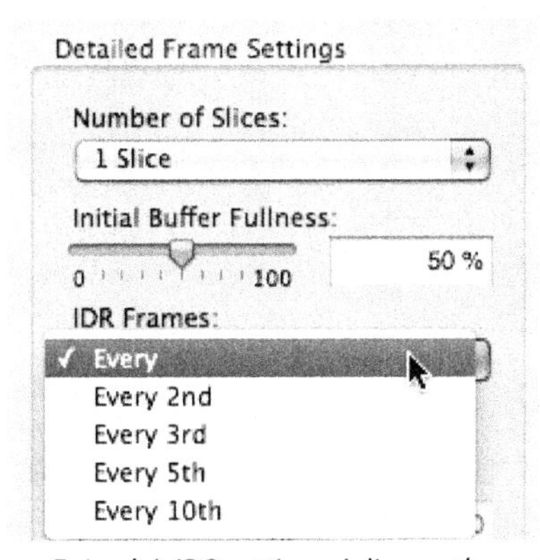

Figure 3-10. Telestream Episode's IDR settings. I discuss the concept of Slices below.

Since one of the key reasons to insert I-frames into your video is to enable interactivity, I use whatever setting results in every I-frame being an IDR frame. With Telestream Episode, I would choose Every as shown in Figure 3-10. If your streaming encoder doesn't offer this option, it's safe to assume that every I-frame is an IDR frame.

Working with B-frames

So the party line with B-frames is an interval of 3 with 5 reference frames. This was kind of a mantra for me—backed up, in the early days of H.264 encoding, by results from encoding tools.

The first shock to this system came in the profile-related tests shown above where the quality of the Baseline encoded video was quite close to video encoded using the High profile, with all options set for maximum quality and B-frame/reference frame settings set to 3 and 5. If you scan back to Table 3-2, you'll see that there are six encoding techniques available in the High profile that are not available in Baseline. If there's so little difference between the Baseline and High profiles with all the techniques working, how important could B-frames be?

Then, while running some tests on some YouTube-encoded files, I noticed that YouTube wasn't using B-frames, as shown in Figure 3-8 and in subsequent tests using a program called Semaphore, which was produced by Inlet Technologies before it was purchased by Cisco (and then no longer sold separately). Not only does Semaphore provide a useful data rate graph, you can scroll through the file and see identify the frame type for each frame (Figure 3-11).

Figure 3-11. Inlet Semaphore identifies this frame as a B-frame.

So, two programs both show that YouTube didn't use B-frames in the file I had downloaded, which certainly seemed to support a case against the value of B-frames. I later corresponded with the compression team at YouTube and was assured that there were instances where YouTube did deploy B-frames and that B-frames "do result in bit rate savings overall." Still, if B-frames were really so efficient, why wouldn't YouTube use them all the time?

B-frames While Standing on One Leg

I'm guessing there is a group of readers out there who don't share my fascination with long compression-related mysteries and thrillers, and another group who does, but perhaps is in a hurry. So let me lay out B-frames definitions and recommendations before returning to our YouTube mystery.

First, the recommendations of 3 B-frames and 5 reference frames stand; if you need some default parameters, they are as good as any. Second, Figure 3-12 contains most of the B-frame related parameters that you'll see in any program; this is from Sorenson Squeeze when using the MainConcept H.264 codec. Note that I copied the Reference Frames control from another portion of the preset screen to create one neat screen for you; if you're looking for that setting in MainConcept Squeeze preset, it's on the upper right.

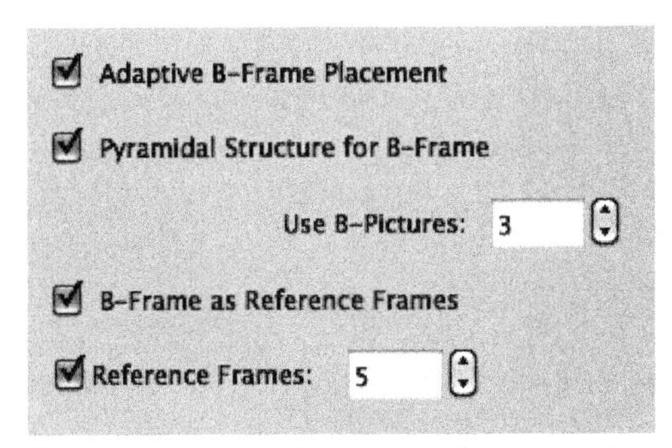

Figure 3-12. Squeeze and the most common B-frame encoding parameters.

So, there are three new concepts shown in Figure 3-12: Adaptive B-frame Placement, Pyramidal Structure for B-fame, and B-Frame as Reference Frames. Here are definitions and recommendations from Rhozet's helpful white paper "Advanced H.264 Encoding with Carbon Coder" (rhozet.com/rhozet_H264guide.pdf).

- ***Adaptive B-frame Placement.*** Allows the encoder to adaptively change the B-frame pattern based on the source content. This is very useful when you are using a longer GOP and trying to hit low bit rates. In general, this increases the encoder's efficiency and should always be checked.

- ***B-Frame as Reference Frames.*** Allows the encoder to use B-frames as a reference to build other frames. This also increases the efficiency of the encoder and should always be checked. Not available in Baseline profile.

- ***Pyramidal Structure for B-frame.*** Allows the encoder to compress B-frames based on other B-frames. Like the above two settings, this increases the efficiency of the encoder and should always be checked. Not available in Baseline profile.

By way of background, in MPEG-2 encoding, B-frames can't be used as reference frames. With H.264, they can, and most authorities agree that this improves overall quality. Note that when a program has two separate controls—one for Reference B-frames, the other for Pyramidal B-frames—the first enables B-frames to be used as reference for P-frames, while the second enables B-frames to be used as reference for other B-frames. I know, I know. Take a deep breath and read the line again, and it will start to make sense.

I discuss the importance of reference frames and examine the best interval below. If you're in a hurry, take Rhozet's recommendations and scoot on down to the discussion on slices. If you're interested in exploring B-frames further and unraveling the YouTube mystery, read on.

B-frames: What's the Right Number?

Let's start with a couple of observations. First, as we learned in our Baseline vs. High profile comparisons, using B-frames isn't going to make a huge difference in any case, but will be most noticeable when using very aggressive encoding parameters. As an example, compare Figure 3-13 and Figure 3-14.

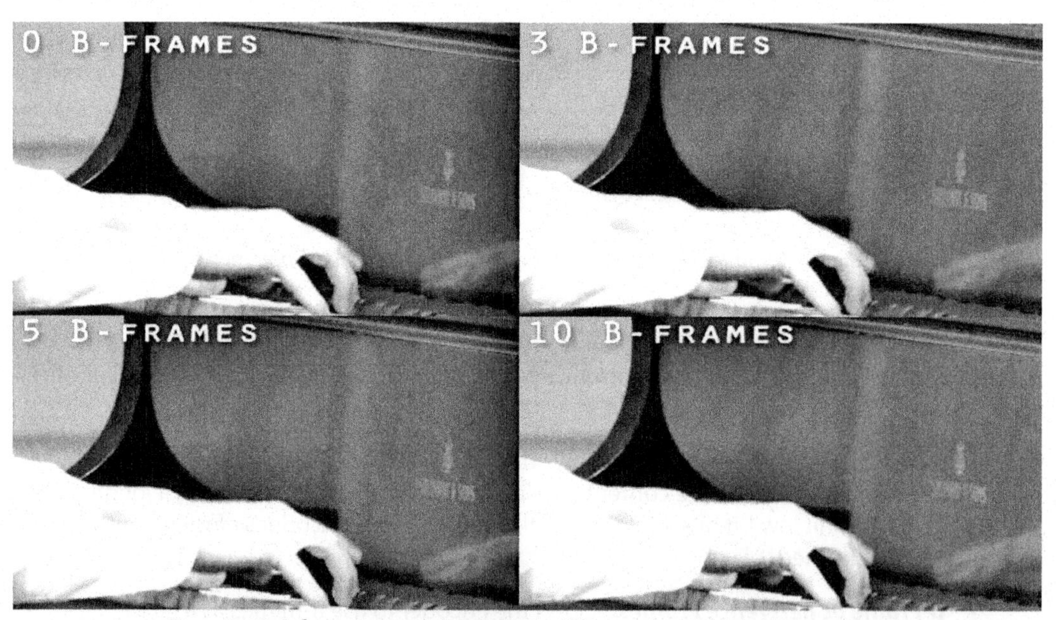

Figure 3-13. B-frame interval showed little difference in this high-bit-rate video.

The files shown in Figure 3-13 are encoded at fairly normal encoding parameters—specifically 640x360x29.97 @ 1200 kbps, for a bits-per-pixel value of about 0.175, which is in between CNN and ESPN. You can see the B-frame interval used in the files on the top left of each quadrant. I've artificially brightened the frame grab so you can more clearly see the Steinway logo on the piano. As you can see, there is very little difference in quality at these encoding parameters, not only in this frame, but also over the entire duration of this test file.

Compare that with Figure 3-14, which I encoded at 640x360x29.97 @ 240 kbps for a bits-per-pixel value of 0.035. This sounds pretty aggressive, but these parameters are used in the adaptive group of one of the three-letter networks with which I have consulted in the past. Figure 3-14 shows the same frame, same brightening, and you can see that the 0-B-frame image on the top left is less clear than others in the group. You can also see that the 3-B-frame image is about as clear as the 10-B-frame image, and that—inexplicably—the 5-B-frame image has less detail than either 3 or 10.

Now would be a good time to remind you that you can download a PDF file containing just the images from the book at **bit.ly/Ozer_multi**.

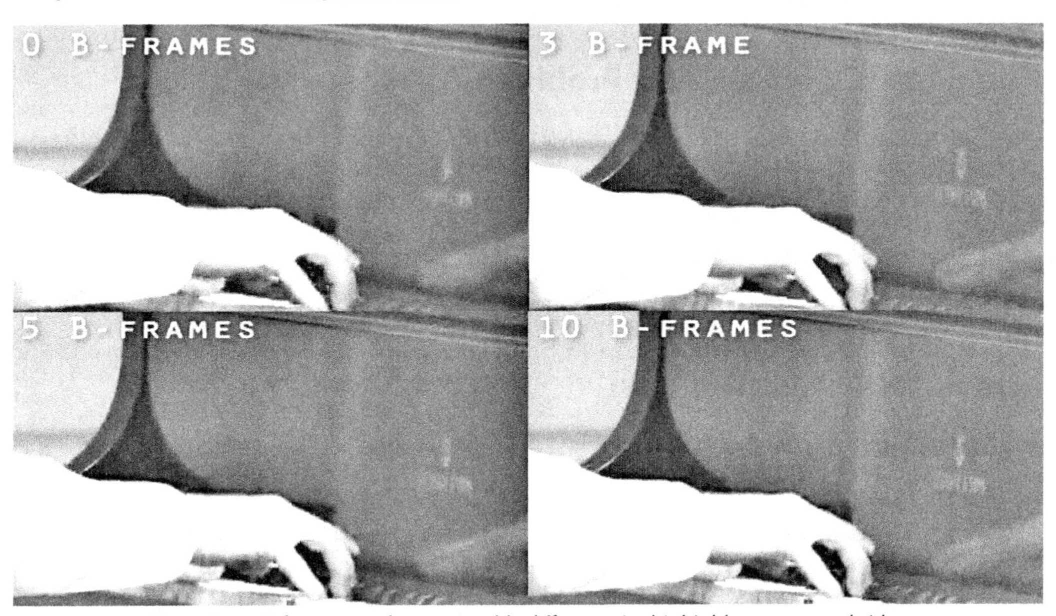

Figure 3-14. B-frames made a noticeable difference in this highly compressed video.

So, my tests indicate that the existence of B-frames and the B-frame interval make more difference at lower bit rates than at higher bit rates. I'm not the only one to draw that conclusion. Here's a snippet from a poster on the Videohelp forum named (gulp!) Poisondeathray (**bit.ly/poisondeathray**), answering the question, "How many B-frames is 'optimal' for h264 encoding?"

> There is no straight simple answer. There are too many variables involved, including specific type of content, level of compression, speed tradeoffs you're willing to make. At lower parts of the compression curve, more B-frames are almost mandatory for maximum compression. … On higher parts of the compression curve, if you give adequate bit rate for that type of content, differences quickly become negligible.

Interestingly, when encoding for an adaptive group, most of the lower-bit-rate encodes are for mobile devices anyway, which means you'll likely use the Baseline profile—which, of course, doesn't support B-frames. On the higher-bit-rate encodes that you produce for iPads and other higher-end mobile devices, as well as OTT and computers, you'll produce at much higher quality, so using B-frames, or the B-frame interval that you choose, will have minimal impact.

Along similar lines, I noticed that B-frames seemed to make a bigger difference with low-bit-rate CBR encodes than VBR encodes. This is most likely because high-motion sequences in CBR files are harder to encode at good quality than in VBR files because the latter allocates more data to harder-to-encode sequences. Basically, the more challenging the encode, the more benefit B-frames can deliver.

Going too High Can Cause Problems

One other phenomenon I noticed in my analysis was that in most instances where B-frames improved quality, the benefit was fully realized with 3 B-frames. Although this wasn't universally the case, I noticed frequent instances where extending B-frames out to 10 frames actually decreased quality. You can see this in Figure 3-15, where 3 B-frames is clearly better than none, but it's also superior to 5 and 10.

This seemed to occur in the scenes with the highest amount of motion, where B-frames, in general, are least effective because there's so little inter-frame redundancy. So going beyond 3 B-frames seldom seemed to produce any benefit and often degraded the results.

Adaptive B-frames Sets the Limit

Interestingly, this is precisely the issue that enabling the configuration option Adaptive B-frames is designed to avoid. Specifically, in the files above, I encoded with Adaptive B-frames disabled, forcing the encoder to encode the file using the requested number of B-frames. However, in most cases, the optimal strategy is to enable adaptive B-frame placement and let the encoding tool decide which frames should be B-frames or P-frames. For example, when I selected 10 B-frames in Squeeze and enabled adaptive B-frame placement, the longest interval of consecutive B-frames in the file was 4, with 2 to 3 in most high-motion sequences.

Why not simply set the B-frame target at 16 and let the encoder limit as it sees fit? This is actually the recommendation made by Encoding.com in an article titled "Advanced Configuration Options for the libx264 Video Codec" (**bit.ly/encoding_h264**), where it states:

> B-frames are a core element of H.264 and are more efficient in H.264 than any previous standard. Some specific targets, such as HD-DVD and Blu-ray, have limitations on the number of consecutive B-frames. Most, however, do not; as a result, there is rarely any negative effect to setting this to the maximum (16) since x264 will, if B-adapt is used, automatically choose the best number of B-frames anyways. This parameter simply serves to limit the max number of B-frames.

Figure 3-15. The frame encoded with 3 B-frames produced the clearest image of the four.

While perhaps this is a safe approach with x264, given that B-frame values in excess of 3 have produced ugly video in some of my tests, I'm most comfortable with recommending 3, with Adaptive B-frames enabled in all instances.

More On Reference Frames

Again, if you're in a hurry, go with the recommendation to use 5 reference frames as a default, and jump over to the discussion on Slices. If you want to dig a bit deeper, read on.

Let's take a closer look at the trade-offs associated with reference frames. Briefly, a reference frame is a frame that the frame being encoded can use for redundant information. If you're encoding frame 10 in a video file, for example, it can use redundant information found in frames before (1-9) or, if it's a B-frame, after (11+) to increase encoding efficiency. You have two basic controls over reference frames: first, setting the number, and second, enabling B-frames to be used as reference frames.

What's the Right Number?

The first question is the optimal number of reference frames. Intuitively, increasing the number of reference frames will increase encoding time, because the encoder has to search through more frames for redundancies. For example, set reference frames at 1, and once the encoder finds a frame with redundancies, it's done. Set it at 16, and the encoder has to continue searching until it finds 16 or reaches its maximum search setting. How does this affect encoding time? Table 3-3 tells the tale.

By way of background, this test involved encoding a 93-second file to 1280x720x29.97 @ 800 kbps on a 3.3 GHz 12-core HP Z800 workstation using Sorenson Squeeze. All times are in min:sec format, and I extrapolated the actual encoding time to that of a one-hour file for perspective. As you can see, increasing the number of reference frames from 1 to 5 increased encoding time by 46%, while jumping from 1 to 16 more than doubled the encoding time.

	Encoding Time	Extrapolated to One Hour	Percentage Increase
1 Reference Frame	1:30	40:54	
5 Reference Frames	2:11	59:32	46%
16 Reference Frames	3:18	90:00	120%

Table 3-3. Encoding times for a 93-second file with different reference frames enabled. All times min:sec.

How did the quality compare? Again, in most test cases there wasn't a huge difference. When I saw a difference, however, most, if not all was realized with 5 reference frames. Going beyond this to 15 reference frames seemed to provide little, if any, benefit.

Figure 3-16. Five reference frames was a clear improvement over 1 reference frame and seemingly superior to 15 reference frames.

Intuitively, particularly in a video with significant motion, most of the redundancies exist in frames proximate to the encoded frame, so it makes sense that 15 reference frames wouldn't deliver significant additional value. Even in a low-motion video, the most proximate frames would still contain the most redundancy, so other than special cases like animation, it's hard to justify the extra encoding time.

B-Frames as Reference Frames/Pyramid Coding

Now that you've seen the genesis of my recommendation of 5 reference frames, let's discuss the issue of using B-frames as references. Here, there is one concern.

Specifically, some authorities relate that using B-frames as reference frames can cause instability on some playback platforms. For example, the "x264 FFmpeg Options Guide" for Linux developers (bit.ly/ffmpeg_linux), comments, "B-references get a quantizer halfway between that of a B-frame and P-frame. This setting is generally beneficial, but it increases the DPB [decoding picture buffer] size required for playback, so when encoding for hardware, disabling it may help compatibility." Given Rhozet's recommendations to always enable both reference and pyramid B-frames, this isn't a problem that Rhozet is seeing, but if you have issues with OTT or mobile device playback, disable both and see if the issues resolve.

OK, we're done with B-frames; let's move on to one other common H.264 encoding parameter you'll see in multiple encoding tools.

Encoding Slices

Some encoding tools offer the option of choosing the number of slices. What the heck are these?

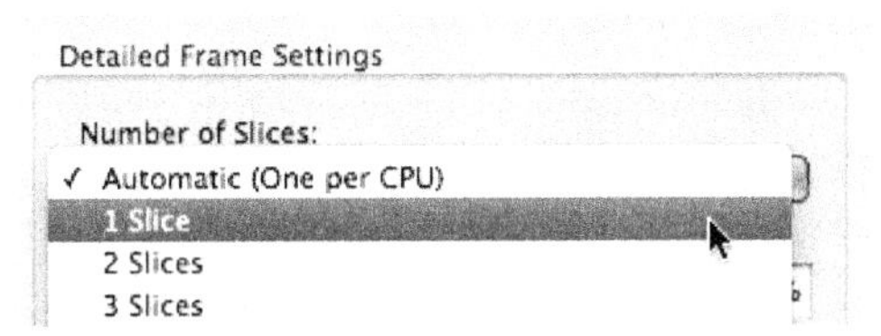

Figure 3-17. Use 1 slice for maximum quality.

Briefly, when you use multiple slices, the encoder divides each frame into multiple regions and searches for redundancies in other frames only within the respective region. This can accelerate encoding on multi-core computers, because the encoder can assign the regions to different cores. However, since redundant information may have moved to a different region between frames—say, in a panning or tilting motion—encoding with multiple slices may miss some redundancies, decreasing the overall quality of the video.

Here's a snippet from the Telestream Episode help file:

> Speed up processing by transcoding parts, slices, of the same frame in parallel. Using more slices may decrease image quality somewhat as redundancies between parts of the frame cannot be fully utilized.

This is illustrated in Figure 3-18. If this video were encoded in a single slice, movement within the quadrants would be irrelevant, and the encoder could find inter-frame redundancies irrespective of where the content is (or was) in the frame. If that distinguished rider on the left (then Governor, now Senator Mark Warner from Virginia) started on the left and moved through the video toward the right in a sequence of five or six frames, there would be significant inter-frame redundancies, which translates to very good quality.

Figure 3-18. This video frame cut into four slices.

With slices, however, the encoder can't refer to information in previous or subsequent frames that wasn't in that slice, so as the rider moves from left to right, significant inter-frame redundancies may go unrealized. For this reason, unless you're in a significant hurry, I recommend setting slices to the lowest value, which is 0 for Squeeze and 1 for Episode.

OK, these are the most common H.264 parameters enabled by the various H.264 encoding tools. In Chapter 7, I'll take a deeper dive into the H.264 encoding parameters used by a range of encoding tools, including (hopefully) one that you might be using to produce your streaming file. One last topic before we move on: Let's take a look at the comparative quality produced by different H.264 codecs.

H.264 Quality Varies by Codec

As mentioned previously, H.264 is a standard that allows multiple parties to develop their own codecs, each with their own sets of strengths and weaknesses. In order to optimize the quality of your H.264 video, you have to choose the right encoding tool, and sometimes the right codec within that encoding tool. You can see an example in Figure 3-19, which shows the same file produced to 1280x720x29.97 @ 800 kbps using three different H.264 codecs. The first shows the Apple codec, the next the x264 codec, and the last the MainConcept codec.

Figure 3-19. The same video produced with different H.264 codecs.

Obviously, the Apple codec is well behind the others in terms of quality. If Compressor is your encoding tool of choice, in Chapter 7 you'll learn how to download and use a free version of the x264 codec from within Compressor. Otherwise, simply choose another tool.

The difference between x264 and MainConcept is an interesting story. Briefly, x264 is an open-source codec released under the terms of the GNU General Public License. This means that anyone can use it, though like all other H.264 codecs, its use may be subject to royalties as discussed earlier. x264 is developed by a group of disparate developers, not a formal company. In contrast, MainConcept is a codec developed by the MainConcept subsidiary of Rovi Corporation. Although I've always found the quality produced by MainConcept and x264 neck and neck, there's a more authoritative source that disagrees.

Specifically, every year, the University of Moscow runs an extensive codec comparison (**bit.ly/umoscow**) with multiple tests. There's a free version of the report that you can download via the supplied link. Figure 3-20 shows a summary chart containing the results; the university concluded that x264 is the best by a significant margin, as it has for the past few years.

One issue I take with the University of Moscow report is that it relies primarily on mathematical comparisons like the Structural Similarity (SSIM) index and Peak Signal-to-Noise Ratio (PSNR). I use the old-fashioned approach, encoding the files and eyeballing the results. Nonetheless, the market has listened, and x264 has been adopted by multiple vendors of desktop software, including Sorenson and Telestream. X264 is also the codec used by free tools like FFMPEG and HandBrake, by cloud encoding vendors like Encoding.com and Zencoder, and by large user-generated content (UGC) sites like YouTube and Vimeo. Note that building your own encoding tool with x264 doesn't create any royalty obligations until you sell more than 100,000 units, so if you download any of these free tools, you're not likely to get a demand letter from MPEG LA.

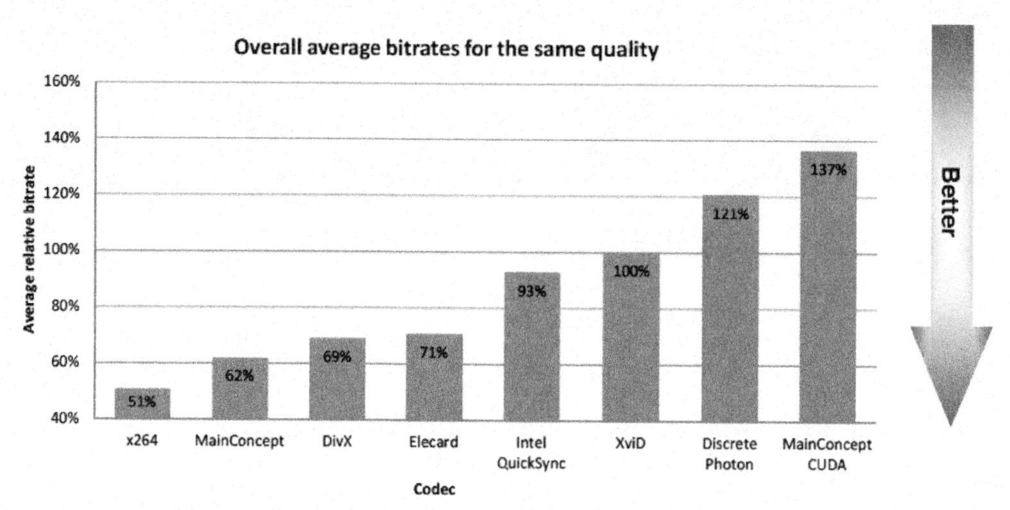

Figure 3-20. University of Moscow found x264 to be the highest-quality codec.

You should take another key point from the chart. That is, hardware-accelerated H.264 codecs like Intel QuickSync and MainConcept CUDA are clearly behind the quality curve and should be avoided unless encoding speed is more important than output quality. In this regard, the need for encoding speed has pushed multiple vendors like Elemental Technologies and Telestream to create their own H.264 codecs that allow them to create enterprise-class encoders with much faster encoding speed than can be realized from workstations encoding with off-the-shelf software encoders.

While the H.264 encoding parameters used in all H.264 encoding tools are necessarily very similar, output quality may not be. When using a tool that offers multiple codecs, like Squeeze and Episode, be sure to use the highest-quality codec available, which I'll cover in Chapter 7. When comparing enterprise-encoding tools, be sure you can assess their respective quality before buying, although all companies in this space have been around for a while and I've found output quality at this level fairly consistent.

Audio Encoding Options

If you're encoding your H.264 video into an MP4 wrapper, chances are you'll want to use AAC audio compression. As discussed above, AAC stands for Advanced Audio Coding, and the codec is defined in Part 3 of the MPEG-4 specification. AAC is the successor to the MP3 format, which is likely the most popular and famous audio codec in the world. Technically, there are three versions of AAC offered by most streaming encoding programs. These are:

- ***AAC-Low Complexity (AAC-LC).*** The most basic and most broadly compatible. In my tests, indistinguishable from HE AAC/HE AACv2.
- ***High Efficiency AAC (HE AAC-LC; 2003).*** Also called AAC+ and aacPlus.
- ***High Efficiency AACv2 (HE AAC-LCv2; 2006).*** Also called enhanced AAC+, aacPlus v2 and eAAC+.

Problems arise when working with various encoding tools because the names of the three versions usually don't match their official designations, though the differences are generally simple to decipher. For example, Figure 3-21 shows the three codec options available in Adobe Media Encoder; while the names don't precisely match, you should be able to figure things out.

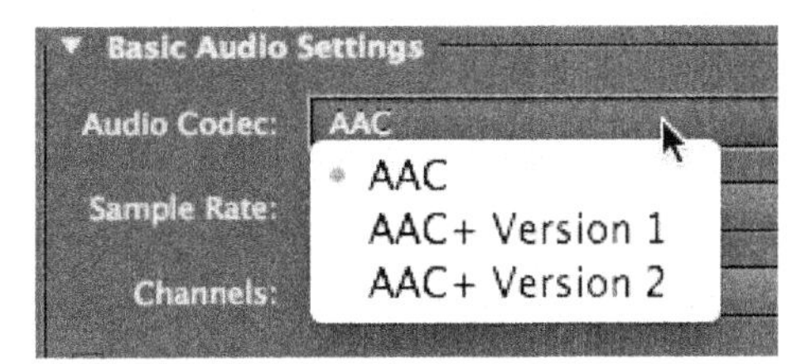

Figure 3-21. The three AAC codec options offered by Adobe Media Encoder.

Which codec should you use? My practice is to use Plain Jane AAC because it's the most broadly compatible. For example, as you'll see in Chapter 4, although decode support for all three codecs is included in the Android operating system, Google recommends using AAC for lowest-common-denominator encodes. In addition, according to their specifications, most older iPod touch and iPhones only play AAC audio, not the two higher versions. In addition, though I'm not an audiophile, I heard very little difference between the three codecs in the comparative tests I've run, so I don't have a lot of motivation to push the envelope.

As in all things compression-related, reasonable minds can differ. For example, according to Encoding.com, "the general rule for the two types of … AAC audio codecs is as follows: For low-bit-rate audio 40 kbps to 64 kbps, use HE AAC @ 44100 Hz; for high-bit-rate audio 80 kbps to 320 kbps, use AAC-LC @ 44100 Hz" (bit.ly/encoding_aac). On the other hand, I checked what a few media companies are doing and noticed that ABC, ESPN and CNN were all using AAC-LC, as was YouTube. These companies push a lot of streams and pay a lot in bandwidth costs. If they're not using the High Efficiency AAC codecs, it's either because the quality difference is minor or the compatibility risks are too great.

So, if you're creating files for multiple-platform deployment, I recommend using AAC-LC unless your tests prove that the higher-quality codecs produce noticeably better quality than AAC-LC. If you go this route, however, be sure to make sure the resulting files play on all target platforms.

In Chapter 4, I detail the playback capabilities of a number of different target platforms, including desktop, mobile and OTT. If you're customizing files for a particular platform, you can learn there which audio codecs these platforms support.

Conclusion

OK, you've learned the basics of H.264 encoding; you know all about profiles, levels and the like. In the next chapter, we'll examine how these parameters should be used when producing for the desktop when delivering via Flash and HTML5, as well as for a range of mobile and OTT devices.

Chapter 4: Configuring H.264 for Desktop, Mobile and OTT Viewers

In Chapter 1, you learned the delivery technologies available for computers, mobile and over-the-top (OTT) devices. All play H.264 files, and in Chapter 3, you learned the basic H.264 configuration options.

In this chapter, you'll learn how to configure your H.264 video for delivery to all three platforms. Along the way, you'll learn:

- The container formats and H.264 requirements for Flash, QuickTime, Silverlight and HTML5 playback
- The playback capabilities of older desktop systems, and how that should affect how you configure your files
- The container formats and H.264 configuration requirements for iOS and Android devices, as well as Windows Phone and Windows RT
- The container formats and H.264 configuration requirements for OTT platforms such as Apple TV, Roku, Boxee, Google TV and Xbox 360.

While you probably won't want to read this chapter from start to finish, I'm hoping the information contained herein will be a valuable resource for your encoding efforts going forward.

Producing H.264 for Specific Environments

As we've learned, you can encode H.264 into an .MP4 file, an .F4V file, an .FLV file, an .M4V file, an .MPG file, a .3GP file and even an .M2T or .MTS file. Fortunately, the H.264-specific parameters that you learned in Chapter 3 are the same when producing for all these wrappers, but you gotta know which wrapper to use for your target playback environment.

To be more specific, by "wrapper," I mean container format, or the format into which the file is written. As you learned in Chapter 3, you can identify a file's container format by the file extension, with .MP4 meaning the standard MPEG-4 container format, .MOV meaning the QuickTime format, and so on. As you'll learn, Flash and Silverlight are very flexible regarding the container format and will play virtually any H.264-encoded file. QuickTime is bit more persnickety, and mobile devices even more so.

Let's start at the beginning, with a quick review that I'll supplement with more format-specific data in the individual sections below. H.264 is a component of the MPEG-4 specification, and MPEG-4 Part 14 defines the media format for MPEG-4 files. Under Part 14, files with an .MP4 extension (e.g. video.mp4) contain video or audio and video, while files with an .M4A contain just audio. Since they follow the guidelines of the MPEG-4 specification, files with these extensions are said to have been produced in the .MP4 "wrapper," or MPEG-4 container format.

Apple was among the first to use the H.264 codec, encapsulating it in a .MOV or .MP4 file for desktop playback via QuickTime, which can play files produced with either extension. QuickTime can also play .M4V files produced for Apple devices like the iPod/iPhone/iPad—although you don't have to use that container format with iDevices, as they will play .MOV and .MP4 files as well. To complete the picture, Apple uses the .M4A extension for audio-only H.264 files (encoded with AAC audio).

Adobe was next to adopt H.264, and Adobe designed the Flash Player to play H.264 files encoded into the .MOV, .MP4 and .3GP formats. The Flash Player doesn't care about extensions; it looks into the file itself to determine the format and whether it's compatible. In the past, .FLV was the default extension for Flash files with the VP6 codec. Since then, Adobe created the .F4V container format to designate files produced for Flash playback using the H.264 codec. Again, however, as with iDevices, Flash will happily play files produced using the standard .MP4 wrapper.

Next into the H.264 sandbox was Microsoft, which did not create its own extension, but vowed to support most commonly used H.264 container formats, including .MP4, .FV4, .3GP and .MOV. Finally, three HTML5-compatible browsers currently play H.264 video—Apple Safari, Google Chrome, and Microsoft Internet Explorer 9—although in early 2011, Google indicated that it would remove H.264 playback from subsequent releases of Chrome (but still hasn't done so). The preferred container format for HTML5 is the standard .MP4 wrapper, since the .MOV extension might call the QuickTime player, and .F4V the Flash Player.

Before getting to environment-specific configuration options, let's look at one issue that will likely affect playback performance irrespective of whether you're using Flash, QuickTime, Silverlight or HTML5. Specifically, the moov atom issue.

Where Dat Moov Atom?

Have you ever posted a Flash or QuickTime video on the Net for progressive playback and noticed that it wouldn't start to play until it was fully downloaded? If so that likely relates to the moov atom. Briefly, the moov atom contains the index information in the encoded file that the player needs to play the file. Here's a quick summary from that Apple website at **bit.ly/moovatom**:

> A QuickTime movie file contains information about the movie, stored in a 'moov' atom—which contains one 'trak' atom for each track in the movie, a 'udat' atom for user data, and so on. This information tells QuickTime what's actually in the movie and where it's stored.
>
> QuickTime needs to load the 'moov' atom into the computer's memory in order to play a movie. When you save a self-contained fast-start movie, QuickTime puts the 'moov' atom at the front of the file [it's usually only 1 or 2 kb], followed by the movie data, arranged in chronological order. When you download the file over the Internet, the 'moov' atom arrives right away, so QuickTime can play the movie data as it comes in over the Net.
>
> If the 'moov' atom is at the end of the file, QuickTime doesn't know what's in the movie or where it's stored, so it doesn't know what to do with the movie data as it comes in, and the movie can't play until the 'moov' atom arrives at the end of the file.

Although the blog post obviously relates to QuickTime, the moov atom affects Flash playback as well. Here's a blurb from Kaourantin.net, the blog maintained by Adobe Developer Tinic Uro (**bit.ly/flashh264**):

> If you use progressive download instead of FMS [Flash Media Server], make sure that the moov atom [which is the index information in MPEG-4 files] is at the beginning of the file. Otherwise you have to wait until the file is completely downloaded before it is played back. You can use tools like qt-faststart.c written by our own Mike Melanson to fix your files so that the index is at the beginning of the file. Unfortunately our tools [Premiere and After Effects etc.] currently place the index at the end of the file so this tool might become essential for you, at least for now. We are working hard to fix this in our video tools. There is nothing we can do in the Flash Player, and iTunes/QuickTime does behave the same way.

Although I haven't experienced these playback delays personally with Silverlight or HTML5, this moov atom location issue likely affects these other formats as well.

Note that you won't encounter this problem when distributing your video via a streaming server because the server can communicate with the player and pass along any necessary playback information. This only occurs when distributing your videos via progressive download because there's no such communication between server and player.

Moving the Moov Atom

What to do if you have a file encoded with the moov atom at the back of the file? Well, you can re-encode the file selecting the Fast Start option, which places the moov atom at the start of the file, or download the QTIndexSwapper from **bit.ly/fixmoov** (Figure 4-1). It's a free Adobe Air application that analyzes the file, determines if the header is in the right place and corrects the problem if necessary.

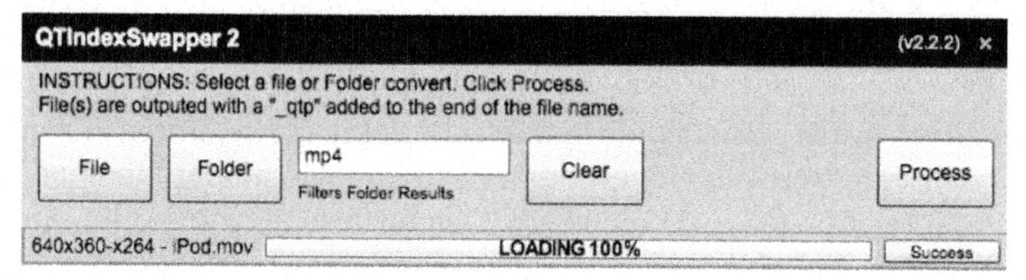

Figure 4-1. Moving the moov atom to the front of the file with QTIndexSwapper 2.

Once you install the application and get it up and running, click File or Folder to add single or multiple files, then click Process. The adjustment only takes a few moments per file, and once you re-upload your files, they should start to play after only a few moments of buffering.

OK, now that we've detailed an issue that can affect all playback environments, let's dig a bit deeper into the specific environments.

Producing for QuickTime Playback

QuickTime Player does best with .MOV and .MP4 files, so use those container formats and extensions when producing files for the QuickTime Player. I couldn't find specs for what profiles and levels the QuickTime Player will play, although through testing, I learned that version 7.6.6 of the old QuickTime player and 10.1 of the newer QuickTime player will both play H.264 video encoded with the Baseline, Main and High profiles. Note that versions prior to 7.2 could not play files encoded with the High profile, but since that version came out in July 2007 and Apple has a very aggressive upgrade strategy, it's likely that the vast majority of QuickTime viewers support the High profile.

In my tests, on an 8-core Mac Pro, the QuickTime Player played a 1920x1080p@20 Mbps file encoded using the High profile at Level 5.1 without problem. So long as your file is High profile or lower, it should play in QuickTime.

As we learned in the previous section, when you're producing a file for progressive playback, use the Fast Start option, avoiding the compressed header to maximize compatibility with other players. If you're producing for the QuickTime Streaming or Darwin servers, use the Hinted streaming option.

In terms of adaptive streaming, Apple has placed its eggs squarely in the HTTP Live Streaming (HLS) basket; while QuickTime has an adaptive streaming function, it requires a QuickTime streaming server and is considered legacy technology by this point.

Producing Flash Playback

The Flash Player supports the Baseline, Main, High, and High 10 H.264 profiles with no levels excluded. Accordingly, when you're producing H.264 video for Flash Player, you can choose the most advanced profile supported by the encoding tool, which is typically the High profile. On the audio side, Flash Player can play AAC Main, AAC Low Complexity, and AAC SBR (spectral band replication), which is otherwise known as High Efficiency AAC, or HE AAC (**adobe.ly/h264encode**).

In terms of container format, the Flash Player should be able to play files in QuickTime, MPEG-4 and 3GP container formats. Here's a blurb from Adobe developer Tinic Uro's blog (**bit.ly/adobeh264**):

> You can load and play .MP4, .M4V, .M4A, and .MOV files using the same NetStream API you use to load .FLV files now. We did not add any sort of new API in the Flash Player. All your existing video playback front ends will work as they are. As long as they do not look at the file extension, that is, though renaming the files to use the .FLV file extension might help your component. The Flash Player itself does not care about file extensions; you can feed it .TXT files for all it matters. The Flash Player always looks inside the file to determine what type of file it is.

Let's be pedantic and identify the key points he's making one by one.

- As I said, Flash Player should be compatible with QuickTime, MPEG-4 and 3GP files.
- Second, file extension doesn't matter. This is very convenient because neither of the two Flash development environments, Flash Builder or Flash Catalyst, can import .MP4, .MOV or .3GP files. However, if you change the extension of the same files to .FLV (or .F4V), they will load right in. The only exception that I've run into is when you have a compressed header, those files won't load.

As mentioned above, .F4V is the primary container format for Flash video containing H.264-encoded video. Exactly what is .F4V? It's a container format based upon the MPEG-4 specification with additional support for metadata and other properties that support playback within the Flash environment. The new container format has four types of files with four different extensions, as shown in Figure 4-2 (**bit.ly/wikiflashvideo**).

File Extension	Mime Type	Description
.f4v	video/mp4	Video for Adobe Flash Player
.f4p	video/mp4	Protected Video for Adobe Flash Player
.f4a	audio/mp4	Audio for Adobe Flash Player
.f4b	audio/mp4	Audio Book for Adobe Flash Player

Figure 4-2. MIME types for Flash-related audio and video files.

If you're producing solely for the Flash environment, .F4V is OK. However, if you're producing for multiple platforms, I recommend using the .MP4 container format, so much so that I penned an article on my website titled "I'm Saying Adios to .F4V" (**bit.ly/adios_F4V**). The pithy conclusion states:

> Long story short, the .F4V format has several disadvantages compatibility-wise, and, at least in its current iteration, no benefits that I could see. It seems as if the format was created when Adobe had designs of Flash being a long-term, vibrant distribution format for general-purpose video. Now, Adobe has ceded the general market to HTML5 while focusing on advanced gaming and premium video. ... Unless and until Adobe adds some functionality to the .F4V format that .MP4 files can't deliver, I'll avoid .F4V.

Long story short, unless you have some strong reason to use the .F4V container format, I would use .MP4.

For adaptive streaming to the Flash platform, you have multiple alternatives, including RTMP-based Dynamic Streaming and HTTP-based Dynamic Streaming, or HDS. We'll cover these alternatives in greater detail in Chapter 5.

Producing for Silverlight Playback

Silverlight should play H.264 files encoded in the .MOV, .MP4 and .F4V container formats, and can play back files encoded using the Baseline, Main and High profiles at all levels. Note the while AAC-LC is supported fully, files produced with both High Efficiency audio codecs will only play back at half fidelity, so 48 kHz files will play back at 24 KHz (**bit.ly/Silverlight_specs**). Since .MP4 files have the broadest playback compatibility, I typically encode files for Silverlight using the .MP4 wrapper.

In terms of adaptive streaming, Smooth Streaming is your only alternative for distribution to Silverlight. I'll cover that in Chapter 5.

Producing for HTML5 Playback

As mentioned, when producing for HTML5, it's best to produce an .MP4 file, because .MOV or .F4V files could trigger playback using either the QuickTime or Flash Player. The H.264 playback capabilities of the three HTML5-compatible browsers—Apple Safari, Google Chrome (for the time being) and Microsoft Internet Explorer 9 and higher—is obviously up to the browser vendor; as far as I could find, none of these vendors have published specifications that detail profile and level-type compatibility.

In my tests, all three browsers played a 1080p file encoded using the High profile, with stereo HE AACv2 audio. So long as you produce a file that doesn't exceed these parameters, you shouldn't run into any playback problems with any H.264-compatible HTML5 browser.

There currently is no browser-based alternative for adaptive streaming, with the Dynamic Adaptive Streaming over HTTP (DASH) in the works and predicted to coalesce into a useful technology in 2013. I discuss DASH in detail in Chapter 5.

Changing the File Extension on an H.264 File

I covered this point above when talking about Flash, but let's quickly reiterate. Suppose you produced an H.264 file in the .MOV format and want to deploy it in the Flash or even Silverlight environment. If you're hand-coding your player, you can keep the file in the .MOV format, and it will play just fine.

On the other hand, tools like Adobe Flash Catalyst or Flash Builder won't import a .MOV or .MP4 file—only .FLV or .F4V—so you have to change the file name so the program will recognize the file. There are no absolutes in video production, only tendencies. However, so long as the file that you've produced doesn't have a compressed header, both Flash tools should be able to import the file after you change the extension to .FLV or .F4V.

Configuring Your H.264 Streams

Now we know the technical aspects of encoding H.264 for desktop playback; let's look at the factors to consider when choosing the resolution of your video. As you recall, back in Chapter 2, we discussed bits per pixel and the qualitative aspect of configuring your video streams. Now we'll take a brief look at the playback capabilities of some legacy systems to get a feel for the types of configurations that will play on these computers.

Flash Playback Statistics

Let's start with a quick look at how H.264-formatted Flash files play back on a variety of desktop and mobile Windows and Mac computers. To perform this test, I encoded a single 29.97 fps 720p test file into six different configurations, starting at 320x240 and ending at 720p. Then I tested playback on nine different computers, playing the file for about two minutes and recording CPU utilization figures from Windows Task Manager or Mac Activity Monitor. You can check out the files yourself at **bit.ly/flashplayback**. I used Mozilla Firefox to play all files on Windows, and Safari on the Mac, checking that each computer had the latest version of both the browser and Flash Player.

Windows Tests

I focused my testing on the slower computers that I have in my office or have access to. On my 24-core HP Z800 or 16-core Mac Pro workstations, I can play multiple streams of 1080p video without a hitch. But few, if any, publishers focus on computers like these since these are the best case. So instead, I tested the older workstations in my office—some that I've had since 2003—to get a sense of how the various test configurations would play on these older computers, which make up a much larger percentage of the installed base that producers do care about.

For Windows desktop computers, I tested a computer with the last Pentium 4 processor (the HP xw4100, circa 2003), a computer with the first dual-core CPU (the HP xw4300, circa 2006), and a computer with the first Core 2 Duo processor (the Dell Precision 390, circa 2006). As you'll see in Table 4-1, the Precision 390 played video up to 720p without much of a hiccup, so it didn't make sense to test any faster, more modern desktops. All three Windows desktops that I tested were configured with NVIDIA graphics cards that accelerated the playback of H.264-encoded Flash, a nice advantage.

Windows	Notebooks			Desktops		
Year Produced	2004	2007	2008	2003	2006	2006
Computer	Dell Latitude D800	HP Mobile 8720p	Acer Aspire One	HP xw4100	HP xw4300	Dell Precision 390
CPU	1.6 GHz Pentium M	2.2 GHz Core 2 Duo	1.6 GHz Atom	3 GHz Pentium 4	3.4 GHz Pentium D	3.93 GHz Core 2 Duo
GPU	NVIDIA GeForce 4200	NVIDIA Quadro FX 1600M	Intel 945 Express	NVIDIA Quadro4 380XGL	NVIDIA Quadro FX 3450	NVIDIA Quadro FX 3500
320x240 File - CPU%	38	12	37	21	4	2
480x360 File - CPU%	63	14	37	28	9	5
640x360 File - CPU%	75	14	35	36	12	6
640x480 File - CPU%	79	17	57	40	15	8
848x480 File - CPU%	NA	16	NA	49	18	10
720p File - CPU%	NA	27	NA	78	41	24

Table 4-1. The percentage of CPU required to play back video at these test configurations on these computers.

Notebooks are a different story, primarily because there are so many different configurations—some built for portability, some for power. Here, I tested a Dell Latitude D800 Pentium M-based notebook from 2004, an HP 8720p mobile workstation from 2007 (with a 2.2 GHz Core 2 Duo) and a netbook, specifically the Acer Aspire One with a 1.6 GHz Intel Atom CPU. I present the results in Table 4-1.

Let's start with the notebooks. My rule of thumb is that if more than 50-60% of CPU is required for video playback, other operations become sluggish, and dropped frames or audio stoppages become likely. Numbers in excess of 70% are big red flags signaling likely playback issues. In this regard, the Latitude D800 struggled to play the 640x360 and 640x480 configuration, and shut down completely on files with a larger configuration. However, both the HP 8710p and Acer Aspire One played both of these files without significant problems.

On the desktop side of the equation, all three computers successfully played files up to 848x480 resolution, though the Pentium 4-based xw4100 had problems with the 720p file. Now let's look at the Mac-based tests.

Mac Tests

I tested three Macs: an oldie-but-still-goodie Dual G5 PowerMac, my daughter's 2.0 GHz Core 2 Duo-based iMac, and a 3.06 GHz Core 2 Duo-based MacBook Pro. Here we see that all three computers sailed through all tests up to and including the 848x480 tests, though the PowerPC started to drop frames at 720p.

Macintosh			
Year Produced	2005	2007	2009
Computer	Dual G5 PowerMac	iMac	MacBook Pro
CPU	Dual .27 GHz Power PC	2 GHz Core 2 Duo	3.06 GHz Core 2 Duo
GPU	NVIDIA GeForce 6800 Ultra	ATI Radeon X160	NVIDIA GeFrce 9600M
320x240 File - CPU%	23	22	17
480x360 File - CPU%	27	33	17
640x360 File - CPU%	31	30	28
640x480 File - CPU%	37	33	24
848x480 File - CPU%	43	34	39
720p File - CPU%	73	53	56

Table 4-2. CPU required to play back video at these test configurations.

If you scan through both sets of results, you'll see that all computers except the old Dell notebook played 640x360 video at less than 40% of CPU, while all but the Dell played the 848x480 stream at less than 60% of CPU. By far the most widely used resolution on the web today is 640x360; the fact that files at this resolution play well on computers dating back to 2003 is probably a consideration.

At the other end of the spectrum, if you're producing only a single stream of video on your site, resolutions like 720p and above will not play well on a large number of older computers. I'm guessing this is one of the reasons that Apple renders all of the product advertisements available on its website at 848x480.

Generalizing These Results

In the interest of full and fair disclosure, I'll admit that I didn't update these test results from when I first produced them back in 2011. How might the Flash-related numbers have changed and how do these results generalize over to HTML5? Let's have a look.

Back in 2010, I wrote an article titled "Flash Player: CPU Hog or Hot Tamale? It Depends" (**bit.ly/CPUhog**), where I compared the CPU utilization of Flash Playback vs. HTML5 using a 720p (1280x720x29.97) YouTube file. I produced the tests because there was a generally accepted sense that Flash consumed much more CPU resources during playback than HTML5. It seemed simple enough to test, so I did, and this article is the most popular ever on my website, with a mention in *The New York Times* (**bit.ly/jans15**) and more than 228,000 page views to date.

How did I test? Although YouTube defaults to Flash playback, there's an HTML5 playback mode that you can opt into (**youtube.com/html5**). You can read about the precise testing procedure I used in the article, but basically, I played the files back using Flash and HTML5 and measured CPU utilization using Activity Monitor on the Mac and Task Manager on Windows.

Back then, I found that performance was very browser- and platform-dependent. For example, on my 3.06 GHz MacBook Pro with a Core2Duo CPU, playback in HTML5 in Safari consumed 12% of CPU, while Flash consumed 32%. In contrast, on the same computer, Chrome consumed 49.79% with Flash, and 49.89% with HTML5. Note that in these tests, WebM wasn't offered on YouTube so Firefox, which can't play back H.264 via HTML5, had no results.

On my Windows notebook of that time (since retired), Flash was much more efficient than HTML in Chrome, though not as efficient as Firefox playing back Flash. Safari was also more efficient when playing back Flash in Windows, but the browser couldn't play YouTube's HTML5-based offering. Internet Explorer 8 was the least efficient, and since it wasn't HTML5-compatible, it couldn't run any HMTL5-related tests.

The bottom line was that on the Mac platform, Apple, with its superior systems knowledge, was much more efficient than any other vendor in both Flash and HTML5 playback. Otherwise, with Chrome, the only other browser that could complete both tests, Flash and HTML5 were neck and neck in CPU utilization. On the Windows platform, Flash was much more efficient than HTML5 on the only browser that could complete both tests, which again was Chrome.

MacBook Pro	Chrome	Firefox	Safari	
Flash	50%	42 %	32%	
HTML5	50%	NA	12%	
Difference	-0.20%		159%	
HP 8710p	**Chrome**	**Firefox**	**Safari**	**IE 8**
Flash	11%	6%	7%	15
HTML5	26%	NA	DNP	NA
Difference	-58%			

Table 4-3. A summary of my CPU utilization tests from 2010.

How have things changed since then? Just so you wouldn't think I'm a total slacker, I redid my tests from 2010, producing the results shown in Table 4-4. Note that I turned in my Core2Duo-based HP 8710p for an 8-core HP Elitebook , so in no way should you compare the results of the older Mac with those of the newer HP.

MacBook Pro	IE9	Chrome	Firefox	Safari
Flash		29%	29%	25%
HTML5		21%	27%	12%
Difference		38%	6%	109%
HP Elitebook 8760w	**IE9**	**Chrome**	**Firefox**	**Safari**
Flash	9%	10%	6%	11%
HTML5	6%	11%	6%	11%
Difference	51%	-4%	6%	-2%

Table 4-4. Updated CPU utilization tests from 2013.

What do these tests show? First, that Flash has gotten much more efficient on the Mac—which, to be clear, was the same system that I tested back in 2010. However, in all cases, HTML5 is more efficient than Flash—significantly so with Chrome and particularly Safari, just like last time. Note that I tested Firefox with WebM playback from YouTube, not H.264. On the Elitebook 8760w, the Flash/HTML5 numbers were similar except for IE9, where HTML5 was about 51% more efficient.

There are a couple of takeaways from this data. First, Flash's CPU utilization has gotten more efficient since my 2011 tests, particularly on the Mac. Second, you can assume that HTML5 playback will be either the same, or more efficient, than Flash, and can figure your HTML5 direct files accordingly.

OK, let's transition over to the mobile side and see how to configure files for playback on the iOS, Android and Windows Phone platforms.

Distributing to Mobile Platforms

According to NetMarketShare, as of January 11, 2013, iOS held an overall 60.13% share in the mobile/tablet operating system market, compared with 24.6% for Android. For obvious reasons, I'll discuss these two in order. While Java ME is third at 10.18%, this OS is typically used on inexpensive "dumb" phones, not smartphones or tablets, so I'll skip it, as well as BlackBerry and Symbian. Although Windows Phone is in sixth place with only 1.05% share, it's garnering lots of attention, so I'll cover that last.

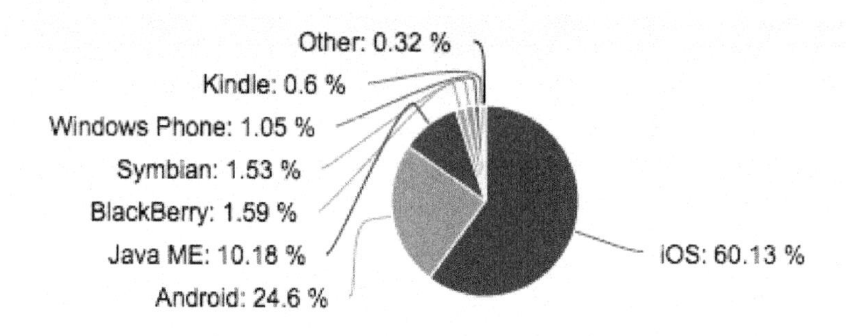

Figure 4-3. Mobile/tablet operating system market share from NetMarketShare as of January 11, 2013.

Distributing to iOS Devices

What do we know about iOS devices? Back in Chapter 1 we learned that they all play H.264-encoded video, whether delivered via cable from iTunes or, for connected devices, via HTML5 in the browser. The only browser-based alternative for adaptive streaming is Apple's own HTTP Live Streaming (HLS), though you can also deliver single files or adaptive streams using Flash technology in an AIR application.

How should you configure your streams for the various devices? Table 4-5 summarizes the playback specifications of all Apple iDevices. Note that all can play video delivered using the MPEG-4 (.MP4), QuickTime (.MOV) and .M4V container formats, and connected devices can decode video delivered using MPEG-2 transport streams (.TS).

I will say that Apple's specs don't always conform to reality, but at least they're conservative. For example, while the spec sheet says the original iPad can only play H.264 video encoded using the Main profile, I've tested 720 video encoded with the High profile and it plays fine. Although my iPhone 4S shouldn't play video with audio encoded using HE AACv2, it plays without problem, and although Apple's maximum audio data rate is 160 kbps, virtually all music videos shipped on iTunes use 256 kbps.

Beyond these mysteries, note that the specs sheets are confusing because they present separate specs for "Audio Playback" and "TV and Video." For example, for the iPhone 4, audio specs include "AAC (8 to 320 kbps), Protected AAC (from iTunes Store), HE AAC" and others. However, in the TV and video section, the spec states "H.264 video up to 720p, 30 frames per second, Main profile Level 3.1 with AAC-LC audio up to 160 kbps, 48kHz, stereo audio." From this, I infer that the device can play standalone audio files up to 320 kbps, but only 160 kbps when delivering audio with the video. However, I've tested multiple Apple devices, and they all can play at least AAC audio at 256 kbps when delivered with video.

	Original iPod (to-5g)	iPod nano/ classic/iPod touch/iPhone to V4	iPhone 4/ iPod touch 4/ iPad 1	iPod Touch 5	iPhone 4S/5 iPad 2/New iPad
Video Codec	H.264	H.264	H.264	H.264	H.264
Profile/Level	Baseline to Level 1.3	Baseline to Level 3.0	Main to Level 3.1	Main to Level 4.1	High to Level 4.1
Max Video Data Rate	768 kbps	2.5 Mbps	14 Mbps	50 Mbps	62.5 Mbps
Max Video Resolution	320x240	640x480	720p	1080p	1080p
Frame Rate	30 fps	30 fps	30 fps	30 fps	30 fps
Audio Codec	AAC-LC	AAC-LC	AAC-LC	AAC-LC	AAC-LC
Max Audio Data Rate	160 kbps	160 kbps	160 kbps	160 kbps	160 kbps
Audio Parameters	48 kHz, stereo	48 kHz, stereo	48 kHz, stereo	48 kHz, stereo	48 kHz, stereo

Table 4-5. Specifications for various iDevices.

Still, when you're distributing to these devices via the web or iTunes, you should conform your streams to the specs shown in Table 4-5, which shows five levels of playback capabilities:

- 320x240 Baseline profile for pre-5G iPods
- 640x480 Baseline profile for later iPods and iPhone and iPod touch devices before 4G
- 720p Main profile for iPhone 4, 4G iPod touch and iPad 1
- 1080p Main profile for the iPod touch 5
- 1080p High profile for the iPhone 4s/5 and iPad 2 and New iPad.

Apple has done a great job defining how to reach these targets in Technical Note TN2444, titled "Best Practices for Creating and Deploying HTTP Live Streaming Media for the iPhone and iPad" (**bit.ly/HTTPLive**). We looked at an abbreviated version in Chapter 2, which lacked information about the profile and stream compatibility. Now that you know what profiles are and the playback capabilities of each device, you're ready for the Full Monty, which is shown in Table 4-6.

For the record, I'll note that there are some differences between Apple's table in Table 4-6 and mine in Table 4-5. For example, mine shows the iPod touch 4 as compatible with streams of up to 720p, which I've personally tested and verified on the Apple spec sheet before penning this chapter. I also verified that the iPad 2 specifications claim compatibility of up to the High profile, which I copied and pasted into a Google spreadsheet that you can access at **bit.ly/idevice_specs**, and you can view the specs yourself at **apple.com/ipad/ipad-2/specs.html**.

16:9 Aspect Ratio	Dimensions	Frame Rate	Total Bit Rate	Video Bit Rate	Audio Bit Rate	Audio Sample Rate	Keyframe**	Restrict Profile to:	Works on iPod Touch Gens 2, 3, 4	Works on iPhone 3G, 3GS, 4	Works on iPad 1, 2	Works on New iPad
CELL	480x320	na	64	na	64	44.1	na	na	*	*	*	*
CELL	416x234	10 to 12	264	200	64	44.1	30 to 36	Baseline, 3.0	*	*	*	*
CELL	480x270	12 to 15	464	400	64	44.1	36 to 45	Baseline, 3.0	*	*	*	*
WIFI	640x360	29.97	664	600	64	44.1	90	Baseline, 3.0	*	*	*	*
WIFI	640x360	29.97	1264	1200	64	44.1	90	Baseline, 3.1			*	*
WIFI	960x540	29.97	1864	1800	64	44.1	90	Main, 3.1			*	*
WIFI	960x540	29.97	2564	2500	64	44.1	90	Main, 3.1			*	*
WIFI	1280x720	29.97	4564	4500	64	44.1	90	Main, 3.1			*	*
WIFI	1280x720	29.97	6564	6500	64	44.1	90	Main, 3.1			*	*
WIFI	1920x1080	29.97	8564	8500	64	44.1	90	High, 4.0				*

Table 4-6. 16:9 recommendations from Apple Technical Note TN2444.

Still, the recommendations in Table 4-6 represent a good starting point for adaptive streaming, which you'll learn to customize in Chapter 5. If you're creating a single stream for web delivery, you can make an informed decision about the configuration necessary to play on the bulk of your most relevant target devices. You'll learn about encoding for iTunes in Chapter 8.

Encoding for Android

What do we know about Android devices from Chapter 1? Again, all Android devices should be able to play H.264-encoded video delivered via HTML5. Though there's a smattering of legacy Flash support in older Android devices, none of the new devices come with the Flash Player, so that's not an option.

For adaptive streaming, HLS is an imperfect choice, available only on Android 3.1 devices and higher, which is still well under 40% of the total installed base. There are also technical issues like aspect ratio problems when switching from landscape to portrait modes, as well as crashing, as reported by my colleague Dan Rayburn (**bit.ly/Rayburn_android**). The optimal approach for adaptive streaming appears to be an Android AIR application distributing Flash video.

> **Tip:** *You can read more about the optimal way to support the Android market in an interview with Jeroen Wijering, the developer of the JW Player, at* **bit.ly/Wijering**.

How should you configure your H.264-encoded video? Unfortunately, distributing to Android devices is more complicated than to Apple devices for several reasons. First, there are multiple hardware manufacturers with multiple devices, with vague or no references to H.264 playback capabilities. For example, my daughter has an HTC Rhyme; the spec sheet on the HTC website has no mention of H.264 whatsoever.

	SD (Low quality)	SD (High quality)	HD (Not available on all devices)
Video codec	H.264 Baseline Profile	H.264 Baseline Profile	H.264 Baseline Profile
Video resolution	176 x 144 px	480 x 360 px	1280 x 720 px
Video frame rate	12 fps	30 fps	30 fps
Video bitrate	56 Kbps	500 Kbps	2 Mbps
Audio codec	AAC-LC	AAC-LC	AAC-LC
Audio channels	1 (mono)	2 (stereo)	2 (stereo)
Audio bitrate	24 Kbps	128 Kbps	192 Kbps

Table 4-7. Google's "Examples of supported video encoding parameters."

The number and diversity of devices, along with the lack of H.264 playback information, makes creating a table like that shown in Table 4-5 an impossible task. Instead, I'll point you to the Android Supported Media Formats page (**bit.ly/androidvideospecs**), where Google details the software-playback capabilities of the Android operating system itself and provides the anemic recommendations shown in Table 4-7. Note that Android devices can play video delivered in the MPEG-4 container format (.MP4) and in MPEG-2 transport streams (.TS).

One would guess that the hardware playback capabilities of most current Android tablets and smartphones far exceed these recommendations. For example, on my Toshiba Thrive tablet, I've played 720p video encoded using the High profile with no problem (although the Toshiba website similarly failed to provide H.264 playback information).

Encoding for Windows Phone

As you probably know, Windows Phone is the latest mobile operating system from Microsoft, now up to version 8. According to Wikipedia, you can't upgrade Windows 7 phones to version 8; you have to buy them with Windows 8 installed.

In Chapter 1, we learned that Windows Phones can play H.264-encoded video delivered via HTML5, and use Smooth Streaming for adaptive streaming. Note that while Windows RT will support Flash (and AIR by mid-2013), Windows phones do not currently support Flash and support for Windows Phone is not on Adobe's Flash Technology Roadmap (**bit.ly/Flash_roadmap**). While Microsoft has its own application development environment and distribution platform, you'll have to use Smooth Streaming; Flash is not an option.

Microsoft's phones are offered by a number of vendors (**bit.ly/win8_fones**) including Nokia, HTC and Samsung. While Microsoft has done a nice job documenting the capabilities of its phones by Qualcomm processor, it's tough to map the processors that Microsoft identifies to those actually used in the phones. I contacted Microsoft for more information and got no help. So, while I'll share the capabilities that Microsoft documents, I can't point you to phones to which they actually relate. I would assume that these are minimum specs and that most phones are more powerful, but you know what they say about assumptions.

Anyway, you can find the specs at **bit.ly/winfone_vidspec**, which I've summarized in Table 4-8. As you can see, there are three categories of phones: the baseline version driven by the Qualcomm 7x27a CPU, and higher-end devices driven by the Qualcomm 8x50 and 8x55 CPUs.

Regarding adaptive streaming, here's an important note from the Microsoft website:

> Streaming Media Element (SSME) allows for dynamic resolution changes. This option is only supported on 8x55 based devices. The 8x50 and 7x27a processors do not support this feature. Only a single bit rate and resolution can be used for playing back video content on those devices by using smooth streaming.

	Qualcomm 7x27a Processor		**Qualcomm 8x50 Processor**	**Qualcomm 8x55 Processor**
Video Codec	H.264		H.264	H.264
Profile/Level	High		High	High
Level	CAVLC – 2.0	CABAC – 1.3	3.0	3.1
Max Video Data Rate	768 kbps	2 Mbps	10 Mbps	10 Mbps
Max Video Resolution	800x480	800x480	720x480@30 fps 720x576@25 fps	1280x720
Frame Rate	30	30	25/30	30
Max Peak Video Data Rate	4 Mbps	4 Mbps	27 Mbps	27 Mbps
Audio Codec	HE AAC v2	HE AAC v2	HE AAC v2	HE AAC v2
Max Audio Data Rate	320 kbps	320 kbps	320 kbps	320 kbps
Audio Params	48 kHz, stereo	48 kHz, stereo	48 kHz, stereo	48 kHz, stereo
Smooth Streaming	Single resolution and bitrate	Single resolution and bitrate	Single resolution and bitrate	Multiple resolutions and bitrates

Table 4-8. Summary of Windows Phone playback capabilities.

If you're encoding for adaptive streaming and need to support these devices, it appears that the only real parameter that you can adjust is the frame rate. This will make it exceedingly tough to produce one set of adaptive files that works well for both mobile and desktop viewers.

Now let's transition over to configuring your files for delivery to OTT devices.

Over-the-Top (OTT) Devices

The last category that we'll cover is over-the-top (OTT) devices like Apple TV, Roku and Xbox 360. By way of background, each device has its own channel-based delivery scheme, so the concept of HTML5, while perhaps involved, is not particularly relevant since most producers aren't seeking to deliver in a browser. What matters are the specs of the files that the device can decode, and the adaptive streaming technology or technologies that they support.

Let's start with Apple TV because we've covered most of the details already.

Apple TV

According to the Apple spec sheet (**apple.com/appletv/specs.html**), Apple TV devices can play "H.264 video up to 1080p, 30 frames per second, High or Main Profile level 4.0 or lower, Baseline profile level 3.0 or lower with AAC-LC audio up to 160 Kbps per channel, 48kHz, stereo audio in .M4V, .MP4, and .MOV file formats." In the Audio Formats section, the specs relate that these devices can play "HE AAC (V1), AAC (16 to 320 Kbps), protected AAC (from iTunes Store)" plus MP3 and other formats.

The only technology that you can use to adaptively stream to Apple TV devices is HLS.

Boxee

If you visit the official Boxee spec sheet (**bit.ly/Boxee_vidspec**), you'll learn that the Boxee Box can play H.264 video delivered in a range of container formats, including MPEG-4 (.MP4) and MPEG-2 transport streams (.TS). If you visit the company's forum (**bit.ly/boxee_guide**), you'll see encoding instructions for several encoding tools, including HandBrake, and you can download presets for both. I downloaded the HandBrake presets and encoded a file using the 1080p version. I then analyzed the result and found that the preset used the High profile with AAC audio at 224 kbps. While it's kind of a clumsy way to figure out which encoding parameters to use, it's the best I could do.

For adaptive streaming, Boxee uses HLS and you can find instructions at **bit.ly/boxee_hls**. Note that according to the Boxee support boards (**bit.ly/boxee_hlsapp**), HLS only works within an application, not in the browser.

Google TV

Google does a nice job detailing the playback capabilities of its device at **bit.ly/googletv_vidspec**, which indicates that these devices can play H.264 video using the High Profile at up to Level 4.1, which includes 1080p video encoded at up to 62.5 Mbps. Google TV plays HLS and Smooth Streaming files, and you can read Google's recommendations on HLS encoding at **bit.ly/googletv_hls**.

Dimensions	Total Bitrate	Video Bitrate	Encoding
640x360	640	600	HiP, 4.1
640x360	1240	1200	HiP, 4.1
960x540	1840	1800	HiP, 4.1
1280x720	2640	2500	HiP, 4.1
1280x720	4640	4500	HiP, 4.1
1920x1080	6040	6000	HiP, 4.1
1920x1080	8196	8156	HiP, 4.1

Table 4-9. Google TV's OTT recommendations.

As you can see in Table 4-9, Google TV adapted its stream recommendations from Apple TN2444, though ignoring the lowest-quality grouping and recommending the High profile for all streams. This reflects the reality that most OTT devices connect via relatively stable connections, and don't need the ultra-low bit rates required for mobile and some computer viewers.

Roku

Roku's comprehensive Encoding Guide (**bit.ly/roku_guide**) details that the Roku box can play 1080p video encoded using the High profile at up to level 4.0, which should translate to a data rate of 50 Mbps. However, the Encoding Guide recommends an average streaming bit rate of 384 kbps to 1.6 Mbps for SD video and 1.6 to 3.2 Mbps for HD video.

	H.264 SD	H.264 HD
Aspect Ratio	4:3	16:9
Resolution	Up to 720x480	720p—1080p
Container	.mp4, .mov, .m4v, HLS: m3u8 and .ts	.mp4, .mov, .m4v, HLS: m3u8 and .ts
Frame Rate	23.976 or 29.976 fps	23.976 or 29.976 fps
Codec	H.264	H.264
Bitrate Control	Constrained VBR	Constrained VBR
Peak Video Bitrate	1.5x	1.5x
Average Streaming Video Bitrate	384 Kbps—1.6 Mbps	1.6Mbps—3.2Mbps
Average USB Video Bitrate	384 Kbps—9 Mbps	384 Kbps—9 Mbps
Keyframe Interval	Under 10 seconds	Under 10 seconds
Audio Codec	AAC LC (CBR)	AAC LC (CBR)
Audio Bitrate	128--256 Kbps	32—256 Kbps
Audio Sample Size	16-Bit	16-Bit
Audio Channels	Stereo	Stereo

Table 4-10. Excerpts from Roku's recommendations.

Since the specs for video delivered via the USB port is 8 Mbps, the low recommendations for streaming probably relates more to concerns over deliverability than the ability to play back higher-bit-rate files. I've tested files encoded at 8 Mbps over the USB, and they played normally.

Interestingly, the specs recommend a keyframe interval of less than 10 seconds and a peak video bit rate of 1.5x the average. Audio should be AAC-LC, although in my tests, the Roku played HE AACv2 with no problem. You can deliver in .MP4, .MOV, or .M4V container formats. I've excerpted some of the specs from the Roku guide and show them in Table 4-10.

Even though Roku supports multiple adaptive specs, its guide (**bit.ly/roku_guide**) makes it clear that HLS is the preferred technique. The guide also identifies the Wowza Media Server as a "very popular, budget-minded choice in the HLS field," with a useful guide to getting up and running with the Roku Streaming Player.

Xbox 360

According to **bit.ly/xbox_vidspec**, Microsoft's game platform will play H.264 video encoded using the High profile to Level 4.1. It recommends a maximum bit rate of 10 Mbps. Audio should be AAC-LC with no bit rate restrictions, and you can deliver the files as .MP4, .M4V, .MP4V, .MOV or .AVI.

I couldn't find any documentation of the Xbox 360's Smooth Streaming playback capabilities. I'm assuming that the box can play any stream that the desktop Silverlight player can play, but that's only an assumption.

Conclusion

OK, now we know how to configure our files for single-file streaming to computers, mobile and OTT devices. Next up is adaptive streaming to the same motley crew.

Chapter 5: Adaptive Streaming

Adaptive streaming technologies make multiple video streams available to your viewers and dynamically switch streams to adapt to changing connection speeds, CPU utilization and other heuristics. Adaptive streaming produces the best possible experience for viewers watching on powerful playback stations over high-bandwidth connections, while delivering a lower-quality, yet still optimum experience for viewers on low-power devices with spotty connections. If streaming video is mission-critical to your organization, you need to be thinking about implementing some form of adaptive streaming.

Between the chapter title and the compelling lead paragraph, it's probably no surprise that this chapter covers adaptive streaming. Here's what you will learn in this chapter:

- How adaptive streaming works
- Available technologies and where they play
- Factors to consider when choosing an adaptive streaming technology
- Technology alternatives for supporting multiple adaptive streaming technologies (e.g. transmuxing)
- How to determine how many streams to offer and how to configure the streams
- How to encode streams to ensure smooth, artifact-free stream switching
- Technology-specific considerations for Dynamic Streaming, HTTP Live Streaming, DASH and Silverlight
- Factors to consider when configuring streams for multiple platforms, including desktop, mobile and over-the-top (OTT) devices.

Technology Overview

Figure 5-1 is a diagram from an Inlet Technologies white paper titled "Powering Smooth Streaming with Inlet Technologies" that shows the multiple components of an adaptive streaming system—in this case, a system using Inlet's Spinnaker encoder to produce the streams from a live event feed and delivering via Microsoft's Smooth Streaming technology. Note that Inlet was purchased by Cisco in 2011, so you can't buy a Spinnaker encoder any longer, but the image is so useful that I decided to use it anyway. Although the diagram details a live event, you can use all adaptive streaming technologies to deliver on-demand video files as well as live.

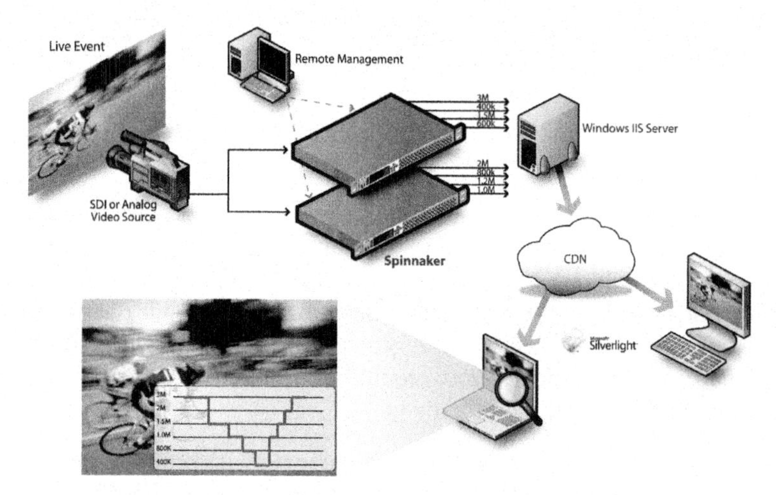

Figure 5-1. A diagram showing adaptive streaming of a live event.

The components are clear in the picture. You encode multiple streams using either a hardware or software encoder, and then upload the streams to a server—which can be a streaming server as shown in Figure 5-1, but can also be just a Plain Jane HTTP web server. Then, the server makes the streams available to the various players over the open IP network and the adaptive streaming system adapts the stream to changing conditions during playback. That's what you see on the bottom left of Figure 5-1, with the data rate graph stepping the viewer through the various streams.

How does the system know when to change streams? This depends upon multiple factors, which can include:

- ***Buffer conditions.*** Each player has a playback buffer—say 2 seconds of video data—that ensures that if the connection stops for intermittent periods, playback will continue. An adaptive streaming technology can monitor the buffer, switching to higher-bandwidth streams when the buffer stays full for a certain duration (indicating a robust connection) or drop to a lower-bandwidth streams when the buffer drops below certain levels.

- ***Client bandwidth.*** For technologies that monitor effective client bandwidth, the system can adjust the stream to match the effective bandwidth.

- ***CPU use.*** Systems can monitor dropped frames, which are an indication of the CPU power available for video decode. For example, even if the buffer were full and bandwidth sufficient for a higher-quality stream, the system wouldn't switch if the player were currently dropping frames, since a higher-bandwidth, higher-quality stream requires more CPU to decode and would cause more dropped frames.

- ***Playback window size.*** Some systems monitor playback size to determine whether to switch to a higher-quality stream. For example, if the video is being played back in a 640x360 window—even if buffer, bandwidth and CPU status indicate the ability to retrieve and playback a higher-bandwidth stream—the system wouldn't switch to a 720p stream because the viewer wouldn't notice the quality difference within the 640x360 viewing window, essentially wasting the additional bandwidth. If the viewer switched to full screen, the system would then switch to the higher-quality stream.

Where does the switching logic reside? On the player, since that's the only program actually running on the viewing station.

What happens when the player determines that it's time to switch streams? This depends upon the type of system. In a server-driven system like Adobe's RTMP-based Dynamic Streaming, the player requests a different stream from the server, and the server is in charge of delivering the higher- or lower-quality stream.

In HTTP-based systems like Apple's HTTP Live Streaming (which also does on-demand), the system depends upon multiple index files that reside on an HTTP server. With HTTP Live Streaming, the index file has an .M3U8 extension, as shown in Figure 5-2. When the viewer clicks on the web page link to play the video, the link points the player to the index file, which the player uses to find the right files to play.

Some HTTP-based adaptive technologies split their files into 2 to 10-second chunks, which are the video chunks on the right in Figure 5-2 (Hi_01.ts, Hi_02.ts, etc.). Others divide longer files into identifiable segments that can be separately retrieved for playback.

The manifest file identifies the available file chunks or segment locations, and their addresses on the HTTP server. The player might start out playing the lowest-quality stream (Low_01.ts), but if it determines that it can handle a higher-quality stream, it would check the manifest file for the next chunk of a higher-quality stream (say, Mid_02.ts) and retrieve and play that chunk. In an on-demand presentation, the manifest file is static, while in a live event, the manifest file is continually updated with the identity and location of all streams and chunks of streams.

For the following discussions of technology alternatives, it's useful to identify the common components of all adaptive streaming technologies.

- ***Encoder.*** The encoder can be live or on-demand and can be proprietary to the adaptive streaming system or third party, but some tool must encode the files to be distributed.

- ***Server.*** The server can be a standard HTTP web server or proprietary RTMP or RTSP streaming server, but the video files must reside somewhere (more on RTMP and HTTP below).

- ***Player.*** Each current adaptive streaming system has a proprietary player, or similar component. Unless that player or component is installed on the target viewer's systems, he or she can't play the files.

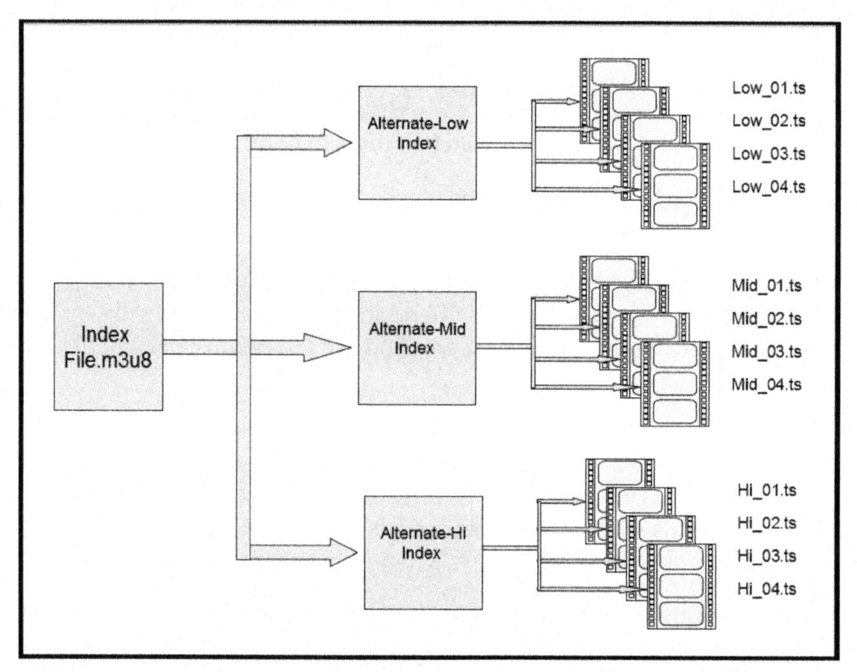

Figure 5-2. The manifest file identifies the different files (or file segments) and their URLs.

With this as background, let's take a quick look at the technology alternatives.

Technology Alternatives

The adaptive streaming market is fast-moving with lots of players in and out. Here are the major players, roughly presented in the order of their entry into the market.

Adobe RTMP-Based Dynamic Streaming

Adobe has two adaptive streaming solutions: the first a Flash Media Server-based solution introduced in 2008. I'll get more into the religious battles surrounding HTTP vs. RTMP in the next section, but this solution requires the Adobe Media Server (3.5 or higher) or equivalent (like the Wowza Media Server) and uses the RTMP protocol for distribution.

When choosing an adaptive streaming platform, player penetration is paramount. The ubiquitous Flash Player is one of the strongest components of Adobe's product offering. You can use either Flash alternative (RTMP or HDS) to deliver to computers running Flash, or to deliver to iOS and Android devices via AIR applications.

Microsoft's Smooth Streaming for Silverlight

In late 2008, Microsoft announced Smooth Streaming for Silverlight, which was essentially the productization of the adaptive-bit-rate streaming technology originally deployed in the 2008 Summer Olympics. More recently, the 2010 Winter Olympics were streamed live via Smooth Streaming, along with NBC's Sunday Night Football offering.

As the name suggests, Smooth Streaming requires the Silverlight player, which is available on about 69% of all Windows computers, but only 49% of Mac computers when I checked on StatOwl (statowl.com) in January 2013. Smooth Streaming delivers via the HTTP protocol, and using that protocol has been the most widely touted competitive advantage over Adobe's RTMP-based Dynamic Streaming. As we've discussed in previous chapters, you can use Smooth Streaming to deliver adaptive streams to computers with Silverlight installed, Windows Phones and Xbox 360 game consoles.

Apple's HTTP Live Streaming

In June 2009, Apple announced its own adaptive-bit-rate streaming technology, the aforementioned HTTP Live Streaming (HLS) which can be viewed on iDevices with cellular or Wi-Fi connections, on Mac computers running QuickTime X, on Android 3.0 and later devices, and on multiple over-the-top (OTT) players including Apple TV (of course), Boxee, Roku and Google TV.

As mentioned in Chapter 1, LongTail Video announced that its pre-fab video player, the JW Player 6, would be able to play back HLS streams (bit.ly/JWP_features). If other pre-fab video players also enable this feature, HLS could be a general-purpose desktop solution as well as the most widely adopted solution for mobile and OTT.

Akamai HD Network

Prior to the launch of Adobe's HTTP Dynamic Streaming offering, content delivery network Akamai launched Akamai HD Network, an HTTP-based service that can deliver adaptive streams to the Flash Player using HTTP on the Akamai network. The Akamai service uses "in the network" repackaging to input traditional RTMP-based streams from streaming producers and convert them into chunks available for HTTP delivery to the Silverlight player or Apple iOS devices. Turner Broadcasting used this service to deliver its streaming presentation of the 2010 PGA Championship in August 2010.

Adobe HTTP Dynamic Streaming (HDS)

In late 2010, Adobe launched HTTP-based Dynamic Streaming (originally code-named Zeri) to provide an Adobe-developed alternative for those seeking an HTTP-based technology. As the name suggests, the technology delivers to the Flash Player using the HTTP protocol, and it doesn't require the Flash Media Server to operate.

At NAB 2011, Adobe announced support for HTTP Live Streaming via the Flash Media Server (**adobe.ly/FlashdoesHLS**), which is now a standard feature of the Adobe Media Server family.

Scalable Video Coding (SVC)

Scalable Video Coding is an adaptive streaming extension to the H.264 standard that was much heralded but never came to market. Technically, SVC's unique advantage is that it can serve multiple streams from a single encoded file, where other adaptive streaming technologies need unique streams for every supported configuration. For example, Major League Baseball streams 11 configurations within its baseball product offering, and has to create 11 unique streams for each game, plus backup streams for fail-over. This requires multiple racks of encoders, and obviously increases both storage and bandwidth cost.

With SVC, only one stream would be required (plus the fail-over), and it's only about 20% larger than the largest stream served. In addition, that single stream can dynamically adjust output resolution, data rate and frame rate to provide a much greater range of output streams. The problem with SVC is that it's currently unavailable, and unlike objects in your rear-view mirror, it isn't closer than it appears. The biggest stumbling block will be the player; to date, none of the major player vendors—Adobe, Apple and Microsoft—have announced support for SVC. Until this happens, SVC is a technology that looks great on paper, but hasn't yet become a real product for general-purpose streaming.

> **Tip:** *For more on Scalable Video Coding, check out "Scalable Video Coding: The Future of Video Delivery?"* (**bit.ly/ozer_svc1**)*, "Seawell and the future of Scalable Video Coding"* (**bit.ly/ozer_svc2**)*, and "Meet MainConcept, the Codec People, and H.264 SVC"* (**bit.ly/ozer_svc3**).

DASH

MPEG Dynamic Adaptive Streaming over HTTP (DASH) is an ISO Standard (ISO/IEC 23009-1) finalized in 2012. As the name suggests, DASH is a standard for adaptive streaming over HTTP that has the potential to replace existing proprietary technologies like Smooth Streaming,

Adobe Dynamic Streaming and HLS. A unified standard would be a boon to content publishers, who could produce one set of files that play on all DASH-compatible devices.

The DASH working group has industry support from a range of companies, with contributors including critical stakeholders like Apple, Adobe, Microsoft, Netflix and Qualcomm. However, while Microsoft and Adobe have indicated that it will likely support the standard, Apple and Google have not given the same guidance. Until these two major players support DASH, it won't realize its full potential.

A more serious problem is that DASH doesn't resolve the HTML5 codec issue. That is, DASH is codec-agnostic, which means that it can be implemented in either H.264 or WebM. Since neither codec is universally supported by all HTML5 browsers, this may mean that DASH users will have to create multiple streams using multiple codecs, jacking up encoding, storage and administrative costs.

Finally, at this point, it remains unclear whether DASH usage will be royalty-free. This may affect adoption by many potential users, including Mozilla, which has already commented that it's "unlikely to implement" DASH unless and until it's completely royalty-free.

DASH was first demonstrated in the 2012 London Olympics (**bit.ly/first_DASH**), although streams weren't interchangeable, so not all players could play all streams produced by all encoders. In February 2012, the DASH Promoters Group, a subset of the DASH Industry Forum, was formed to promote DASH and advance its adoption (**bit.ly/DASH_PG**). Most significantly, the group is working towards developing DASH264, which is "a set of interoperability guidelines, associated test vectors and software tools for building interoperable streaming solutions. DASH264 will provide a rich set of test cases, test vectors and software tools that enables service providers, operators, encoding and client vendors to build and deploy MPEG-DASH-compliant solutions."

As of the time of this writing (March 2013), even though many products and companies claim to support DASH, there were no commercial implementations in operation. The DASH industry forum claims that 2014 "will be the mass deployment year of MPEG-DASH technology" (**dashif.org/faq**).

Perhaps so. But if you check the membership list in the DASH Industry Forum, you'll note that both Apple and Google are conspicuously absent. No surprise; standards are often enacted to level the playing field, and typically that isn't in the best interest of market leaders like Apple in particular. Specifically, Apple owns HLS, which plays on all iOS and Apple TV products, the Flash Player (via the JW Player), Android 3.0 and above devices, and all OTT Players except for Microsoft's Xbox 360. Why would Apple adopt a standard that reduced the value of HLS? Maybe Apple will come around by 2014, but if not, it's hard to call any deployment without Apple's support "mass."

In addition, the problem that DASH is attempting to solve just isn't that problematic. That is, as you'll read about in the next section, there are multiple options available today for converting one set of adaptive streams into another format, for example, inputting RTMP-based Dynamic Streaming and outputting HLS (or DASH, or Silverlight) streams in real time. This makes supporting multiple adaptive technologies as simple as clicking a checkbox in a streaming server. So if DASH does ever come around in a big way, it's not really solving a huge problem for most streaming producers.

Interestingly, although DASH is often seen as the Great White Hope for HTML5-based adaptive streaming, it's unlikely that we'll see it supported in HTML5-compatible browsers in the short term. For example, when I interviewed Long Tail Video president Jeroen Wijering in November 2012, he commented, "As far as I know, no browser vendor has DASH support on its roadmap."

For this reason, it's likely that initial support for DASH will come from players like the Silverlight or Flash Players, or plug-ins thereto. So to use the Great White Hope of HTML5-based adaptive streaming, you'll likely have to (wait for it) install a plug-in or a player from Adobe or Microsoft. Ironic, since the key benefit of HTML5 was supposed to be the end of proprietary players.

> **Tip:** *For more on MPEG-DASH, check out "What is MPEG-DASH?" (**bit.ly/whatisDASH**), "MPEG-DASH: The File Format of the Future?" (**bit.ly/DASH_future**), "CFF: One Format to Rule Them All?" (**bit.ly/DASH_cff**), "Apple to Adobe & Microsoft: With Friends Like You, Who Needs Enemies?" (**bit.ly/DASH_friends**), "MPEG-DASH's Future: Unified Format or DASHed Hopes?" (**bit.ly/DASHed_hopes**), and "MPEG-DASH: Struggling for Adoption?" (**bit.ly/DASH_struggling**).*

Supporting Multiple Adaptive Streaming Technologies

As we've discussed multiple times, there is no single adaptive-streaming solution that will reach all your target viewers. As a practical matter, this means that most producers will have to support either Flash Dynamic Streaming or Silverlight for the desktop, along with HLS for mobile and OTT. Prior to around 2010 or so, this necessitated two essentially different encoding/delivery mechanisms to support streaming to the desktop and iDevices. Fortunately, that dynamic has changed, primarily because the all relevant targets can play video encoded in the H.264 format.

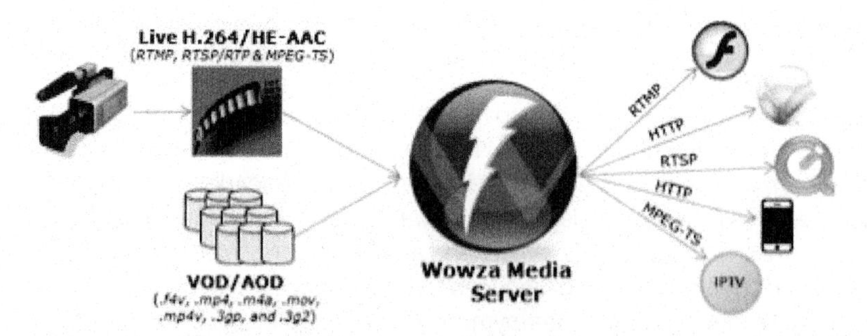

Figure 5-3. The Wowza Media Server can input one stream and deliver it to multiple devices using different technologies.

One of the first solutions on the scene was from Wowza Media, which sells a streaming server that competes with Flash Media Server. As shown in Figure 5-3, the Wowza Media Server can take one input stream and automatically convert it to all formats necessary to deliver to Flash

(RTMP and HTTP), Silverlight, QuickTime and Apple iDevices. There's a very cool demo of that capability at **bit.ly/wowza_demo**.

Essentially, Wowza is inputting an H.264 stream (or streams), re-wrapping it (or them) for the different container formats and protocols involved, and creating any necessary manifest files. As I've mentioned before, this process is called transmuxing. For example, while QuickTime, Flash and Silverlight most frequently work with files wrapped in the MPEG-4 container format, HTTP Live Streaming works with files encoded in the MPEG-2 container format. The Wowza server inputs the file in the MPEG-4 container format, converts to the MPEG-2 container format, creates all the necessary file chunks and the manifest file, and delivers them to HLS-compatible players.

> **Tip:** *If you'd like to see a definition of multiple encoding-related terms like encoding, transcoding, transrating and transizing, see my article "What is Encoding and Transcoding?" at* ***bit.ly/encoding_defs***.

As mentioned, in September 2010, Akamai launched new features in its HD Network that performed a similar function. Here's a blurb from the press release:

> Our focus is to ensure each viewer gets the highest-quality experience possible leveraging the Akamai HD Network, while making it easier for customers to reach these growing platforms with more of their content. We accomplish this through Akamai's "in the network" packaging. In a nutshell, companies provide Akamai with standard H.264 and MPEG-4 video on-demand content, and we deliver it to iPhone, iPod touch and iPad. It's that simple.

The third alternative is from Microsoft, which provides the ability to re-wrap adaptive Silverlight streams for delivery to iDevices. Specifically, in Expression Encoder 4, you generate a Smooth Streaming MPEG-4 bit stream and transmit that Microsoft's Internet Information Services 7 server with the IIS Live Smooth Streaming Feature. Like Wowza and Akamai, the server then re-wraps the transport stream from MPEG-4 to MPEG-2, chunks the streams as necessary, and creates the required .M3U8 manifest files that tell the iDevices playing the stream where to find and retrieve the video chunks.

Other alternatives include the aforementioned Adobe Media Server and Real Network's Helix family of servers. I'm sure all of these servers will support DASH when it becomes generally available. Still, since transmuxing-capable media servers like Wowza do all the heavy lifting for you, supporting multiple adaptive streaming formats is actually fairly trivial.

Choosing a Technology

Now that you know the available technologies, how do you choose between them? Let's consider the two most relevant technology characteristics that most users are most likely to consider: supported platforms and protocol.

Supported Platforms

Probably the most important characteristic of any technology that you choose is whether it can reach your target viewers, with or without a required player download (Table 5-1). Both flavors of Flash offer the broadest reach to desktops, and can also reach iOS and Android devices via AIR. However, no OTT platforms currently play either version of Adobe's adaptive streaming technology.

Smooth Streaming has garnered very little third-party support, so it is largely a Microsoft-only technology—at least as it relates to native support. Given the relatively tiny market share for Windows Phones, it's probably only relevant to those publishing to large numbers of Xbox 360 viewers. As mentioned previously, if you're up for producing an app, you can distribute Smooth Streaming to most mobile and OTT platforms.

Although it was initially an Apple-only technology, HLS has garnered lots of support from other mobile and particularly OTT platforms. It's really the only "must support" technology in the group. If other pre-fab players duplicate HLS support via the Flash Player, like LongTail Video did with its JW Player, HLS could be the only single-technology solution; it's going to be interesting to see what happens in this regard in 2013.

	RTMP Dynamic Streaming	Smooth Streaming	Apple HTTP Live Streaming (HLS)	HTTP Dynamic Streaming (HDS)	DASH
Desktop Player	Flash, AIR	Silverlight	IOS 3.0, Mac OSX, Flash	Flash, AIR	None
Desktop Platforms	Win/Mac/Linux/ Solaris	Win/Mac/Linux	iOS 3.0/OS X 10.6/Windows	Win/Mac/Linux/ Solaris	NA
Player Penetration	Windows – 96% Mac – 87%	Windows – 73% Mac – 49%	Windows – 96% Mac – 99%	Windows – 96% Mac – 87%	NA
Device Support	Android/iOS (via AIR app)	Windows Phone	IOS/Android 3.0+	Android/iOS (via AIR app)	NA
OTT Support	None	Google TV, Roku, XBox 360	Apple TV, Boxee, Google TV, Roku	None	NA

Table 5-1. Supported platforms for adaptive streaming solutions.

I include DASH for completeness and to set up a simple analysis when it does become available. That is, until it's available on all relevant desktop, mobile and OTT devices, it will still be just one of a number of technologies that producers will have to consider.

Today, however, the overwhelming number of producers will have to support HLS and one of the two flavors of Flash. There are two flavors of Dynamic Streaming, RTMP and HTTP, and we'll distinguish them in the next section.

Protocol (HTTP vs. RTMP)

The RTMP vs. HTTP argument first appeared when Microsoft launched Silverlight and promoted its HTTP delivery technology as the primary advantage over RTMP-based Dynamic Streaming. Once Adobe launched HTTP-based Dynamic Streaming, of course, Microsoft's argument was blunted; if Flash users wanted HTTP, they could just switch over.

Now those implementing Flash, or considering whether to convert over from RTMP to HTTP, face the most significant RTMP vs. HTTP decision. So let's explore the differences.

RTMP-based technologies require both a streaming server (like the Flash Media Server) and the player, and consistent communications between the two during the course of media playback. The connection is termed "stateful" because the player continually communicates its states via playback controls like play, pause or stop. In contrast, HTTP-based technologies don't require a streaming server—just the manifest file mentioned above and the player. The connection is called "stateless" because there is no persistent server/player communication.

These designations indicate a number of preliminary differences between the technologies. For example, regarding RTMP-based technologies, we know that:

- Since they require a streaming server, and servers cost money, they are likely more expensive.
- The stateful connection and server means that more playback information is captured, which likely means superior playback analytics.

There are several other key differences between the technologies, which include:

- HTTP is natively supported by caching devices on the web. This should make delivery more efficient than RTMP, which isn't natively supported by these devices. It also makes HTTP-based streaming more scalable.
- The availability of local caching should also improve quality of service for viewers served via the cache as opposed to those served by RTMP-based delivery, which is not likely to be cached.
- HTTP packets are not blocked by corporate firewalls, which should enable HTTP-based technologies to reach more viewers than RTMP-based technologies can.

In addition, some content delivery networks, most notably Highwinds, have discontinued their RTMP-based services. Over time, those using RTMP delivery for large-scale events may find their distribution options limited.

There are counter-points to all of these arguments. ESPN uses RTMP-based Flash; would it do so if HTTP were cheaper or delivered better quality? *The New York Times* uses RTMP-based Flash; would it do so if there were firewall issues? For smaller users who may not be using a CDN and aren't experiencing any firewall or similar issues, there may not be a lot of motivation to switch. After all, if it ain't broken, why fix it?

Still, there's a general perception out there among producers I trust, like Major League Baseball's Joe Inzerillo, that HTTP is better at delivering high-quality streams than RTMP. It's clearly the direction in which the industry is heading. So, if you're implementing Flash-based adaptive streaming today, HTTP is the most logical choice. If you're currently using RTMP and are considering converting over to HTTP, it's probably not a bad decision either. However, if it's working just fine, I probably wouldn't fix it until it starts breaking. On the other hand, if you're starting to distribute Flash adaptively today, HTTP-based Dynamic Streaming is the way to go.

Implementing Adaptive Streaming

With this as background, let's look at how to configure the files you'll produce to distribute. Whichever technology you use, you'll have several similar decisions to make, including the number of streams and their configuration. So let's start there.

In gathering these guidelines, I've consulted multiple sources, including my own consulting projects, and input from a number of actual users for an article for *Streaming Media Magazine* titled "Adaptive Streaming in the Field" (**bit.ly/adaptive_field**). I also checked a number of other articles, white papers and vendor recommendations. A complete list is available at the end of this chapter.

If you're serving multiple platforms, including computer, mobile and OTT, the most efficient approach is to derive a single set of streams that you can deliver to all targets. The benefits of this approach are clear: you have fewer files to encode, and because you're using a single set of files, you can transmux as necessary to reach your targets.

There are two potential objections to this approach. First, if you create a single set of files targeting mobile, computer and OTT, you'll have to encode using the Baseline profile for the streams targeting the oldest mobile devices, and the Main profile for some newer mobile devices and tablets. Since many of these streams will be played by computers and OTT devices that can play video encoded using the High profile, you're potentially reducing the quality of the computer and OTT experience by using a Baseline- or Main-encoded stream.

We discussed this in a section titled "Profiles and Levels" in Chapter 3, where I shared some test results that showed the difference between the Baseline and High profile isn't as significant as you might think. In this regard, you should work through the procedure detailed below for choosing the stream count and configuration, and then encode some test streams to those parameters using both the High and Baseline profile. If the quality difference between the two is significant, encoding different sets of streams for the different targets might make sense. If not, then it probably doesn't.

The second objection to serving multiple platforms with a single set of streams is a perception that streams must be customized for mobile delivery. For example, Table 5-2 shows the specs used by Turner Broadcasting for its excellent NBA League Pass service (see my review at **bit.ly/Turner_NBA**). If you look at the mobile stream configurations, you'll note that the 416x234 resolution used for the first three mobile video streams comes straight from Apple's Technical Note TN2224 (**bit.ly/HTTPLive**), shown in Table 5-3. You'll also see that Turner used three different sets of streams for all targets, not even sharing the 720p file distributed to the Web and OTT.

Web	Type	Format	Resolution	Vid BR	Frame Rate	Aud Codec	Sample Rate	Aud BR	Channels
mosaic	Flash	h264	480x270	452 kbps	29.97	HE-AAC	44.1 kHz	48 kbps	Stereo
low	Flash	h264	768x432	836 kbps	29.97	HE-AAC	44.1 kHz	64 kbps	Stereo
med	Flash	h264	896x504	1436 kbps	29.97	HE-AAC	44.1 kHz	64 kbps	Stereo
high	Flash	h264	960x540	2436 kbps	29.97	HE-AAC	44.1 kHz	64 kbps	Stereo
full	Flash	h264	1280x720	3436 kbps	29.97	HE-AAC	44.1 kHz	64 kbps	Stereo
Mobile	**Type**	**Format**	**Resolution**	**Vid BR**	**Frame Rate**	**Aud Codec**	**Sample Rate**	**Aud BR**	**Channels**
Audio	HLS	---	---	---	---	HE-AAC	44.1 kHz	40 kbps	Stereo
iPhone	HLS	h264	416x234	110 kbps	10	HE-AAC	44.1 kHz	40 kbps	Stereo
iPhone	HLS	h264	416x234	200 kbps	15	HE-AAC	44.1 kHz	40 kbps	Stereo
iPhone	HLS	h264	416x234	400 kbps	29.97	HE-AAC	44.1 kHz	40 kbps	Stereo
iPhone	HLS	h264	640x360	600 kbps	29.97	HE-AAC	44.1 kHz	40 kbps	Stereo
iPad	HLS	h264	640x360	1200 kbps	29.97	HE-AAC	44.1 kHz	40 kbps	Stereo
iPad	HLS	h264	960x540	1800 kbps	29.97	HE-AAC	44.1 kHz	40 kbps	Stereo
Connect Devices	**Type**	**Format**	**Resolution**	**Vid BR**	**Frame Rate**	**Aud Codec**	**Sample Rate**	**Aud BR**	**Channels**
OTT	HLS	h264	1024x576	1200 kbps	29.97	HE-AAC	44.1 kHz	96 kbps	Stereo
OTT	HLS	h264	1024x576	2500 kbps	29.97	HE-AAC	44.1 kHz	96 kbps	Stereo
OTT	HLS	h264	1280x720	3500 kbps	29.97	HE-AAC	44.1 kHz	96 kbps	Stereo

Table 5-2. Videos produced by Turner for its NBA League Pass service.

The problem with the 416x234 resolution is that it's probably useless for web distribution, where encoding and delivery is most efficient when the resolution of the video matches the resolution of the browser playback window. I could be going out a limb here, but I'm guessing few producers use this resolution for browser-based playback. In contrast, 480x270 resolution, which is the next step up in resolution on the Apple Tech Note, would be much more universal.

What's interesting is that the 416x234 resolution has no apparent relation to any iOS device screen resolution. The smallest relevant display is the 480x320 resolution on the iPhone 3/iPod touch 3 and earlier devices (see a summary of iDevice specs at **bit.ly/idevice_specs**). With the iPhone 4/iPod touch 4, resolutions jumped to 960x640, so any lower-resolution video should be fine.

Note that many producers ignore Apple's recommendations and choose their own resolutions. For example, MTV uses 448x252 for its mobile streams, while a large broadcast network (and former consulting client) uses 256x144. So there doesn't appear to be anything magical in Apple's recommendations that would warrant creating custom resolutions for low-end iDevices.

I will say that most larger consulting clients do opt to customize the resolution and profile of their streams for the platforms they serve. In contrast, most smaller clients have used a single set of streams for all targets, which has worked well—although they're pushing far fewer streams than these larger clients.

As an aside, Turner could have many unknown reasons for using the 416x234 resolution, and I wasn't privy to the team's considerations. Certainly it makes a lot of sense to go with Apple's recommendations when you're launching a new service. I just want to point out that you don't need to adapt them as if they were set in stone.

Recommended Procedure

When choosing both the number of streams and their configuration, here's my recommended approach. I'll work through the entire process, then circle back for some additional thoughts on resolutions and data rates.

Choose Mobile First

Start by choosing the number of mobile streams you feel are necessary to effectively reach your mobile viewers, and the data rates and resolutions for their devices. To accomplish this, identify the lowest video data rate you'd like to support, reflecting your views about the lowest connection speed used by a relevant viewer group. For Apple in Tech Note 2224 (Table 5-3), that's 200 kbps—although I've had clients who produced video as low as 110 kbps. Then identify the resolution/frame rate combination that delivers optimal quality at that video data rate.

16:9 Aspect Ratio

	Dimensions	Frame Rate	Total Bit Rate	Video Bit Rate	Audio Bit Rate	Audio Sample Rate	Keyframe**	Restrict Profile to:	Works on iPod Touch Gens 2, 3, 4	Works on iPhone 3G, 3GS, 4
CELL	480x320	na	64	na	64	44.1	na	na	*	*
CELL	416x234	10 to 12	264	200	64	44.1	30 to 36	Baseline, 3.0	*	*
CELL	480x270	12 to 15	464	400	64	44.1	36 to 45	Baseline, 3.0	*	*
WIFI	640x360	29.97	664	600	64	44.1	90	Baseline, 3.0	*	*
WIFI	640x360	29.97	1264	1200	64	44.1	90	Baseline, 3.1		
WIFI	960x540	29.97	1864	1800	64	44.1	90	Main, 3.1		
WIFI	960x540	29.97	2564	2500	64	44.1	90	Main, 3.1		
WIFI	1280x720	29.97	4564	4500	64	44.1	90	Main, 3.1		
WIFI	1280x720	29.97	6564	6500	64	44.1	90	Main, 3.1		
WIFI	1920x1080	29.97	8564	8500	64	44.1	90	High, 4.0		

Table 5-3. Configure the lowest-quality streams for mobile.

If possible, try to choose a resolution that matches a window size used for browser-based playback on your website. You'll see why in the following section.

Then Browser-Based Streams

Then choose the configurations for streams to be played within your website, either by computer- or tablet-based viewers. Here your primary concern is producing at least one stream for every video window size on your website. You can see this in Table 5-4, which shows the encoding parameters used by MTV for its web properties. The two lowest-quality streams are for mobile and very low-bandwidth connections, while the highlighted streams are for playback in a browser window.

As you can see, MTV has a stream for each window size used on its websites. When MTV shared these configurations, it was using RTMP-based Flash streaming, which enabled it to prevent stream from switching until a viewer started watching in a larger window.

Scenario	Format	Frame Size	Total Bitrate	Audio Bitrate	bits/pixel *frame @ 30 fps	bits/pixel *frame @ 24 fps
Mobile & constrained (low)	baseline, mono, 10 fps	448x252	150	48	0.09	0.09
Mobile & constrained (high)	baseline, mono	448x252	450	48	0.12	0.15
Sidebar placements	main profile, stereo	384x216	400	96	0.12	0.15
Small in-page	main profile, stereo	512x288	750	96	0.15	0.18
Medium in-page	main profile, stereo	640x360	1200	96	0.16	0.20
Large in-page	main profile, stereo	768x432	1700	96	0.16	0.20
Full size in-page	main profile, stereo	960x540	2200	96	0.14	0.17
HD 720p (full screen)	high profile, stereo	1280x720	3500	96	0.12	0.15

Table 5-4. Encoding configurations used by MTV Networks.

For example, if the viewer were watching within a medium in-page window (640x360), MTV would never send a higher-quality stream to that viewer unless he or she increased the size of the playback window. That makes wonderful sense, because scaling downward can hinder player performance, and since the quality of the streams are close to identical (as measured by the bits-per-pixel values), the viewer wouldn't notice the difference anyway. I'm not sure this logic is available in HTTP-based technologies, but if it is, it's worth programming in.

What to do if you only have one window size within the browser—say, 640x360—with the next jump to full screen? First choose the highest quality you want deliver at the resolution, using the bits-per-pixel value. For example, at around 1.2 Mbps, the bits-per-pixel value would be around 0.174, which should be perfect for all but the hardest-to-encode video. Obviously, if the video looks perfect at 1.2 Mbps, there's no need to go any higher with a 640x360 stream.

Then choose a number of streams that provides good coverage from your highest-quality ***mobile*** stream to the stream you just configured. For example, if your highest-quality mobile stream were 640x360@640 kbps, you might include another stream between the two at 900 kbps. This would deliver a meaningful quality difference between all three streams. On the other hand, in TN2224, Apple recommends jumping from 640x360@600 kbps to 640x360@1200 kbps, so an intermediate stream may not be essential.

Note that more streams are not necessarily better; more streams mean that the streams are closer together, minimizing the quality difference while increasing the frequency of stream switching, which can disrupt viewing. The ideal scenario is when the viewer quickly identifies the optimal stream and continues to watch that through the end of the video.

For example, in his article "Live dynamic streaming With Flash Media Server 3.5" (adobe.ly/kapoorlivefms), Adobe's Abhinav Kapoor recommended:

> If your target viewer covers a broad spectrum of bandwidth capabilities, it is best to keep a wide range of stream bit rate encodings while keeping a large enough difference between successive bit rates. Too many bit rates too close to one another could result in too many stream switches, even with smaller bandwidth fluctuations. Besides the slight overhead in switching, the viewer's experience with too-frequent quality fluctuations may not be pleasant. Meanwhile, too few streams with a huge gap in bit rates would not provide the best quality or the most optimal stream for a particular bandwidth.

Here are some other factors to consider:

- You'll need more streams for HD content than for SD. Most clients encoding SD video (often converted from DVD) used three or four streams, although I did have one client in the entertainment space with eight streams for SD content. Most clients producing HD content range from five to eight streams.

- You'll need more streams for subscription content than for free or advertising-supported content. For example, the last time Major League Baseball shared its adaptive streaming schema with me, it was encoding 11 streams. I've seen the stats on other subscription services, and some use up to eight streams for their content. Note that the extra streams are usually at the high end to really satisfy users watching in their living room on large flat-panel displays, either via an OTT device or computer.

OTT and Full Screen Playback Last

Lastly, configure the streams bound for OTT and full-screen playback on computers or mobile devices. In this regard, cost sets your maximum. For example, looking at Apple's highest-quality stream, at 8,564 kbps for 1080p video, an hour of video would consume around 4GB. According to Dan Rayburn's latest blog on the subject (bit.ly/CDN_pricing), CDN pricing for customers buying from $100,000 to more $1 million a year in bandwidth ranged from a low of $0.01 per GB to a high of $0.12.

At these prices, it would cost between $0.04 and $0.48 to stream an hour of video at Apple's highest recommended rate. However, I've seen legacy bandwidth pricing for more modest-sized commitments as high as $1.10/GB, which would boost the per-hour transfer cost of this 1080p configuration to $4.40, which is pretty pricey.

When configuring your highest-quality stream, choose the highest data rate you can afford, given your monetization strategy and cost structure. Since your top-quality stream has to look very good, you'll have to adjust video resolution accordingly. For example, if you can only afford 3 Mbps at the top end, encode at 720p, not 1080p.

Then, if you need to add a stream to bridge between the highest browser-based stream and the maximum OTT stream, add that in, using the same logic we just applied for the browser-based streams. For example, if your highest-quality stream is 720p@3 Mbps, a 2 Mbps stream at the same configuration would serve as a nice bridge from the 640x360@1200 kbps stream and the 720p@3 Mbps stream.

Other Considerations

Here are some other issues to consider when configuring your streams.

Low resolution or low frame rate? For low-bit-rate files, I prefer to drop frame rate rather than resolution; you can read all about why in an article on my website titled "Configuring low data rate adaptive streams" (**bit.ly/configadaptive**). Briefly, when producing at low data rates, you have several options to preserve quality, including lowering the resolution, the frame rate or both. All options have some negatives. For example, lower resolutions preserve frame quality but can look pixelated when scaled for display. Higher resolutions avoid scaling artifacts, but frame quality can suffer. Dropping the frame rate preserves frame quality, but reduces smoothness.

Figure 5-4. Dropping resolution (on the right) results in an easier-to-encode file, but loss of detail throughout. For a better look, download the PDF of all figures in the book at **bit.ly/Ozer_multi**.

To help a client decide which strategy to pursue, I encoded files at the following configurations, all at 300 kbps and otherwise identical encoding parameters:

- 640x480x15 fps
- 640x480x30 fps
- 400x300x15 fps
- 320x240x15 fps
- 320x240x30 fps.

You can view all streams at bit.ly/configadaptive, and Figure 5-4 shows a comparison of the 640x360@15 fps and 320x240@30 fps streams. Although the latter was obviously smoother, this was offset by a persistent fuzziness and loss of detail. Ultimately, the client decided to encode all streams at 640x480 and adjust the frame rate downward rather than dropping the resolution and scaling upward for display. When making choices like this for low-bit-rate files, you should perform the same kind of comparison and see which streams look best.

If you must scale to fit a window size, scale to a higher resolution, not lower. For example, while the smallest window size the Apple schema has to support is 480x320 (iPhone 3/iPod touch 3 and earlier), the smallest stream is 416x234, which gets scaled upward to 480x320, not downward. This is better because most graphics processing units (GPUs) have built-in functions that can scale to higher resolutions and filter the results very efficiently, which means high-quality video without excessive power consumption or dropped frames. However, most GPUs are very inefficient when scaling down, which means more power consumption or slower playback, plus lower quality. Accordingly, always have at least one stream exactly the same resolution as your smallest target window, or smaller.

Stream #	Picture Size	V	A	AV
1	256x144	150	64	214
2	256x144	250	64	314
3	512x288	450	64	514
4	512x288	600	64	664
5	512x288	800	64	864
5	512x288	1200	64	1264
6	768x432	1400	64	1464
7	1280x720	1700	64	1764
8	1280x720	2500	64	2564
9	1280x720	3500	64	3564
10	1920x1080	4200	64	4264
11	1920x1080	5300	64	5364

Table 5-5. Adobe's Levkov recommends clustering your streams around certain frame sizes.

Cluster streams at target resolutions. If you only have a few display resolutions, cluster your encoding targets at these resolutions. For example, in his white paper "Video encoding and transcoding recommendations for HTTP Dynamic Streaming on the Flash Platform" (**adobe.ly/Levkovhttp**), Adobe Flash technology evangelist Maxim Levkov points out that streams of identical resolution switch most smoothly, and he recommends clustering streams at specific window sizes as shown in Table 5-5.

Don't worry about mod-16. It's interesting that Levkov's recommendations don't include 640x360, which is the most widely used resolution on the planet. Of course, that's because 640x360 isn't "mod-16," so 16 doesn't divide evenly into the height and width. We discussed this back in Chapter 2 in a section titled "What About Mod-16," where I concluded, "When configuring your video, try using a mod-16 configuration, or mod-8 (where both height and width are divisible by 8) if mod-16 won't work. You should never use a resolution that's not at least mod-4." With his recommendations, Levkov is presenting the politically correct, mod-16-compliant answer.

Don't encode at resolutions larger than your source. This should go without saying. Scaling to higher resolutions for encoding doesn't add quality; you'll get a higher-quality result and save bandwidth by using the GPUs on your target playback device to scale the video.

More on Choosing Data Rate

Intuitively, for your lower-quality streams, deliverability is the key factor. That is, as we've discussed, you'll choose the lowest data rate you want to support and configure your video appropriately. At higher resolutions, maintaining quality is the most important consideration.

For example, if you were supporting window sizes of 480x270, 640x360 and 853x480 with 30 fps, low-motion video, you might consider data rates of 448,948 bps, 691,200 bps, and 1,063,860 bps from Table 5-6, which is a portion of Table 2-7 shown back in Chapter 2. As you may recall, this table used the bits-per-pixel value and the Power of .75 rule to calculate data rates that maintained consistent quality at the various resolutions.

30 fps – 16:9		Low Motion		High Motion	
Width	Height	Data Rate (bps)	Bits per Pixel	Data Rate (bps)	Bits per Pixel
320	180	244,376	0.14	366,564	0.21
480	270	448,948	0.12	673,421	0.17
640	360	691,200	0.10	1,036,800	0.15
853	480	1,063,860	0.09	1,595,790	0.13
1280	720	1,955,009	0.07	2,932,513	0.11
1920	1080	3,591,581	0.06	5,387,371	0.09

Table 5-6. Recommended data rates and bits-per-pixel values for video produced at 30 fps.

Interestingly, these recommendations map relatively closely to Apple's 480x270 recommendation of 464 kbps and 640x360 recommendation of 640x360. If you were delivering two streams at each resolution, consider using the low-motion data rate recommendation for the lower value and the high-motion data rate recommendation for the higher value.

Another consideration is the difference between the data rate of the respective streams. Specifically, the difference between the stream data rate should be sufficient to produce meaningful quality differences between the streams and avoid overly frequent stream switching. For example, in Table 5-5, stream 3 (512x288@450) and stream 4 (512x288@600) are too similar for my liking—the quality difference would be minimal and the data rates are too close. I would probably skip stream 4 for that reason.

To further explore the analysis that I would apply, unless the content was ultra-high-motion, I'd also recommend against the second stream 5 in Levkov's table, since the 1,200 kbps data rate represents a bits-per-pixel value of a typically wasteful 0.271. In contrast, the first stream 5—at 800,000 bps—represents a bits-per-pixel value of 0.181, which should be sufficient for all but the highest-motion video. If you scan back to Table 5-4, you'll note that MTV encodes its 512x288 video at 750 kbps, which supports this conclusion.

To summarize, you want the data rates far enough apart to represent meaningful quality differences and to avoid too many stream switches. The ideal scenario is one where the viewer quickly identifies the optimal stream and continues to watch that through the end of the video.

Other Configuration Items

Beyond these basic stream configuration options, you'll need to make other adaptive streaming-specific configuration decisions. Among the most important is choosing the H.264 profile for your encoded files.

Choosing The H.264 Profile

By this point, you've already made the decision to either use one set of files for all targets, or customize your streams for each target. Either way, let's revisit what we've learned so far about configuring streams for different targets. Back in Chapter 3, you learned what profiles are and how they affect compatibility with your target playback platforms. In Chapter 4, you learned more details about the H.264 compatibility of your respective targets. To summarize what we've learned to date:

- ***Mobile.*** For Apple devices, follow the profile-related recommendations shown in Table 5-3. For Android, the most conservative approach would be to use the Baseline profile for all streams to maintain compatibility with software-only Android playback. If you're configuring one set of streams for all three targets, these mobile configuration options are the lowest common denominator and dictate how to encode the entire adaptive group.

Regarding Android, I would assume that most Android devices share similar hardware playback capabilities as Apple devices of the same form factor and approximate release date, so I wouldn't create a set of Baseline-only streams for Android. Rather, the schema shown in Table 5-3 is probably safe for Android, and probably a good schema for efficient, single-adaptive-group encoding for all computer, mobile and OTT targets.

Note that Windows phones result in several issues. First, the phones with specs maxed out at 720p, and several CPU configurations require that all files have the same resolution. Basically, if you're configuring adaptive streams for Windows Phones, you're almost certainly going to have to create a separate set of streams specifically for those devices.

- ***Computer Playback.*** If you're customizing streams for computer playback, use the High profile, but make sure you have some configurations at 720p for older viewing platforms.

- ***OTT Playback.*** If you're customizing streams for OTT playback, use the High profile and exclude ultra-low-bit-rate streams.

Once you choose your profile, implement the other H.264 configuration options as discussed in Chapter 3.

VBR or CBR

The knee-jerk response to the issue of variable bit rate (VBR) vs. constant bit rate (CBR) encoding would be CBR to simplify the data transfer, particularly over constrained connections. That is, when encoded using VBR, the size of a 5-second file chunk could vary pretty dramatically, which could make it challenging to deliver that chunk in time to ensure smooth playback.

In addition, you should consider the potential for a VBR bit stream to disrupt the stream-switching algorithms implemented by the adaptive streaming schema. Specifically, buffer status is one of the factors monitored to determine if a stream switch is necessary. If a high-motion scene within a video were encoded at 2x the normal data rate, during transmission to the viewer, which would take longer than normal, the buffer could drop below a threshold level, triggering a stream switch that would not have been necessary with a CBR stream.

Most technical authorities that address this issue recommend CBR or very tightly constrained VBR. For example, Adobe's Levkov recommends 2-pass CBR when available, and also states, "VBR maybe used as well; however, exceeding 10% above the minimum set bit rate is not recommended." I read this to say that you can use constrained VBR with the maximum set to 10% over the target.

On the other hand, as we've discussed, VBR should produce higher-quality video than CBR in videos with diverse content, including sequences that are hard and easy to compress. For this reason, MTV uses 2x-constrained VBR, while Harvard University uses CBR for low-bit-rate streams, and VBR for higher-bit-rate streams.

With my consulting clients, I typically recommend either CBR or 1.25 to 1.5x constrained VBR, with lots of testing to make sure the latter approach doesn't induce data-rate-related stream switches. I see no reason to change that recommendation here.

> **Tip:** *Note that if you're producing video for delivery via an app distributed in the Apple App Store, you should use CBR, or your app may be rejected. See* **bit.ly/AppleTN2235** *for details.*

Key Frame Interval

As discussed above, stream switching must occur on a key frame, so use the same key frame interval in all encoded streams. All resources and users I reviewed advise that the typical 10-second interval was just too long, with Apple recommending a key frame interval of 3 seconds, Adobe recommending between 2 and 5 seconds, and most users I checked with in the 2-to-3 second range. I typically use an interval of 3 seconds for all of my projects, so that's what I recommend. Otherwise, when setting key-frame-related configurations, make sure that:

- ***You disable scene change detection.*** This ensures that key frames are at the specified interval. Some encoders do offer a "fixed key frame" option, which forces key frames at the specified interval even if others are inserted in between. If that is available and you're sure you're forcing the key frames at the requested interval, you can enable key frames at scene changes.

- ***You choose a key frame interval that divides evenly into chunk size.*** As we've discussed, HTTP adaptive streaming technologies send videos in chunks, which can range from 2 to 10 seconds. Remember to consider the duration of the media segment when choosing a key frame. Here's a blurb from the Wowza Media Segmenter documentation that provides specific direction:

  > Chunks must start on a key frame. So it is best to use a key frame interval that is factor of the chunkDurationTarget setting. For instance if chunkDurationTarget is set to 10 seconds then use a key frame interval of either 2, 2.5, 5, or 10 seconds.

  In non-Wowza-speak, make sure whatever key frame interval you choose, it divides evenly into the selected segment size.

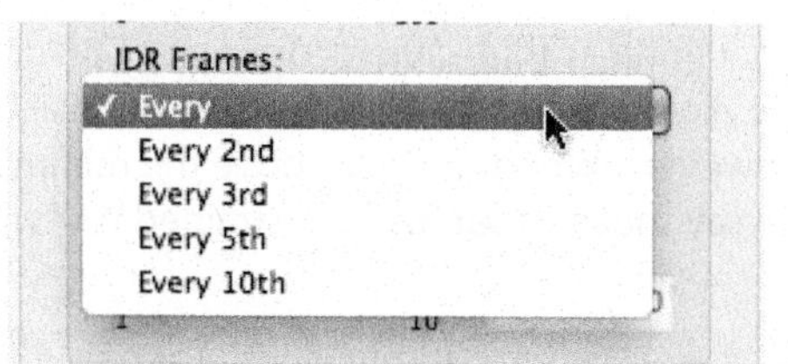

Figure 5-5. Make all I-frames IDR frames when encoding for adaptive streaming.

- ***All key frames are IDR frames.*** We covered the concept of IDR frames back in Chapter 3 in a section titled "Instantaneous Decode Refresh (IDR) Frames." Briefly, as defined in the

Telestream Episode help file, "An IDR frame is an I-frame whose preceding frames cannot be used by predictive frames." For general-purpose streaming, I recommended that every I-frame be an IDR frame, and that's particularly true for adaptive streaming. If your encoding tool includes a configuration option like that shown in Figure 5-5, make sure every I-frame is an IDR frame.

If your encoding tool doesn't offer an I-frame-related configuration option, in my experience, this means that the tool automatically makes every I-frame an IDR frame, so you're in good shape.

Audio Parameters

Here, the most conservative approach is to use the same audio parameters for all video files at the top and bottom of the adaptive group. If you scan Table 5-3, you'll note that Apple recommends using the same audio parameters for all video files. Similarly, in his article "Live dynamic streaming with Flash Media Server 3.5" (**adobe.ly/kapoorlivefms**), Adobe's Kapoor warns, "When there is a switch between two streams with different audio bit rates, there may be a slight audible 'pop' sound. If you are concerned about this switching artifact, then keep the audio bit rates constant between all the streams to have a completely seamless audio switch."

Kapoor recommends that if you do offer different streams, switch from stereo to mono using the same per channel sampling rate and bit rate. For example, if your stereo stream is 128 kbps (44,000 sample rate, 16-bit sample size), use a mono stream that's 64 kbps with the same sample rate and sample size.

In my practice, when audio quality isn't paramount, as in talking head videos and the like, I use 64 kbps mono for all streams. If audio quality is an important component of the overall experience, I recommend 96 kbps stereo, which are the parameters used by MTV for all computer- and OTT-bound streams. For those producers who want to scale audio quality with video, I recommend that they use the approach recommended by Kapoor and switch from mono to stereo using the same per-channel sampling rate and bit rate.

OK, these are the general decisions you'll have to make for all adaptive streaming technologies. During my research, I did encounter some technology-specific recommendations that I'll share with you here.

HTTP-Based Flash Dynamic Streaming (HDS)

In October 2010, Adobe Flash Media Encoding evangelist Maxim Levkov released his detailed white paper "Video encoding and Transcoding Recommendations for HTTP Dynamic Streaming on the Flash Platform" (**adobe.ly/Levkovhttp**). Levkov is both very knowledgeable and experienced, and he made some very specific recommendations that include:

- ***Regarding I-frames.*** Using an I-frame interval of 3 to 8 seconds for long form content (1 minute or longer), with all I-frames designated as IDR frames, which I discuss above and scene change detection disabled.

- ***Other recommendations.*** Levkov also recommended using:

    - A B-frame interval of 2 maximum.

    - 2 to 4 reference frames, "closer to 2."

    - Two-pass CBR recommended when available (i.e. when not live). Levkov does state, "VBR maybe used as well; however, exceeding 10% above the minimum set bit rate is not recommended." I read this to say that you can use constrained VBR with the maximum set to 10% over the target.

    - A consistent frame rate across all files.

    - Buffer length of 1 to 2 seconds.

    - For encoding applications that give you access to the Video Buffer Verifier (VBV), use a buffer of 70% initially, and 100% final.

- In very detailed tables at the end of the white paper, Levkov recommends using 16-bit, 44 kHz stereo audio at 64 kbps for all encoded streams. These tables also recommend using the Main and High profiles, and CAVLC rather than CABAC.

Most of the consulting projects I've worked on were for RTMP-based Flash and mobile distribution, and the general recommendations I've made above and in Chapter 3 have worked well. I haven't worked on any that involved HTTP-based Dynamic Streaming, so I don't have any actual experience to share here.

Unfortunately, Levkov's paper is long on detail but short on reasoning and disagrees with several of my recommendations. For example, he doesn't state why you should use CAVLC rather than CABAC, why the B-frame interval should max out at 2 or why reference frames should be limited to 2. When I asked questions about these details, his responses didn't clarify things for me.

All that said, if I were tasked with creating presets for a client distributing via HTTP Dynamic Streaming, I would follow Levkov's recommendations for three reasons. First, he may know something that I don't. Second, I wouldn't expect any significant quality differences to arise from using 2 B-frames rather than 3, 2 reference frames rather than 5, or CAVLC rather than CABAC. So there's little downside from following Levkov's recommendations, and little upside for using my own. Third, if my presets didn't work and Levkov's did, I'd have some explaining to do.

So, if you get tasked with creating encoding presets for HTTP Dynamic Streaming, I would follow Levkov's recommendations presented above.

Smooth Streaming with Silverlight

Here's some additional input for those working with Smooth Streaming. First, Microsoft Expression Encoder is the go-to tool for encoding for Silverlight and Smooth Streaming. As you would expect, Expression Encoder has multiple presets for Smooth Streaming which obviously reflect Microsoft's view of the optimal number of streams and their configuration. I've copied these parameters for the 1080p Smooth Streaming template onto a shared Google spreadsheet that you can view at **bit.ly/expression_1080p**.

I also wanted to share some points made by Microsoft's Alex Zambelli (who has since moved on to iStream Planet) in a *Streaming Media* webinar on March 24, 2011. You may have to register to get them, but his slides (and those used by other speakers) should be available online at **bit.ly/smtranscoding**. Here are the major points.

- ***Use only a single audio bit rate.*** I didn't know this, but Zambelli stated that the Silverlight client only supports a single audio bit rate with Smooth Streaming, so don't change audio parameters when you're producing your streams. Note that the Expression Encoder template uses 160 kbps stereo in all streams, which is way too high for most adaptive groups.
- ***Use a single resolution if streaming to Window Phone 7.*** I mentioned this when discussing Windows Phone back in Chapter 4. If you're streaming to a Windows Phone, use a single resolution for all streams.
- ***Use a 2-second key frame interval.***
- ***Disable scene-change detection.***
- ***Set client side buffer at 2-3x key frame interval.***
- ***Drop to 15 fps at lower than 300 kbps rather than reducing the resolution.*** Zambelli said that this delivered superior quality over streaming at a lower resolution.

Conclusion

Encoding for adaptive streaming, particularly for multiple target platforms, is a complex task with lots of inputs and moving parts. My best advice is to be conservative and to choose the parameters that give your implementation the best chance of working smoothly. For example, use CBR rather than VBR and one audio configuration for all files. Once you get things working reliability, you can experiment with additional configurations that may improve audio or visual quality. And, of course, test all streams on as many target platforms as you can access to ensure that your video files play.

Chapter Appendix: Resources

Adobe Flash Dynamic Streaming: RTMP

- Abhinav Kapoor, "Live dynamic streaming with Flash Media Server 3.5" (**adobe.ly/kapoorlivefms**)

- Larry Bouthillier, "How to do Dynamic Streaming with Flash Media Server" *Streaming Media* website (**bit.ly/fmsdynamic**)

- Akamai white paper: "Akamai Flash Streaming Solutions Whitepaper" (**bit.ly/akamaiwhitepaper**)

Apple HTTP Live Streaming (HLS)

- Apple Tech Note: "Best Practices for Creating and Deploying HTTP Live Streaming Media for the iPhone and iPad" (**bit.ly/bestpracticehttplive**)

- Apple Tech Note: "HTTP Live Streaming Overview" (**bit.ly/httpliveoverview**)

- Apple Tech Note: "Using HTTP Live Streaming" (**bit.ly/usinghttplive**)

- Akamai White Paper: "Akamai HD for iPhone and iPad Encoding Best Practices" (**bit.ly/akamaiwhitepaper**)

Adobe Flash Dynamic Streaming: HTTP

- Maxim Levkov, "Video encoding and transcoding recommendations for HTTP Dynamic Streaming on the Flash Platform" (**adobe.ly/Levkovhttp**)

- Akamai white paper: "High Definition for Flash over HTTP" (**bit.ly/akamaiwhitepaper**)

- Adobe website: "Tutorial: On-demand HTTP Dynamic Streaming" (**bit.ly/ondemanddynamic**)

Smooth Streaming

- Akamai white paper: "High Definition for Silverlight" (**bit.ly/akamaiwhitepapers**)

- Alex Zambelli, "IIS Smooth Streaming Technical Overview" (**bit.ly/smoothstreamzambelli**)

- Case study data collected from Microsoft regarding the Olympics Broadcast and NBC's Sunday Night Football broadcast.

Chapter 6: Choosing an On-Demand Encoding Tool

You've learned a lot about encoding video files; now let's take a high-level look at on-demand encoding tools, which are where you put much of this theory to practice. Briefly, there are five major categories of software encoding tools: free, bundled, desktop, enterprise and cloud. In this chapter, I'll review the members of each category and point you towards reviews I've written about these tools in the past.

I won't extensively compare the tools as these comparisons have short shelf lives and limited value within months. Rather, I'll focus on the categories themselves so you'll understand what each category does well, and when you need to consider moving to the next level.

Specifically, in this chapter, you will learn:

- The questions that you should ask yourself before choosing an encoder
- The major categories of encoding tools, including free encoders, bundled encoders, desktop encoders and enterprise encoders
- The three major categories within the enterprise encoder market (workflow systems, Swiss Army Knife transcoders and high-volume encoders)
- When you should consider moving from one category to the next
- When you should consider moving to the cloud.

Note that in Chapter 7, I'll review how to encode using many of the tools mentioned herein. Here, just sit back and enjoy the narrative as we tour the encoding landscape.

What You Should Know Before You Go Shopping

There are multiple encoding workflows for most encoding tasks, and the encoder you purchase must neatly fit into that workflow. As an example, if you use a media server like the Wowza or Adobe Media Servers to transmux .MP4 files for various adaptive targets, you don't need an encoder that can spit out chunked video files and metadata for HTTP Live Streaming (HLS), HTTP Dynamic Streaming (HDS) or Dynamic Adaptive Streaming over HTTP (DASH).

If you just need an encoder to produce a high-quality file to upload to your online video platform (OVP) or user-generated content (UGC) site, that's another kettle of fish altogether. So here are the questions you should ask before you go shopping.

Question 1: What output formats must the encoder support?

Start by defining the delivery requirements for the file(s) you are encoding, including codec, and single-file and adaptive delivery container formats. Be as specific as possible. For example, do your Flash programmers need an .F4V file or can they work with .MP4? Does your authoring program need program or elementary streams? For adaptive streaming, does the encoder need to create file chunks and manifest files or just the raw encoded files?

Question 2: What audio/video input formats must the program support?

Identify all audio/video input formats that the program must support. Will you be starting with Apple ProRes or Avid DNxHD input, or raw camera formats? Different programs, particularly in the enterprise class, have very different capabilities in this regard.

Question 3: What features do you need surrounding your video?

Beyond the target platforms and the number of streams, you should also consider the features surrounding your video. For example, will closed caption support be a requirement, or digital rights management (DRM) or advertising insertion? If so, few low-end encoders can handle these requirements, so you'll have to step up to the enterprise class.

Question 4: What's your overall encoding workflow?

Again, do you need your encoder to produce a single file for uploading to an OVP, or all files necessary for multi-format adaptive delivery? Virtually all encoders can handle the single-file requirement, fewer can output streaming-ready files for adaptive delivery, and fewer still can add captions or DRM to the adaptive streams. Map out the precise formats your encoder will need to produce and you'll be a long way toward choosing the optimal tool. Let's start with a quick look at several free tools.

Free Encoding Tools

I'm not sure this is a big category for most readers, who in most instances will at least have one bundled encoder, like Compressor (Final Cut), Adobe Media Encoder (Premiere Pro) or Squeeze

(Avid). But if you're interested in a free tool, I thought I would point you to my article on OnlineVideo.net titled "H.264 Encoding: Four Free (or Cheap) Encoders Compared," available at **bit.ly/freeh264enc**, which includes a video summarizing my findings. The tools I compared include HandBrake, Miro Video Converter, MPEG Streamclip, and Apple Compressor 4—although for this chapter, I'll talk about Compressor in the next category of bundled encoders.

Here's a summary of each of the three tools.

Miro Video Converter

Miro Video Converter (**mirovideoconverter.com**) is a template-driven tool with very limited configuration options. I concluded, "For general-purpose use, Miro is a non-starter, since it allows no configuration options whatsoever. If simple is your most important criteria, however, Miro is it."

HandBrake

HandBrake is one of those rare tools that's functional and free (**handbrake.fr**), and it's cross-platform—another positive. Performance and quality were both top-notch, and if you're looking for an easy way to rip and convert DVDs to .MP4 files, HandBrake is a great option. Overall, I concluded, "The negatives? The program's interface can feel disjointed, with resolution controls completely separate from encoding controls, and HandBrake crashed several times during my tests, though it completed all tasks. And, it only produces H.264 files, so had limited use as a general-purpose encoder."

There's no adaptive streaming support or support for higher-end functions like captioning or DRM. If you need to produce a single file for uploading to an OVP, however, you could do worse than HandBrake.

MPEG Streamclip

MPEG Streamclip hasn't been updated since 2008, a big negative, but it supports x264 with full access to all configuration options—a nice positive (**www.squared5.com**). What makes this program indispensable, however, is the rare ability to trim H.264 and many other files without re-encoding, a feature I'll explore in more detail in Chapter 15. For a sneak peek at that functionality, check out a video titled "Trim Without Re-encoding with MPEG Streamclip" at **bit.ly/trimh264**. Otherwise, I prefer HandBrake for casual encoding.

Overall, if you need an encoder on a computer that's not your primary workstation, HandBrake is the best choice. When you need to trim a file without re-encoding, try MPEG Streamclip.

Let's move onto the next category of bundled encoders.

Bundled Encoders

Although Apple's Compressor is sold separately, I'm including it in this group because it's the clear choice for Final Cut Pro X (FCPX) producers. The other prominent entry is Adobe Media Encoder (AME), which is included with Premiere Pro when sold on a stand-alone basis or in a suite. Both programs offer the fastest, simplest workflow for encoding content on the editing timeline—no intermediate file to create; just choose an export preset and start encoding. You can even keep editing in the suite's respective editors, since both encoding tools render in the background.

Bundled Encoders	Adobe Media Encoder	Apple Compressor
Platforms	Windows/Mac	Mac
Price	Bundle only	$49.99
Preset Customizability	Limited	Limited
Adaptive Streaming Support	Single file presets	Single file presets
Scalability	No	Yes-Clusters
Parallel Encoding	Yes	With Qmaster
Watch Folder Operation	Yes	No
FTP Retrieve/Deliver	No/Yes	No/No
Intro/Outro/Watermark	No/No/No	No/No/Yes

Table 6-1. Features of the bundled encoders.

Working through Table 6-1, Compressor costs $49.99, and is available only on the Mac; neither is a noteworthy negative for most Final Cut Pro producers. Preset customizability is relatively limited in both tools. For example, neither lets you control an entropy-encoding technique (CABAC or CAVLC) or set B-frame interval, the number of reference frames, or any search-related options. This doesn't mean that quality suffer—AME actually outputs very high-quality H.264 video—but if you're looking for tweaking controls, neither program gives them to you. Although each program provides adaptive streaming presets to encode individual files in an adaptive streaming set—Apple for HLS and Adobe for Flash and HLS—neither tool creates the required manifest files for any adaptive streaming format.

Each product provides some bright spots in the performance realm. For example, Compressor can encode multiple files in parallel via Qmaster and share encoding tasks over multiple Compressor nodes on a network via clustering. AME can accelerate encoding with NVIDIA-based CUDA acceleration, and can encode a single file to multiple outputs in parallel, but offers no option to scale encoding over multiple computers.

File input/output is limited with both tools. AME can retrieve files for encoding from watch folders, but not FTP—although it can deliver encoded files via FTP. Compressor doesn't support watch folders or FTP input or output, removing FTP delivery functionality in 4.0.4. Compressor can insert a watermark over videos, but neither program can insert intros and outros before or after a video. Of course, you can always accomplish all three of these functions in the bundled video editor.

You'll know that you've outgrown your bundled program, particularly Compressor, when you require additional output formats or more codec configuration options. As you'll see, the three desktop programs we'll consider also have several unique advantages not offered by either bundled program.

Resources

Here are some resources for the Adobe Media Encoder:

- "Tutorial: Adobe Media Encoder CS6—Update" at **bit.ly/Ozer_AME6_1**
- "Video tutorial: Producing H.264 video for Flash distribution with the Adobe Media Encoder" at **bit.ly/Ozer_AME_2** (although I'm producing to the .F4V container format, which I no longer recommend)
- Overview of the encoding workflow at **bit.ly/Ozer_AME_3**

For an introduction to producing H.264 video with Compressor, check out **bit.ly/Compressor_tutorial**.

Stepping Up to Desktop Encoders

The third class of encoding tools are desktop encoders, which are typically designed for stand-alone use by a single user. Table 6-2 contains a summary of the tools and their respective features.

Starting at that top, Sorenson offers three versions of Squeeze 8.5: Squeeze Standard ($649), Squeeze Pro ($899), and Squeeze Premium ($1,995). Telestream offers three versions of Episode: Episode ($495), Episode Pro ($995), and Episode Engine ($3,995). I'll talk briefly about the differences between the versions here, and cover them completely in Chapter 7.

As mentioned, encoding customization is another key differentiator between bundled and desktop versions, and all three desktop encoders offer exceptional H.264 customization. Although it's not highlighted in the features table, note that both Squeeze and Episode offer both the x264 and MainConcept codecs for maximum flexibility and quality.

On the adaptive streaming front, both Expression Encoder and Squeeze offer single presets with multiple outputs within the user interface, and they produce the chunked output and metadata files required for the supported adaptive streaming formats. While AME and Compressor can output single files, they can't produce the final files necessary to distribute in adaptive-bit-rate format. With version 6.2, Episode Pro supports HLS and Smooth Streaming, but only via their command-line interface or application programming interface (API), not via program controls.

In terms of scalability, Expression Encoder operates on a totally stand-alone basis, so multiple copies on networked workstations buys you nothing. In contrast, Squeeze can send files to an installation of Squeeze Server for encoding, which is great for spreading the encoding load. Like Compressor, multiple copies of Episode on a network can share jobs, which is another convenient way to spread the load over multiple encoders.

Desktop Encoders	Microsoft Expression Encoder	Sorenson Squeeze	Telestream Episode
Price	$199	$649, $899, $1,995	$495, $995, $3,995
Platforms	Windows	Windows/Mac	Windows/Mac
Preset Customizability	Extensive	Extensive	Extensive
Adaptive Streaming Support	Smooth Streaming	HLS, HDS, Smooth, DASH	HLS, Smooth via command line
Scalability	No	Via Squeeze Server	Share via clustering
Parallel Encoding	No	Yes	Varies by version
GPU Acceleration	Yes	Yes	No
Watch Folder	No	Yes	Yes
FTP Retrieve/Deliver	No/No	No/Yes	Yes/Yes
AME/Final Cut Pro Integration	No/No	Yes/Yes	No/No
Intro/Outro/Watermark	Yes/Yes/Yes	No/No/Yes	Yes/Yes/Yes
Split and Stitch	No	Yes	Engine
YouTube Delivery	No	Yes	No
S3 Delivery	No	Yes	Yes
Notifications	No	Yes	Via API

Table 6-2. Features of desktop encoders.

In terms of compression efficiency, Expression Encoder encodes serially, while all versions of Squeeze offer split and stitch and parallel encoding. Briefly, split and stitch breaks a long file into multiple components for faster encoding on a multiple-core computer. Episode, Episode Pro, and Episode Engine can respectively encode one, two, and an unlimited number of files simultaneously, but only Episode Engine offers split and stitch. Also on the performance front, Expression and Squeeze can also encode using GPU acceleration with the MainConcept codec, although the quality may suffer using this technique.

Squeeze offers the ability to access Squeeze presets from inside Final Cut Pro 7 and Adobe Media Encoder. Both Squeeze and Episode enable email or text message notification upon encoding success or failure, though Episode's feature is via the API, while Squeeze's is through program controls. Other features in Table 6-2 are self-explanatory.

Desktop Encoder Resources

Here are some resources to learn about Sorenson Squeeze:

- "Sorenson Squeeze 8: A Video Tutorial" at **bit.ly/Squeeze_tutorial**
- Sorenson video on presets at **bit.ly/Squeeze_presets**
- My review of Squeeze 7 at **bit.ly/Squeeze7**

Here are some resources to learn about Telestream Episode:

- Introduction to Episode at **bit.ly/Episode_ozer**
- Telestream tutorials at **bit.ly/Episode_demos**

Tip: *Sorenson released Squeeze 9 in April 2013. Key new features include an updated user interface; new HTML5 presets that encode files to both WebM and H.264 formats and supply the required HTML codes for embedding; closed caption support by converting CEA-608 and CEA-708 captions into TTML output; a new filter that simplifies adding pre- and post-roll advertisements.*

Differentiating Enterprise from Desktop Encoders

Several encoding vendors, most notably Sorenson and Telestream, offer both desktop and enterprise encoders. What differentiates these desktop and enterprise products? And what are the characteristics of a true enterprise system? Let's review.

- ***Shared usage and an application programming interface (API).*** Where Sorenson Squeeze runs via an interface or watch folders, Squeeze Server runs via watch folders and an API, which lets customers integrate operation with their digital asset management systems and content management systems. Virtually all enterprise encoders feature such an API, while few desktop encoders do—although some, like Telestream Episode, offer a command line interface that is automatable.

- ***Advanced features like closed captions, digital rights management, and advertising insertion.*** If you need these features, you won't find them on any desktop encoding tool (in March 2013, anyway; they are coming). So you will need to step up into the enterprise class.

- ***Highest possible performance.*** All enterprise encoders offer parallel encoding, split and stitch and other techniques to speed encoding on the multi-core computers or dedicated hardware they typically run on.

- ***Scalability.*** All enterprise encoders offer some form of scalability, usually via server farm operation with a single point of control and failover should any modules go down. This level of controlled scalability isn't available with any desktop tools.

- ***All relevant outputs.*** There are some exceptions, but for the most part, enterprise encoders output all relevant single-file and adaptive formats.

- ***Access to quality control.*** Although it's not universal, most enterprise encoders either have a sister product that performs quality control or integration options with third-party QC programs. If quality control is essential to your environment, you'll need an enterprise-class encoding tool.

- ***Programmability.*** As we'll discuss in a moment (and as you'll see in more detail in Chapter 7), some enterprise encoders are workflow systems that incorporate features like quality control, and branched workflows based upon file characteristics like resolution or aspect ratio. If you're working with a wide variety of inputs, outputs and file destinations, workflow systems can make overall operation more efficient and reliable, and much less prone to error.

Unfortunately, there are many vendors in the enterprise encoding space, each with multiple products. This makes a features-table evaluation unworkable. For this reason, I'll first identify the three main categories of enterprise encoders, and then detail some high-level buying considerations.

Class 1: Swiss Army Knife Transcoders

Swiss Army Knife transcoders can input and output every format known to man (or woman), including camera formats, intermediate formats, and specific formats for cable playout servers and other non-traditional outputs. Swiss Army Knife transcoders that we know and love include Amberfin iCR 1101, Ateme TITAN KFE , Thomson ViBE Convergent Video System, Harmonic Rhozet Carbon Coder, Telestream Episode Engine, Sorenson Squeeze Server, Digital Rapids Transcode Manager, and Rovi TotalCode. Every large organization needs one of these encoders to simply manage the various conversions necessary for day-to-day operation.

The downside? Typically performance. Usually encoding speed falls well behind the performance of products in our next category of encoder, which are high-volume encoders.

Class 2: High-Volume Encoders

These products typically have limited input and output capabilities, but the operations they do perform, they perform at warp speed. Products in this class include the Elemental Server, VBrick 9000 Series, Envivio 4Caster, Digital Rapids StreamZ, ViewCast Niagara 9100 and Haivision Makito. Typically, these products are either sold as a hardware/software combination or come with very stringent hardware requirements.

If you're a high-volume operation, you probably need one of these around as well, but they offer limited functionality when it comes universal file transcoding, and they don't offer the capabilities of our third class of enterprise encoding, which are workflow systems.

Class 3: Workflow Systems

Workflow systems don't want to encode your files; they want to change your life. They exist so you can program in every file input and output and standardize and automate repetitive operations. For example, workflow systems can:

- Make encoding decisions based upon metadata.
- Interrogate incoming files and place them different encoding buckets; for example, placing 4:3 files in one encoding bucket and 16:9 files in another.

- Perform incoming QC checks to kick out files that don't meet set requirements or have flaws like inadequate sound or too many black frames, or letterboxing.

- Perform post-encoding quality checks and re-encode a file if it doesn't meet certain quality metrics. (For an overview of how quality control systems work, check out **bit.ly/vScvK1**.)

These tools are ideal for environments like a post house, which might need to encode files received from a variety of remote producers and deliver them to a range of outputs, from broadcast to mobile streaming. They're also ideal for content producers who need to produce and deliver their content to multiple destinations in a range for formats.

In these instances, you should consider a workflow system like Telestream Vantage Workflow Portal, the Harmonic Rhozet Workflow System, Digital Rapids Kayak, or AmberFin iCR Works (iCR 5102). These programs are typically modular in nature and sit above encoding and quality-control modules that are available separately. Interestingly, a workflow system can coexist with Swiss Army Knife transcoders and high-volume encoders, directing the work to the optimal tool to get the job done.

Figure 6-1 shows the architecture of the Harmonic Rhozet Workflow System. You see the WFS controller sitting on top, with Swiss Army Knife transcoder Carbon Coder (now ProMedia Carbon) performing all the encoding chores. In early 2013, Harmonic added a high-volume encoder, ProMedia Xpress, to the family; you'll read about that in Chapter 7. Rhozet QCS, for Quality Control System, tests files input to the system before encoding, and then after encoding and before distribution.

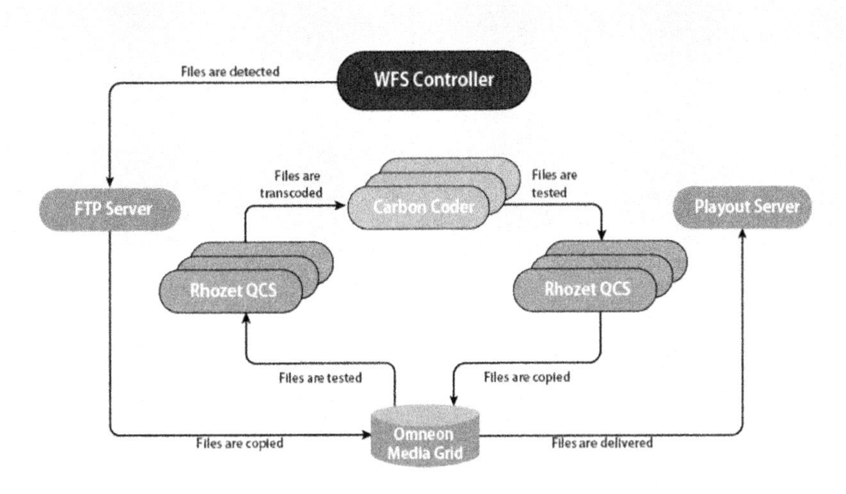

Figure 6-1. The Harmonic Rhozet Workflow System.

Working through the workflow shown in Figure 6-1, the workflow system detects and retrieves input from the remote FTP server (or watch folder), checks the files upon input, then encodes, checks post-encode quality and sends the file to the Playout Server (or web server or content management system). If a file fails inspection, or any other problem occurs, the system will send an email to the designated party, or post a report to website.

Similarly, Telestream's Vantage Workflow Portal sits atop Vantage Transcode (formerly FlipFactory, a Swiss Army Knife transcoder) and Vantage Analysis, Telestream's quality control program. New to the family is Vantage Lightspeed, a high-volume encoder you'll read about in Chapter 7.

These integrated workflows can maximize encoding efficiency and identify problem footage before delivery, both of which are exceptionally valuable features. The obvious downside is cost. For example, for all three modules (workflow, encoding, and QC) in the Rhozet system, you'll pay more than $30,000, and you'll pay more than $20,000 for the Telestream system. In contrast, Sorenson's Squeeze Server is less than $5,000, and Episode Engine costs less than $3,995. Or you can buy stand-alone encoders such as Telestream Vantage Transcode for $5,500 and ProMedia Carbon for $6,000.

The other challenge is complexity. Figure 6-2 shows a workflow built in Telestream Vantage, and they can get quite complicated. But, like anything else, after you create a few workflows, it gets much, much simpler.

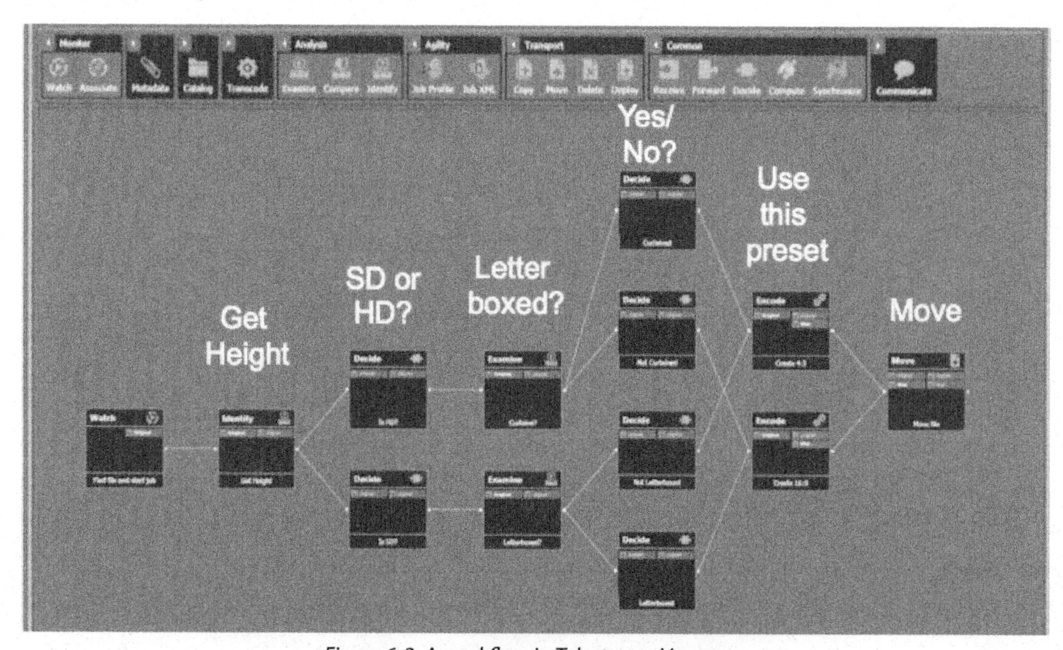

Figure 6-2. A workflow in Telestream Vantage.

So, these are three classes of enterprise encoder. Job No. 1 when buying an enterprise encoder is to choose the class that you need. Then you can at least identify the most relevant alternatives and make a better-informed decision.

Choosing an Enterprise System

When buying an enterprise system, to ensure complete integration, it's important to identify all the required input and output points that the enterprise encoder will have to tie together. Figure 6-3, which is a grab from a Telestream Vantage product webpage, provides an example of some of the typical input sources and output targets.

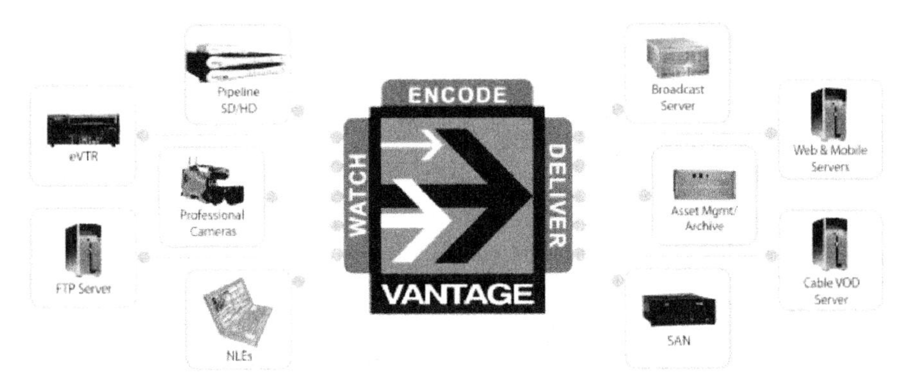

Figure 6-3. All the tools with which Telestream's Vantage encoder coexists.

Then, think about the optimal APIs for integration with existing programs. Depending upon the enterprise system, vendors offer REST, SOAP, .NET, WCF, and other integration options. In addition, since you're not working with the system via a user interface, you need to ensure that the notifications supported by the system are sufficient for your needs.

The next major architectural difference relates to how each system manages server-farm operation. There are two basic approaches: peer-to-peer, where the various encoding nodes function separately, and server-controlled operation. Generally, peer-to-peer operation is cheaper because there is no server component to buy, but load balancing is often less sophisticated because there may be no server component to monitor actual throughput or other indications of encoding efficiency. There's also no single point of failure, since if any individual encoder goes down, the rest simply pick up the load.

In contrast, server-based farm operation has a server component distributing encoding chores. Since the server component can track the performance and capabilities of the available encoding resources, load balancing can be more efficient. However, the server component is an extra cost, and it represents a single point of failure, so you may have to buy two servers for complete failover support. When comparing systems, you need to understand how the server farm works as well as the pricing model.

While metadata is typically an afterthought for most bundled or desktop systems, it becomes paramount with enterprise systems because the "packages" delivered to various distribution platforms require accurate and complete metadata. So be sure all of your short-list candidates meet your needs in this regard. The same holds true for features such as closed-caption support, which will only become more important over time, or the ability to ingest broadcast footage and remove interstitial commercials or slugs.

As a group, enterprise encoding systems are very complex, and each system has its own set of unique advantages. When buying an enterprise system, it's critical to start with a specific, detailed list of your requirements. With the foregoing as background, you should have a solid base to begin your comparisons, but expect it to be an iterative process that will involve one or two trial installations before picking a winner.

I've spoken several times about choosing an enterprise encoder at Streaming Media Conferences, and you may find the handout from my last presentation helpful, as it presents quality, performance, pricing and other comparative data. You'll get a lot of valuable data plus a framework to use when making your encoder selection. You can download the handout at **bit.ly/enterprise_enc**.

A couple of observations from the handout. First, when assessing price, be sure to include all necessary modules in your analysis, which is harder than you might think. For example, some systems need a software controller to operate a server farm, and some don't. Some need additional modules for failover and redundancy, and some don't.

Also, when comparing a hardware-based system like Elemental Server to a software-based system like Rhozet ProMedia Carbon, be sure to factor in the price of the computer you'll need to run the software encoder on. In addition, if you're a high-volume operation, be sure to factor in the increased production a product like Elemental Server delivers over most software-based encoders. For many operations, three Elemental Servers can outperform five instances of any software-based enterprise encoding tool, so you should compute overall price accordingly.

Second, on quality, every encoder claims to offer the best quality—usually because engineering teams tell their marketing folks, "Of course, our quality is the best," and the marketing folks happily spread the word. I've tested lots of these systems, and the qualitative difference is seldom commercially relevant.

Third, performance is a huge differentiator. So do download the handout and future handouts to see how the systems compare.

Enterprise Encoder Resources

Here are some other resources you may find valuable:

- "Elemental Server Review: The One to Beat" at **bit.ly/Ozer_elemental**
- "Telestream Episode Engine Review: Room for Improvement" at **bit.ly/Ozer_Engine**
- "Review: Rhozet Carbon Coder 3" at **bit.ly/Ozer_carbon** (This one is quite old, but the overall workflow hasn't changed so it's a good introduction to the product.)

- Future review: Google "Vantage Lightspeed" and "Jan Ozer" to find a review of Telestream's new high-volume encoder that I was working on while finishing this book. Ditto for "ProMedia Xpress" and "Jan Ozer."

Now let's turn our attention to the cloud.

Cloud Encoding

The economics for on-demand cloud encoding are clear: Rather than invest in your own encoding hardware and software and develop internal encoding expertise, you can save money by outsourcing both to a cloud facility. Perhaps you'll pay for a few hours' consulting time to identify the best encoding presets, but it's cheaper than employing your own compressionist, and you never have to worry about buying, upgrading and supporting your own encoding hardware or software.

The 500-pound gorilla for cloud encoding is upload time. Mezzanine files are large and upload times can be quite lengthy. Of course, if the bulk of the files you encode are uploaded to you—rather than generated internally, as with user-generated content sites or service bureaus—upload time isn't an issue, and cloud encoding is a natural. Ditto if you store your content in the cloud already.

Other sweet spots exist at both ends of the volume spectrum. If volume is low, upload time isn't so much of an issue and it makes sense to offload encoding chores to a service. If volume is very high, encoding speed is typically critical, and even with the upload time, the ability to tap into hundreds of cloud instances may speed encoding overall.

Or, if encoding chores are limited but episodic, like having to re-encode a huge library for a new target device or format, you can send a hard drive to Amazon and let it copy the files to an S3 bucket, solving the upload time dilemma. Obviously, as upload speeds increase and connectivity prices decrease, the 500-pound gorilla will shrink in importance.

Here are some questions to ask when considering a cloud encoding service for on-demand encoding:

1. ***Purchase or software as a service (SaaS)?*** While all cloud transcoders are available as a service, some are also available for purchase and installation in your own private cloud or leased cloud facility. If you're interested in purchasing, get that question on the table early.

2. ***Input and output format support?*** There will be a great deal of uniformity here, but this is worth checking up front, particularly if you're seeking to distribute beyond desktop and mobile devices into OTT and other markets.

3. ***How easy is the system to use?*** As with live transcoding, the ease of use varies dramatically by vendor. Some enable you to create presets in familiar desktop applications and upload files à la YouTube; some require that you support their API for any operation at all.

It's obviously best when the technical expertise required to work with an encoding service matches your internal capabilities.

4. *How extensive is the API and how easy is it to use?* Most high-volume users of cloud encoding services automate operation via the cloud encoder's application programming interface (API), which differ in scope and ease of use. Review the API to make sure it includes all critical elements, like notifications for all relevant encoding events and precise error codes that detail any problems to make them easy to resolve. Have your technical personnel review the API to estimate how long it will take to integrate, and to identify any potential feature gaps.

5. *What's the price and how simple is the pricing model?* Again, simpler is better. Most of the larger vendors are moving towards simplified fees based on the size of files imported and created, with performance aspects like encoding speed dictated by service-level agreements and volume commitments. Other vendors look at the actual CPU instances that are created for a particular job, which forces the client to make the tradeoff between encoding speed and encoding cost.

6. *How stable is operation?* Although 99.9999% uptime probably doesn't matter for smaller clients, large media companies need guaranteed throughput and therefore redundancy. For this reason, some of the larger cloud encoding vendors operate on the Amazon Cloud and Rackspace, as well as having their own private encoding facilities.

Regarding the robustness of the encoding facility itself, if you have a library of problem files that have crashed your existing encoding programs, try uploading them to your candidate cloud facilities to see how they are handled. Errors happen, but look for clear error messages and notification capabilities that detail what's happening and why.

7. *What's the guaranteed speed and throughout?* One key benefit of the cloud is the ability to scale capacity to meet demand and then shut off capacity when not needed. If you're a news organization or UGC site with an emphasis on encoding speed, scrutinize your service-level agreements to ascertain the throughput commitments the cloud encoding service is willing to make. This may include guarantees relating to queue time and encoding time.

If speed is critical to your operation, try uploading several large files to different suppliers and timing the upload, queue and encoding process. Check each service two or three times at various hours during the day, and you should to get a feel for comparative performance.

8. *How's the service?* Ascertain the hours and cost for real-time telephone support. Consider calling the support centers to get a feel for the technical level of the support personnel. Most are trained to solve simple problems, like login or server address problems, but may not be sufficiently knowledgeable for more complex issues, like the optimal encoding configuration for your source encoder.

9. *Does the service offer the necessary features?* If you'll need closed captioning, DRM and advertising insertion, check early, because not all cloud vendors offer these services. Beyond this, if a substantial component of your video library exists in the cloud, you may require basic editing features like trimming, splitting and concatenating video files, along with text and image overlay. You'd also want these features available via the API so you can automate them.

Another class of features centers around translation. If you're a multinational organization distributing videos worldwide, you may need translation assistance to create voice-overs in other languages or via captioning. The ability to integrate these services into the encoding workflow is much more efficient than handling it as a pre- or post-process.

Yet another class of features centers around format support. Virtually all cloud services can accept H.264 files, but what about Avid DNxHD or Apple ProRes? Even fewer will accept QuickTime reference movies, which may be the most efficient way to upload multiple iterations of the same basic content.

Finally, if you need both live and on-demand encoding services, it's usually simplest and cheapest to get them both from the same vendor. So if you're in the market for both types of services, check to see if your candidate vendors support them both.

Conclusion

So now you know the different categories of encoder, and hopefully have a good idea about which category best suits your needs, the products within that category and features to consider when choosing among them. In Chapter 7, you'll learn how to operate many of the products discussed herein.

Chapter 7: Encoder-Specific Instruction

> ***Note:*** You can download a PDF file with all figures from this book at bit.ly/Ozer_multi. Since this chapter contains lots of images, now would be a good time to do so.

Up until now, we've learned a bunch of theory; now it's time to translate that into practice. Specifically, in this chapter, you'll learn how to encode into H.264 format using a variety of desktop and enterprise encoders. I'll describe encoders I have access to and know reasonably well; if I don't cover your tool, I apologize in advance.

For each desktop encoder, I'll cover a range of topics, including the availability of common configuration options like bit rate and key frame control, and your best options for encoding to H.264 format for both single-file and adaptive delivery. Basically, it's a brain dump that transfers everything I know about making these encoders work most efficiently.

With the enterprise encoders, which are much more complex, I'll demonstrate the high-level workflows so you can get a feel for how the programs operate. These are really cool tools, but really deep, so a quick flyover is all we have time for. The encoders that I'll cover include:

- ***Desktop encoders.*** Adobe Media Encoder, Apple Compressor, Sorenson Squeeze and Telestream Episode.
- ***Enterprise encoders.*** ProMedia Carbon, Telestream Vantage Transcode Multiscreen, Harmonic ProMedia Xpress, and Elemental Server.

Note that when discussing the presets used for encoding for iDevices, I refer to an iTunes usage survey in Chapter 8 that you won't be aware of if you're reading this book linearly. If you're doing a lot of encoding for iTunes or iDevices, you should skip ahead and scan the survey-related data presented in the next chapter.

Again, I'm sorry if I'm not including the encoder you're using.

Adobe Media Encoder

The Adobe Media Encoder (AME) is included with each copy of Adobe Premiere Pro and Adobe After Effects, and all bundled Creative Suites that contain these products. It's a very functional product that produces very high-quality H.264 video using the MainConcept H.264 codec.

H.264 codec. MainConcept

Adaptive-bit-rate strategy. None. To create an adaptive group, you apply multiple presets to the source file. AME can only output .MP4 files, which will work for RTMP-based Dynamic Streaming. If you're producing files for HTTP Live Streaming, HTTP-based Dynamic Streaming or Smooth Streaming, you'll have to use another tool to create the file chunks and metadata files.

Encoding efficiency. When you apply multiple presets to a single file, AME will encode all files at once, in parallel. If you load multiple source files into the batch window, AME will encode them in order. If there are multiple presets applied to a source file, AME will encode them in parallel. If only a single preset is applied, AME will encode that source file using that preset before starting any subsequent encodes.

Other notable features. AME has a very usable watch folder feature; you can watch a tutorial describing that feature at **bit.ly/AME_watch**.

General encoding controls. Sufficient.

H.264 controls. Very limited; choose Profile and Level; no B-frame, search or other H.264 enabled controls.

iTunes presets. See Table 7-1. Profiles appear to be correct for all presets (but double-check anyway). The AME data rate is a bit low for the first three categories, but close to the 1080p rate used for HD TV episode downloads. Note that Adobe uses the same data rate for 23.976, 25 and 29.97 fps, despite a 20% difference in the respective bits-per-pixel value due to the additional frames. If your 29.97 video doesn't look great, consider increasing the data rate. Conversely, if your 23.976 video looks pristine, drop the data rate by 20% and see if the quality remains sufficient. Less is always more when it comes to download time and file size on your iDevice.

	Profile	**Video**		**Audio**	
Preset		**Survey Average**	**AME Preset**	**Survey Average**	**AME Preset**
320x240x29.97	Baseline	532 kbps	400kbps	95 kbps	160 kbps
640x360x29.97	Baseline	1.14 Mbps	750 kbps	113 kbps	160 kbps
1280x720x29.97	Main	2.9 Mbps	2.2 Mbps	169 kbps	160 kbps
1920x1080x29.97	High	5.08 Mbps	5 Mbps	158 kbps	160 kbps

Table 7-1. AME's iDevice settings.

Getting acquainted. If you're totally unfamiliar with the AME, there are a ton of free video tutorials that you can view, many produced by yours truly. For example, you can check out "Tutorial: Adobe Media Encoder CS6—Updated" at bit.ly/Ozer_AME6_1. There's a more H.264-specific tutorial at bit.ly/Ozer_AME_2, although I'm producing to the .F4V container format, which I no longer recommend. My bank account compels me to mention that I produced a training series on Adobe Media Encoder for Video 2 Brain, which has several free videos, including an overview of the encoding workflow, which you can watch at bit.ly/Ozer_AME_3. If you're new to the program, I suggest that you watch at least the first two videos to get acquainted with the program before reading any further.

Encoding with AME. AME ships with a ton of presets segmented by container format and target. While device presets are appropriately encoded into the .MP4 container format, all Flash videos are produced into either the .FLV container format, for videos encoded using the VP6 codec, or the .F4V container format, for H.264-encoded videos. Since these files wouldn't load or play on many devices, unless your playback target is exclusively Flash, it's best to use the .MP4 format—which plays everywhere, including on the Flash Player.

Creating a preset is easy. To open an existing preset, right-click a preset in the Preset Browser and choose Preset Setting.

Figure 7-1. Opening a preset so you can customize the settings.

Then choose the H.264 format as shown in Figure 7-2.

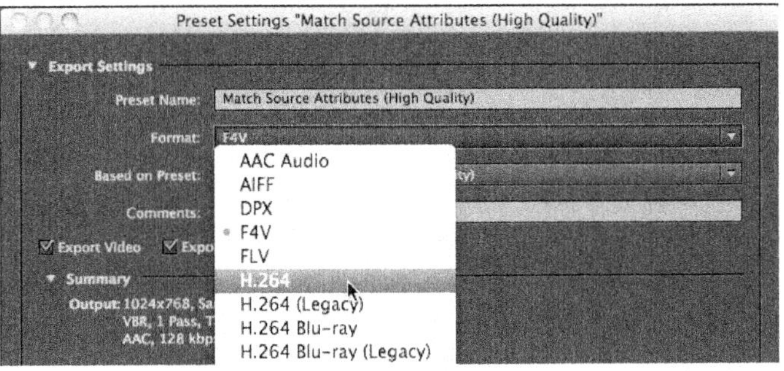

Figure 7-2. Choosing the H.264 format.

Then modify the preset as desired, select a new name, and click Save A Copy on the lower right, which saves the new preset into the User Presets & Groups folder.

As noted, H.264 configuration options are limited; you can choose the Profile and Level, but that's it. If you attempt to enter in configuration options that exceed the selected Level, AME will open an error message telling you that you've done so. If you're encoding for devices, you should back down on your configuration options so as not to exceed the capabilities of the target device. If you're encoding for computer or OTT playback, just choose a higher level.

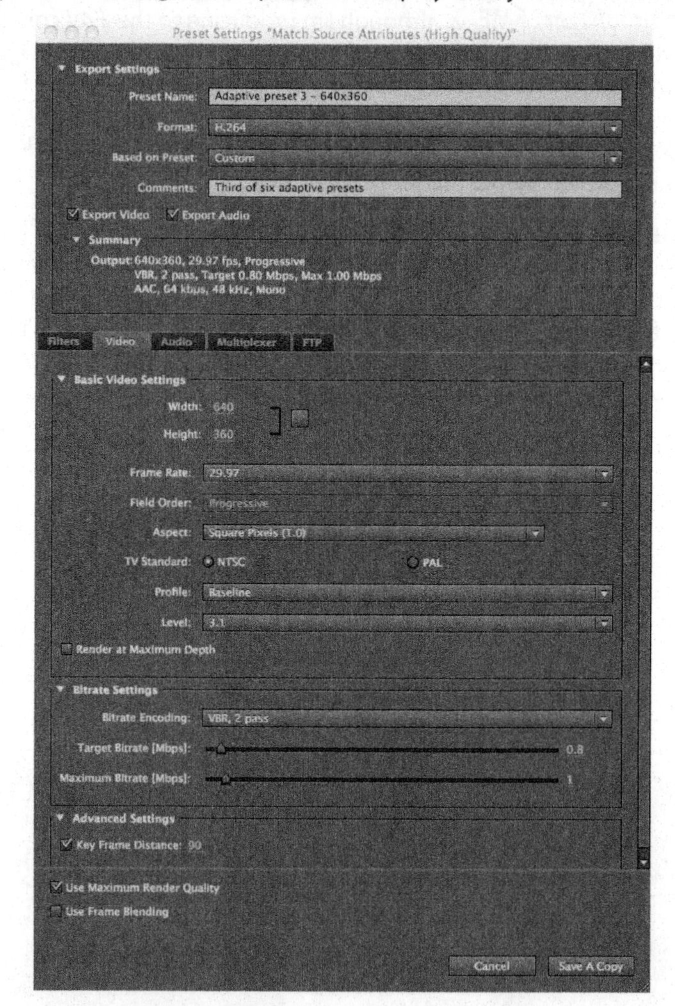

Figure 7-3. AME video encoding parameters.

H.264 Output

Note that your profile selection also controls the B-frame interval and the entropy encoding technique applied to the file. Details are shown in Table 7-2, and there is no way to change these results.

Profile	B-frames	Reference Frames	Entropy Encoding
Baseline	NA	3	CAVLC
Main	3	3	CABAC
High	2	3	CABAC

Table 7-2. H.264 details by H.264 Profile with AME.

Quirks, Tips and Tricks

Here are some quirks, tips and tricks regarding AME's operation.

- ***Use Maximum Render Quality checkbox.*** You should always enable the Use Maximum Render Quality checkbox for final output (as opposed to draft output) because it can noticeably improve output quality. You can see a tutorial discussing that at bit.ly/maxrenderquality. If you have an NVIDIA card with CUDA acceleration in your computer, you shouldn't notice any significant increase in encoding time, but you almost definitely will if you have a graphics card from a different vendor. For details on this, check out bit.ly/AME_graphics.

- ***Use Frame Blending.*** Check this only when changing the frame rate during encoding.

- ***Render at Maximum Depth.*** I haven't seen this control have any impact on quality and never enable it.

- ***Data rate controls.*** Not a huge deal, but the data rate controls are in Mbps. So, in Figure 7-3, you have to enter 0.8 to encode at 800 kbps.

- ***Minimum data rate.*** The minimum video data rate supported by the H.264 codec in AME is 190 kbps.

Producing Audio

Click the Audio tab in the Preset Settings dialog to open those parameters. Most controls are straightforward; I always select High Audio Quality and Bit Rate to take precedence over sample rate.

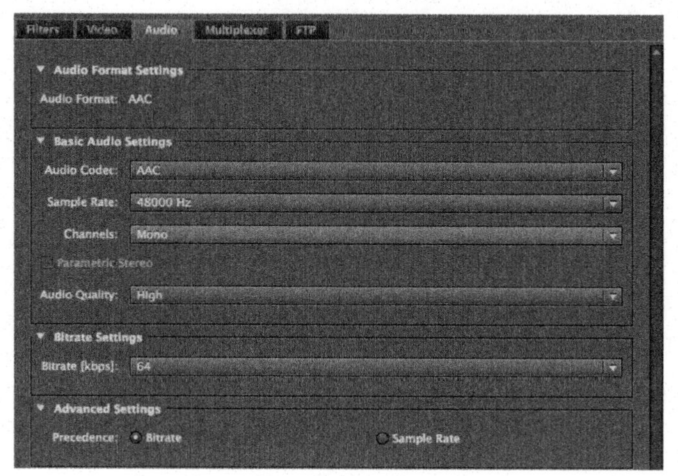

Figure 7-4. AME audio encoding parameters.

OK, that's it for AME; next up is Apple Compressor.

Apple Compressor

Apple Compressor is a very popular encoding tool that provides great control over output, but is generally hard to use and offers minimal H.264-related controls. More seriously, the Apple codec produces the lowest output quality of all software-only H.264 codecs that I've tested, so it should be avoided whenever possible. You can see an example of this in Figure 7-5, where I compared the Apple codec with the x264 codec you'll learn to use below. You can read all about it at **bit.ly/x264vApple**.

H.264 codec. By default, the Apple codec. You can load QuickTime-compatible x264 codecs as discussed below, and I recommend doing so.

Adaptive-bit-rate strategy. None. To create an adaptive group, you apply multiple presets to the source file. Compressor will output multiple .MOV files that should work for RTMP-based Dynamic Streaming. If you're producing files for HTTP Live Streaming, HTTP-based Dynamic Streaming or Smooth Streaming, you'll have to use another tool to create the file chunks and metadata files.

Figure 7-5. Apple's codec is a clear step behind x264 in this 720p file compressed at 800 kbps.

Container format strategy. Compressor can create .M4V, .MOV and .MP4 files. When creating a setting, choose H.264 for Apple Devices to create an .M4V file, MPEG-4 to create a file an .MP4 file, and QuickTime Movie to create an .MOV (Figure 7-6).

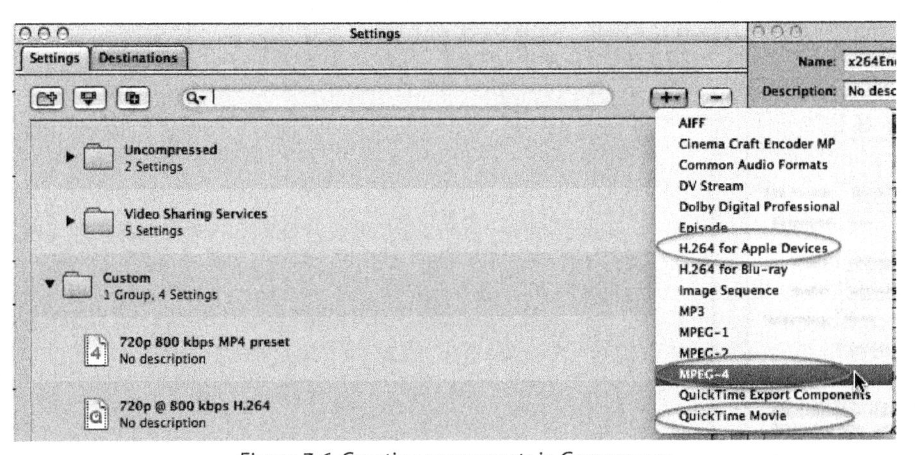

Figure 7-6. Creating new presets in Compressor.

Each selection has a completely different interface with different capabilities, which are shown in Table 7-3. All let you configure the basics—like resolution, aspect ratio and data rate—and use the same Apple H.264 codec by default, so output quality is identical. Otherwise, there are some very key differences. Here are some highlights:

	Choose Profile	High Profile	Choose VBR/CBR	Single/ Multipass	Keyframe interval	B-frame on/off	Access x264
H.264 for Apple Devices (M4V)	No	Yes	No	Yes	Frame Sync	No	No
MPEG-4 (MP4)	Yes	No	No	Yes	Yes	No	No
QuickTime Movie (MOV)	No	No	Yes	Yes	Yes	Yes	Yes

Table 7-3. Capabilities of different H.264 encoding interfaces.

H.264 for Apple Devices. This interface outputs .M4V files. While you can encode using the High profile, you can't select it directly; you have to choose the right iDevice preset, which you'll learn below. Otherwise, you have minimal control over how the file is encoded, including the inability to specify VBR or CBR, or enable or disable B-frames. You also can't access the x264 codec from this interface.

MPEG-4 Interface. This interface outputs .MP4 files and lets you choose the profile, but only the Baseline or Main profiles are accessible, not the High profile. Otherwise, you have minimal control over how the file is encoded, including the inability to specify VBR or CBR, or enable or disable B-frames. You also can't access the x264 codec from this interface.

QuickTime Movie. The QuickTime interface exports .MOV files and provides the most controls, but always encodes using the Main profile; you can't choose the Baseline or High profile from this interface when encoding using the Apple codec. Since the x264Encoder we'll discuss later in the chapter is a QuickTime component, this is the only interface you can use to access that codec. The primary downside of the QuickTime Movie interface is that it outputs .MOV files, not .MP4.

> **Tip:** *The .MP4 container format is based largely upon the QuickTime container format. I've had near universal success producing .MOV files with the QuickTime Movie interface and then simply changing the extension from .MOV to .MP4. Changing the extension in this manner should work universally for Flash playback, since the Flash Player ignores the extension and directly interrogates the file header. However, if you plan to use this technique for broader distribution, you should check to see if files with changed extensions work in all target devices and any tools along the way (like transmuxing servers).*

Encoding efficiency. Apple Qmaster is a feature that should enable more efficient parallel encoding, which speeds the encoding of single and multiple files. However, since the changeover to Compressor 4, I've never been able to get it to work reliably. You can find a PDF file detailing how to set up Qmaster at **bit.ly/qmaster_setup**.

In addition, all Macs with Compressor on a local area network (LAN) should be able to send jobs to other Compressor-equipped Macs on the same LAN. This is a great way to use computer resources that might otherwise lay fallow, particularly when you're on a deadline. Operationally, you accomplish this by creating a cluster of Macs on the LAN and then sending encoding jobs to machine configured to accept and process them. You can learn how to set up and use a cluster at **bit.ly/create_cluster**.

Other notable features. Compressor allows you to create droplets, which are an efficient way to encode files without working through the Compressor interface. Specifically, a droplet is an icon on the desktop tied to a Compressor preset (Figure 7-7). Drag a source file onto the icon, and you start an encoding run. Larry Jordan has a nice tutorial on creating droplets on the Peachpit website, which you can access at **bit.ly/Larry_presets**.

Figure 7-7. A droplet on my desktop.

General encoding controls. Very extensive but equally confusing. Very few users I know of are comfortable working through the Compressor's various screens.

H.264 controls. With the Apple codec, flexibility depends upon whether you're producing .M4V, .MP4 or .MOV files.

iTunes presets. Although there are only four settings in the Apple Devices folder (on the left in Figure 7-8), controls in the Inspector window let you choose from among 10 different presets.

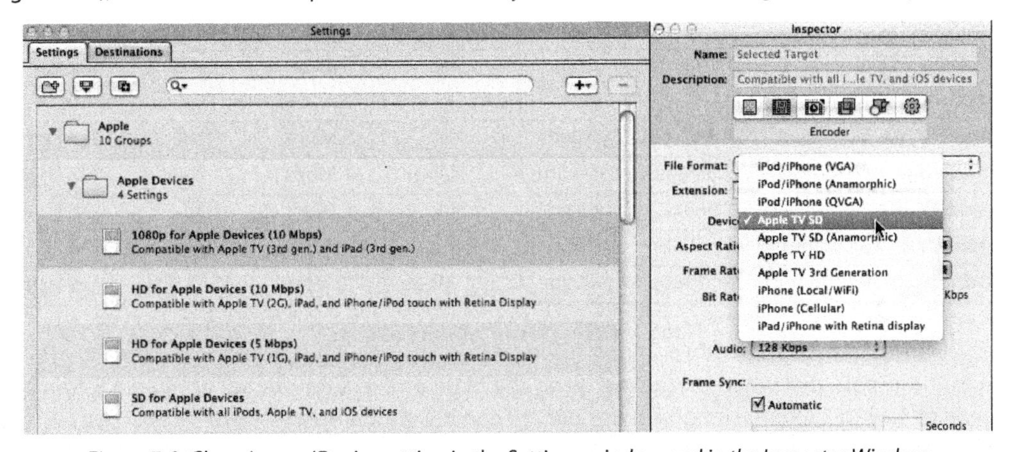

Figure 7-8. Choosing an iDevice setting in the Settings window and in the Inspector Window.

Table 7-3 contains a summary of the encoding parameters used in each setting. Profiles appear to be correct for the first few presets. The iPhone (Local/Wi-Fi) is funky; it makes no sense to use the Baseline profile with video produced at 854x480, since all devices that can load 854x480 files can play the Main profile, but I'm not sure where you would use that preset anyway. The Apple TV 3rd Generation 720p preset is correct for that device, but exceeds the specs of the first 720p-capable devices (iPhone4/iPod touch 4/iPad 1)—although I know my iPad can play 720p files encoded with the High profile. I'm not sure I would recommend doing so, of course, but there is no way to change the profile in the preset.

Apple's data rates are OK for the lowest-quality files, but start diverging from survey results at 640x360 resolution. Since there were no survey results for the iPod/iPhone (Anamorphic) preset, I compared it with the average of the music videos and SD TV shows, and it was right in the ballpark: 1.5 compared to 1.54. That's why there's an asterisk next to the 1.54.

There were no survey or other equivalents to the next few presets, so I skipped those, while 720p and 1080p were clearly out of whack (I compared the 1080p with HD TV episodes, hence the two asterisks). I would definitely adjust these down to survey-levels before encoding. Note that all audio except the iPhone (Cellular) preset is produced at 128 kbps, which should be fine for most non-music video or network TV productions. The bottom line is that when the preset values for data rate diverge significantly from the survey average, I would use the survey average.

	Rez	**Profile**	**Video**		**Audio**	
Preset			**Survey Average**	**Apple Preset**	**Survey Average**	**Apple Preset**
IPhone (Cellular)	176x99	Baseline	56 kbps	NA	NA	24 kbps
iPod/iPhone (QVGA)	320x180	Baseline	532 kbps	600 kbps	95 kbps	128 kbps
iPod/iPhone (VGA)	640x360	Baseline	1.14 Mbps	1.5 Mbps	113 kbps	128 kbps
iPod/iPhone (Anamorphic)	640x480	Baseline	1.54 Mbps*	1.5 Mbps	113 kbps	128 kbps
Apple TV SD (Anamorphic)	720x480	Main	NA	3 Mbps	NA	128 kbps
Apple TV SD	853x480	Main	NA	3 Mbps	NA	128 kbps
iPhone (Local/WiFi)	854x480	Baseline	NA	1 Mbps	NA	128 kbps
Apple TV HD	960x720	Main	NA	5 Mbps	NA	128 kbps
Apple TV 3rd Generation	720p	High	2.9 Mbps	10 Mbps	95 kbps	128 kbps
iPad/iPhone Retina	1080p	High	5.1** Mbps	10 Mbps	158 kbps	128 kbps

Table 7-4. Compressor's iDevice settings.

Interestingly, note that when producing using the Main profile, Compressor applied a B-frame interval of 1, with 2 reference frames. As near as I can tell, no Compressor preset ever uses CABAC entropy encoding; every file I analyzed used CAVLC. I can't tell if the B-frame strategy or the entropy encoding strategy are intentional or due to limitations in Compressor. They diverge from my recommendations, but I stand by my recommendations and point out Compressor's defaults in the interest of full disclosure.

Getting acquainted. If you're totally unfamiliar with Compressor, there are a ton of free video tutorials that you can view, including one you can access on YouTube at bit.ly/Compressor_tutorial. If you're a total Compressor newbie, I suggest that you watch this video before reading any further.

Encoding with Compressor. Not to be judgmental, but in its quest to simply the interface, Apple has made working with H.264 very obscure. Let's work through the three different interfaces and you'll see that examples abound.

H.264 for Apple Devices Interface

Figure 7-9 shows the H.264 for Apple Devices preset. Here are some tips for using this interface.

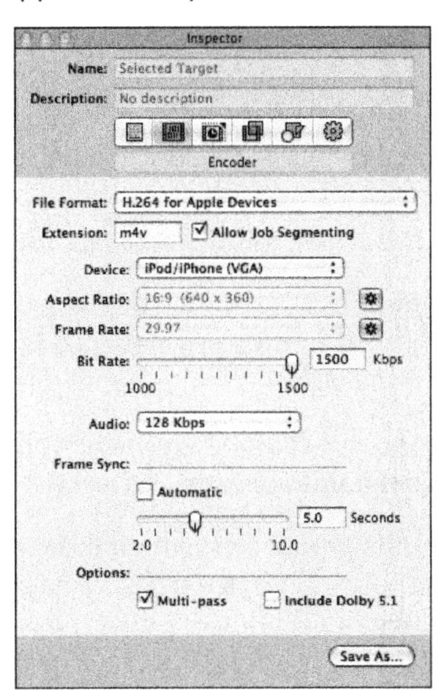

Figure 7-9. H.264 controls in the H.264 for Apple Devices encoding interface.

- ***H.264 profile.*** You can't specify H.264 profile directly, although changing presets in the Device list box will change the profile.
- ***Other controls.*** As mentioned above, there are no controls for setting VBR or CBR, and you set the key frame interval via the Frame Sync control.
- ***Producing audio.*** All audio produced via this interface is AAC-LC audio, and your only option is data rate in the Audio drop-down list.

MPEG-4 Encoding Interface

Figure 7-10 shows the MPEG-4 encoding interface. Here are some tips for using this interface.

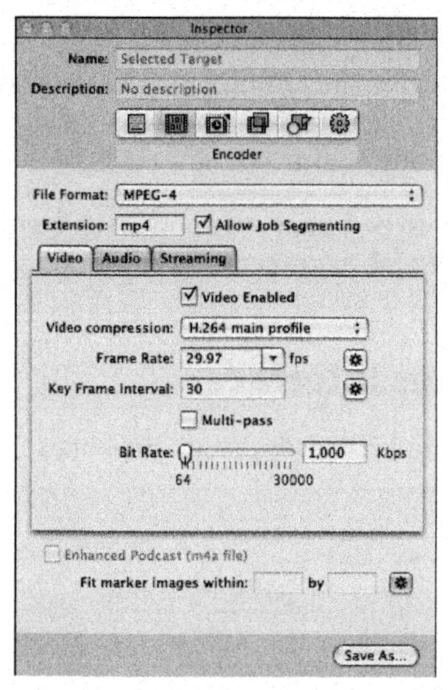

Figure 7-10. H.264 controls in the MPEG-4 encoding interface.

- ***H.264 profile.*** In the Video compression list box, you can choose the Main or Baseline profiles, but not the High profile. In my tests, files encoded using the Main profile used no B-frames and CAVLC rather than CABAC.

- ***Other controls.*** As mentioned above, there are no controls for setting VBR or CBR, and you can't enable or disable B-frames directly.

- ***Producing audio.*** All audio produced via this interface is AAC-LC audio. Click the Audio tab to view these controls, which should be familiar. There is a quality list box, which I always set to High (Figure 7-11).

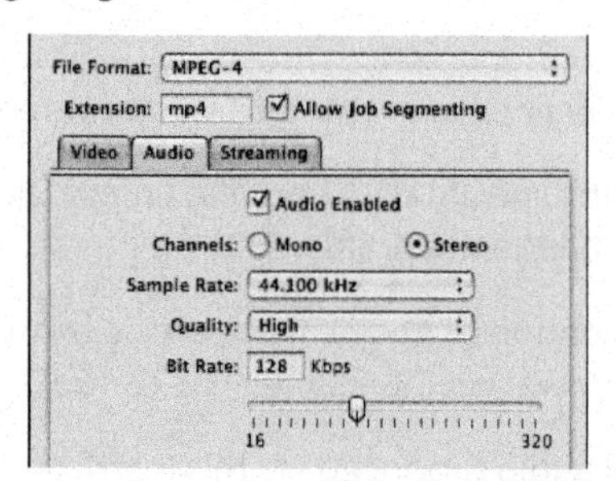

Figure 7-11. Audio controls in the MPEG-4 encoding interface.

- ***Streaming controls.*** Click the Streaming tab to open those controls. The Streaming hints enabled checkbox is not checked in any of Apple's streaming-related presets, and I recommend the same.

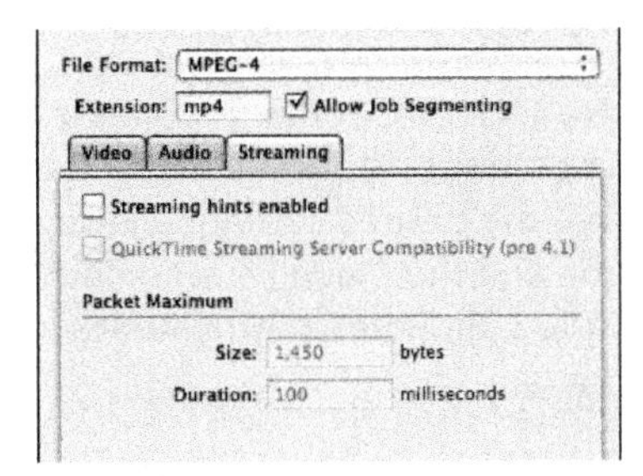

Figure 7-12. Streaming controls in the MPEG-4 encoding interface.

QuickTime Movie Encoding Interface

Figure 7-13 shows the H.264 controls in the QuickTime Movie encoding interface, which you open by clicking the Video Settings button shown on the lower left and selecting H.264 as Compression Type if necessary. Here are some tips for using this interface.

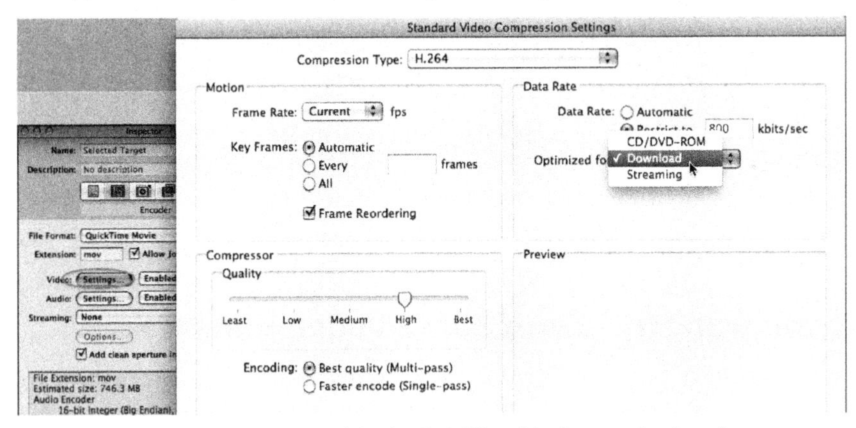

Figure 7-13. H.264 controls in the QuickTime Movie encoding interface.

- ***H.264 profile.*** Compressor uses the Main profile for all files encoded from this interface. To produce Baseline files, use a different interface.

- ***B-frames.*** You enable B-frames by checking the Frame Reordering checkbox, and Compressor uses a B-frame interval of 1.

- ***Key frame interval.*** Apple recommends setting the key frame interval to Automatic when encoding with H.264, which lets Compressor choose the optimal key frame interval (**bit.ly/QT_h264_keyframes**). The obvious exception to this is when creating presets for adaptive streaming, in which case you should choose an appropriate setting.

- ***Producing audio.*** To produce AAC audio from this interface, click the Audio Settings button to open the Sound Settings dialog shown in Figure 7-14. Choose AAC from the format list box, then choose the desired settings. I always set quality at Best. Average Bit Rate is the default mode and the mode recommended in Apple Technical Note TN2271 (**bit.ly/TN2271**) titled "Bit Rate Control Modes for AAC Encoding."

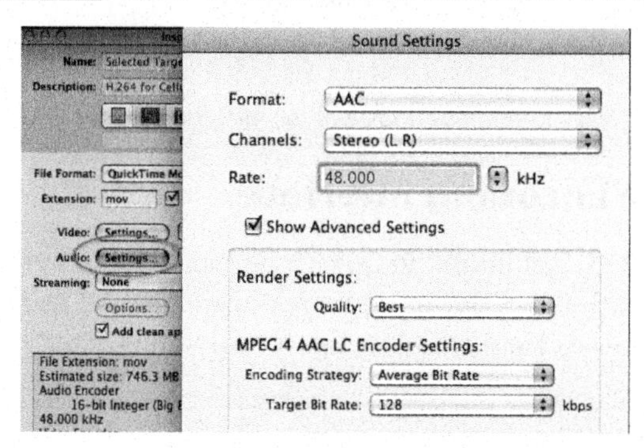

Figure 7-14. Audio controls in the QuickTime Movie encoding interface.

Here's what Apple says about the Average Bit Rate mode: "A target bit rate is achieved over a long-term average [typically after the first few seconds of encoding]. Unlike CBR mode, this mode does not provide constant delay when using constant-bit-rate transmission, but this mode provides almost best global quality while still being able to strictly control the resulting file size."

Downloading and Installing the x264 Component

As shown in Figure 7-5, the Apple H.264 codec can't match the quality of the x264 codec, so I avoid its use. Instead, I recommend the x264Encoder, a QuickTime component available from a website named MyCometG3, which was run by Japanese developer Takashi Mochizuki. Note that Mochizuki discontinued development on the component at the end of 2011, so the code is frozen as of that date. This means that as the x264 development community advances the codec further, you won't get the benefits of these advances.

That being said, the existing download available at MyCometG3 is current as of December 30, 2011. As far as I know, this is the most advanced x264-based implementation available that you can operate within Compressor. I used this update to encode the x264 files shown in the figures above, so obviously this component produces much higher quality than the Apple codec.

When you click the download link on MyCometG3 (bit.ly/X264encoder), you'll download a DMG file that you can double-click to open. Within the contents, you'll find a file named x264Encoder.component. Copy that into your Library/QuickTime folder, and the next time you run Compressor (or QuickTime, for that matter) the x264 codec will be available as shown below. Although you probably won't need it, there's a YouTube video showing you where to copy the file at youtu.be/sZI0ET3r4O4.

Using the x264Encoder

To use the x264Encoder within Compressor, you can change the codec in an existing preset or create a new preset; I'll show the latter technique. In Compressor's Settings window, click the Create a New Setting list box and choose QuickTime Movie, which opens a new setting in the Inspector window.

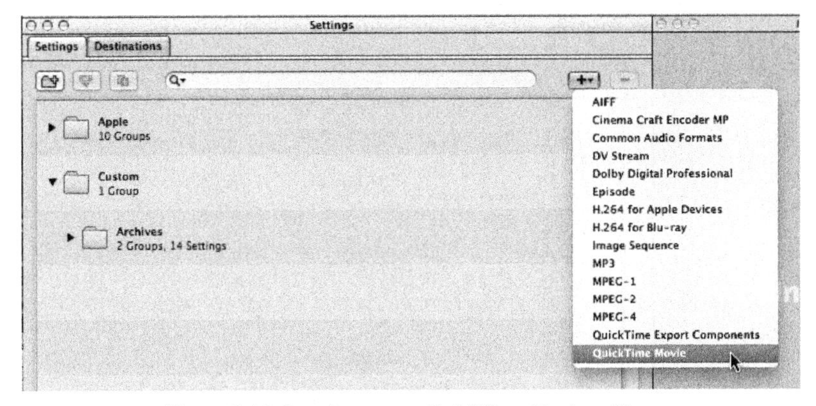

Figure 7-15. Creating a new QuickTime Movie setting.

In the Inspector window, name the setting something memorable. Then, click the Video button to open the Standard Video Compression Settings dialog. Click the Compression Type list box and choose the x264Encoder.

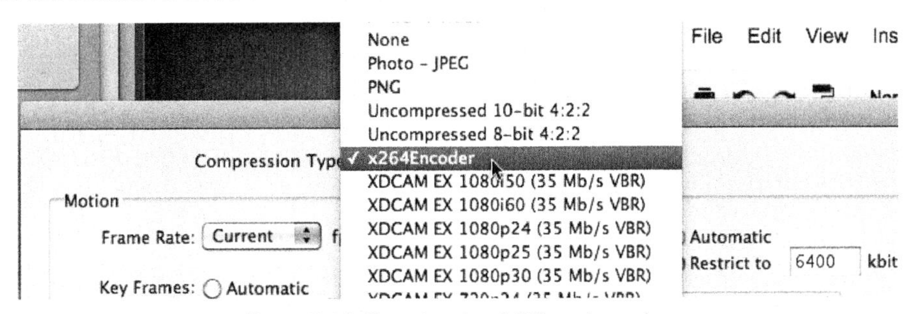

Figure 7-16. Choosing the x264Encoder codec.

x264 Presets and Tuning

Next, click the Options button to open the libavcodec settings dialog. While the settings are very comprehensive, there are only one or two that you need to adjust, which I'll come back to in a moment. First, click Load preset on the bottom left to load an encoding preset.

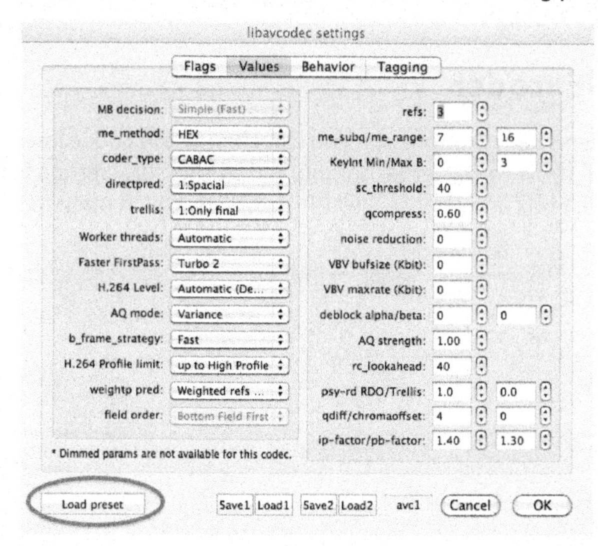

Figure 7-17. The libavcodec settings dialog with the myriad options enabled by the x264Encoder.

Briefly, the developers of x264 created the 10 presets shown in Figure 7-18 to enable x264 users to simply and easily balance encoding time against quality. You'll see the same presets whether you access x264 via plug-ins like x264Encoder, free programs like HandBrake, or professional encoding tools that offer x264 encoding like Sorenson Squeeze or Telestream Episode. As with these tools, when you choose a preset, the x264Encoder applies a certain set of parameters that you can later customize. If you're interested, you can find a complete breakdown of the encoding parameters used in each preset at **bit.ly/x264_explained** and some performance (but not quality) comparisons at **bit.ly/x264_speed**.

I produced the file comparison shown in Figure 7-5 using Medium preset, which is the default preset, and one that for most tools seems to represent the best balance of encoding speed and quality. Note that if you choose slower presets you increase the risk of creating files that may not play on your target platforms.

As an example, if you encode using the Slower preset, the encoder uses 8 reference frames. If you scan through the readme file included with the x264Encoder download, you'll note that files with more than six reference frames may not play in older versions of QuickTime. To avoid potential problems like these, it's best to use the Medium preset unless you're certain you can correct any potential problems before encoding.

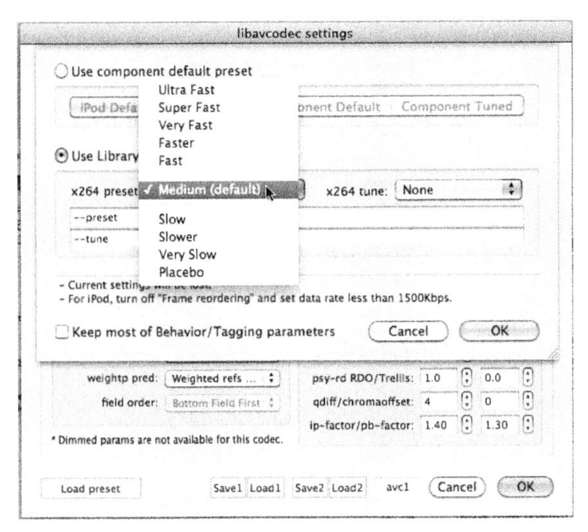

Figure 7-18. Choosing an x264 encoding preset.

In addition to choosing a preset, you can also choose an x264 tuning option as shown in Figure 7-19. On its website, Zencoder, a cloud encoding facility that uses the x264Encoder, explains the tuning options at **bit.ly/x264_tuning** as follows:

- ***Film.*** Optimized for most non-animated video content (not only feature films).
- ***Animation.*** Optimized for animation. Note that most 3D animation behaves more like film and not like hand-drawn animation, so only use this for hand-drawn animation (anime, classic Disney, etc.)
- ***Grain.*** Optimized for film with high levels of grain.
- ***PSNR.*** Uses peak signal-to-noise ratio to optimize video quality.
- ***SSIM.*** Uses structural similarity to optimize video quality.
- ***Fast Decode.*** Reduces encoding complexity to allow for easier decoding.
- ***Zero Latency.*** x264 will keep an internal buffer of frames to improve quality; this setting removes that buffer, but reduces quality.

In general, if your video falls in one of the first three categories or is a Touhou game, you should apply the fine-tuning option. Otherwise, I generally use None, though some notable compressionists, like Amazon's Ben Waggoner, recommend the Film tuning option. If you have some representative footage you can test, you should test with Film and None and see if there's a difference with your footage. I ran these comparisons with my standard test files and saw no difference.

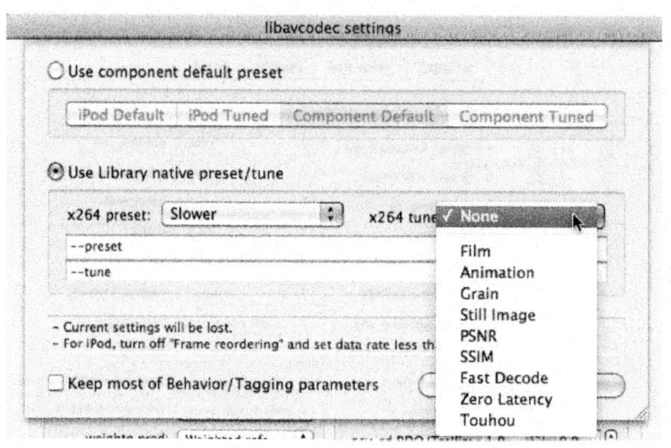

Figure 7-19. Choosing a tuning option.

Once you choose your preset and tuning option, click OK to close that dialog and return to the libavcodec settings screen shown in Figure 7-17 (and again in Figure 7-20, below, for convenience). Here's where you can customize one or two specific parameters for your target playback platforms. For example, if encoding for older mobile devices, you should limit the H.264 profile to the Baseline profile. Unless you absolutely know what you're doing, I would leave all other parameters at those selected by your preset.

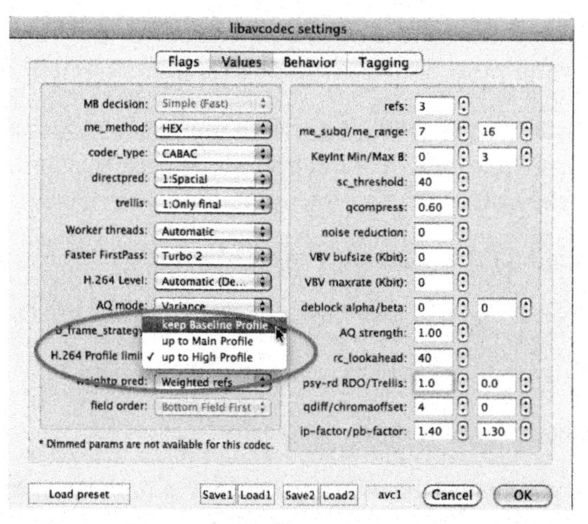

Figure 7-20. Customizing H.264 encoding settings.

When you're done, click OK to return to the Standard Video Compression Settings screen. In general, you can configure and save these and all other Compressor settings as you have in the past.

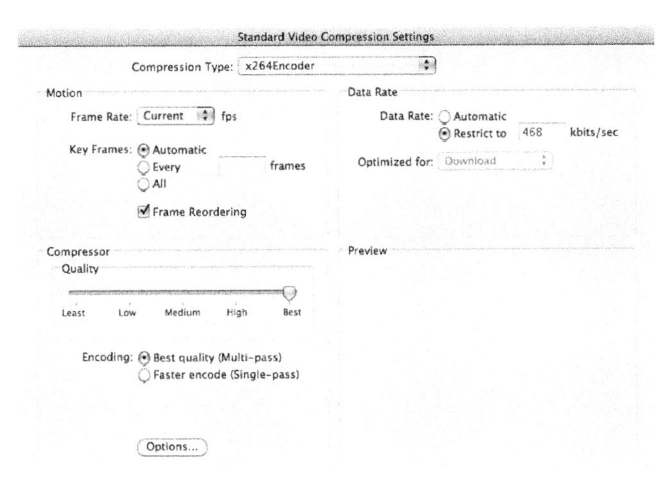

Figure 7-21. Finalizing your video encoding parameters.

For those seeking additional information on the x264 encoding options, here's a list of x264 resources—although if you Google "x264," you can find thousands more.

- ***The x264 MeWiki.*** A comprehensive guide to x264 encoding options (mewiki.project357.com/wiki/X264_Settings)

- ***Diary Of An x264 Developer.*** The blog of Jason Garrett-Glaser, the primary developer of x264 (x264dev.multimedia.cx). Very topical and useful.

- ***Doom9 Forum.*** Doom9 has some of the most active forums on x264. If you have a particular question or issue, Google that issue plus "Doom9," and you'll probably find a discussion on point.

Any time you change codecs or encoding parameters, there's a potential compatibility risk. Remember, however, that since YouTube, Zencoder, Sorenson, Telestream, and many other companies use the x264 codec, there's nothing inherently wrong with the codec; if there is a problem, it almost certainly relates to your settings. You can minimize the impact of any incompatibility by testing the first few files you encode on all relevant target platforms.

Note that potential incompatibilities are most likely on devices that have the least tolerance for out-of-spec video encoding. Test tablets and other mobile devices first. Note that there's a list of x264-related resources below in the Squeeze section titled "Tuning Your x264 Encodes."

That's it for Compressor; now let's move to Sorenson Squeeze.

Sorenson Squeeze Desktop

Sorenson Squeeze is one of the best and most popular desktop encoding programs on the market. In addition to working on a stand-alone basis, Squeeze can send jobs to high-volume encoder Squeeze Server, and serves as the preset creation tool for Squeeze Server. There are three versions of Squeeze Desktop: Standard, which costs $649; Pro, which costs $899; and Premium, which costs $1,995. I've cribbed a chart from the Sorenson website as Figure 7-22 to explain the differences (**bit.ly/squeeze_ver**).

	Squeeze Standard Learn More | Buy | Upgrade	Squeeze Pro Learn More | Buy | Upgrade	Squeeze Premium Learn More | Buy | Upgrade
Windows Server Certified	○	○	✓
Premium Support + Rep	○	○	✓
1 Year of Upgrades	○	○	✓
Avid DNxHD Encoding	○	✓	✓
Apple Pro Res Encoding (OSX Only)	○	✓	✓
5GB of Storage	✓	✓	✓
Adaptive Bitrate Support	✓	✓	✓
NLE Plugins	✓	✓	✓
Squeeze Server Integration	✓	✓	✓
Industry-Best Squeeze Engine	✓	✓	✓

Figure 7-22. Squeeze versions.

It appears that if you don't need Avid DNxHD and ProRes output, you can save $250 and get the Standard version. At the other end of the product spectrum, Squeeze Premium is certified on Windows Server 2008, a more industrial-strength OS than Windows 7 or 8 or Mac OS X.

Note that Squeeze Server is a totally separate product that also runs on Windows Server 2008, but also offers a REST-based API for easy integration into a larger publishing workflow. Squeeze Server also offers database functionality—either with the bundled MySQL or with SQL— which enables server farm operation with queueing, job priority, status, archiving, and more.

> **Tip:** *Sorenson released Squeeze 9 in April 2013. Key new features include an updated user interface; new HTML5 presets that encode files to both WebM and H.264 formats and supply the required HTML codes for embedding; closed caption support by converting CEA-608 and CEA-708 captions into TTML output; and a new filter that simplifies adding pre- and post-roll advertisements.*

This version also debuted an updated MainConcept codec that's supposedly equal to or better than the x264 codec. I'll be running my own tests on the product after this book is published, and if you Google "Jan Ozer" and "Squeeze 9," I'm sure you'll find the review.

Fortunately, neither the interface nor the encoding controls changed significantly so the advice contained herein should be just as applicable to Squeeze 9 as it is for Squeeze 8.

Container formats. Squeeze can produce most relevant container formats and groups its presets by container format.

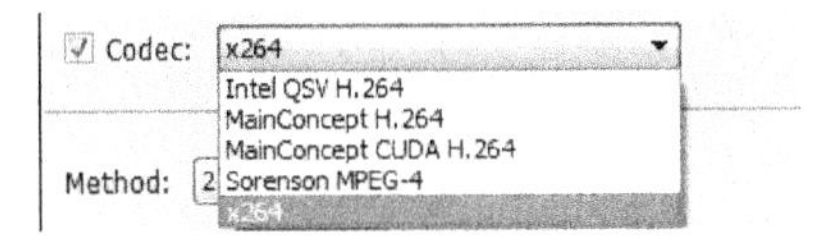

Figure 7-23. Squeeze's H.264 codecs.

H.264 codecs. As you can see in Figure 7-23, Squeeze has multiple H.264 codecs, including the Intel QSV H.264 codec, the MainConcept H.264 codec, the CUDA H.264 codec, and the x264 codec. Sorenson MPEG-4 is the MPEG-4 codec, not H.264. I've seen older Squeeze installations that contained the Apple codec as well, which produces the worst quality I've seen and should be avoided.

The MainConcept H.264 and x264 codecs are software-only codecs, while the Intel QSV codec is accelerated by some Intel CPUs, and the MainConcept CUDA codec is accelerated by NVIDIA graphics cards with CUDA acceleration. Both are much faster than the two software-only codecs, but this comes at a substantial price in terms of quality.

Specifically, Figure 7-24 shows a quality chart produced by the University of Moscow in its annual rating of H.264 codecs that you may recall from Chapter 3. As you can see, Intel QuickSync is much less efficient than the MainConcept and x264 codecs, while the MainConcept CUDA codec was dead last. I don't recommend using either hardware-based codec and won't demonstrate their use in this chapter, although I'll show how the MainConcept CUDA codec compares with the software-only MainConcept codec near the end of the section on Squeeze.

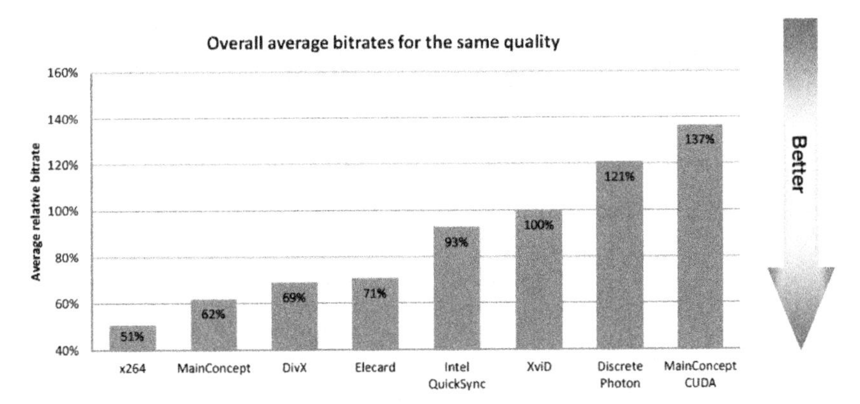

Figure 7-24. Results from University of Moscow encoding trials.

Adaptive-bit-rate strategy. Squeeze has the most advanced adaptive-bit-rate strategy of any desktop encoding tool and can produce both chunked files and metadata files for HLS, HDS, DASH and Smooth Streaming as shown in Figure 7-25.

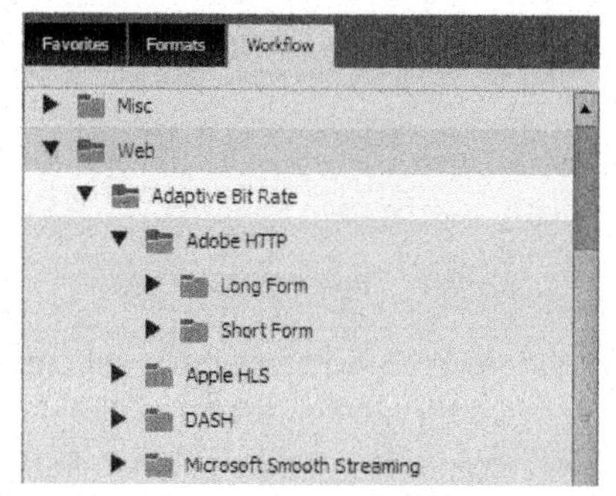

Figure 7-25. Squeeze's adaptive bit rate presets.

Sorenson provides separate presets for Long Form and Short Form encodings, with long form for videos more than five minutes long. Both apply a key frame interval of two seconds, but the long form uses a fragment duration of six seconds while the short form use a fragment duration of two seconds. With longer productions, which should soon identify the optimal stream and stop switching, the six-second fragment size cuts the number of files by 66%. With shorter productions, the premium is on more nimble switching, which the two-second fragment size enables.

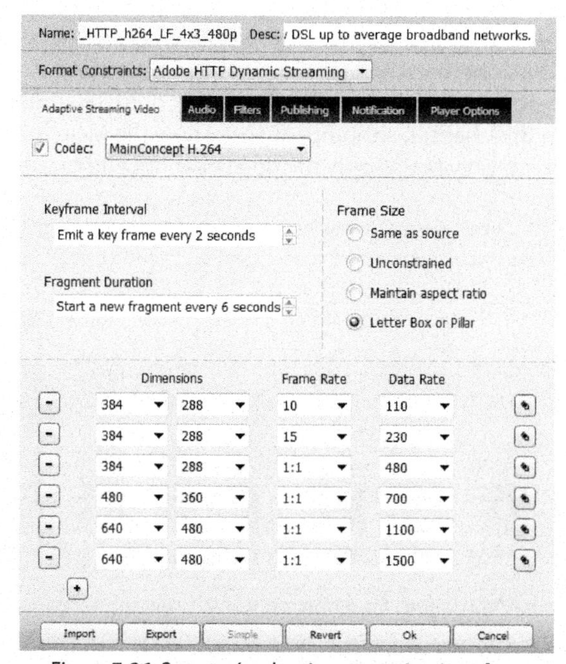

Figure 7-26. Squeeze's adaptive streaming interface.

Now would be a good time to gaze in wonder at Squeeze's adaptive streaming presets (Figure 7-26). They're structured so that they all use the same key frame interval, which is absolutely necessary, and the same audio parameters, which is the conservative approach. Unfortunately, if you've decided to adjust audio parameters over the adaptive group, you can't use this tool.

As you can see in Figure 7-26, you build your streams in the Presets window, adding streams using the plus icon on the lower left, and deleting them via the dash icon to their left. To access any of the individual encoding parameters for any single stream, click the pencil icon to its right. This opens the Squeeze Presets window that you'll see below. You can change any encoding parameters like Profile and Entropy Coding for any stream, so you can use Baseline profile for the lower-quality streams, Main for mid-quality streams and High for the top-quality streams.

When you press the Squeeze It! button, Squeeze encodes the files and builds the fragments and the metadata files. Boom, you're done. It's a slick, easy-to-use approach that ensures a consistent key frame interval and chunk size over all files, which is obviously essential.

Encoding efficiency. Squeeze can encode multiple files simultaneously and will do so when encoding a single source file to multiple presets, or multiple source files encoded to a single or group of presets, adaptive or otherwise. You can see Squeeze happily chugging away and encoding six different files over two different source files in Figure 7-27.

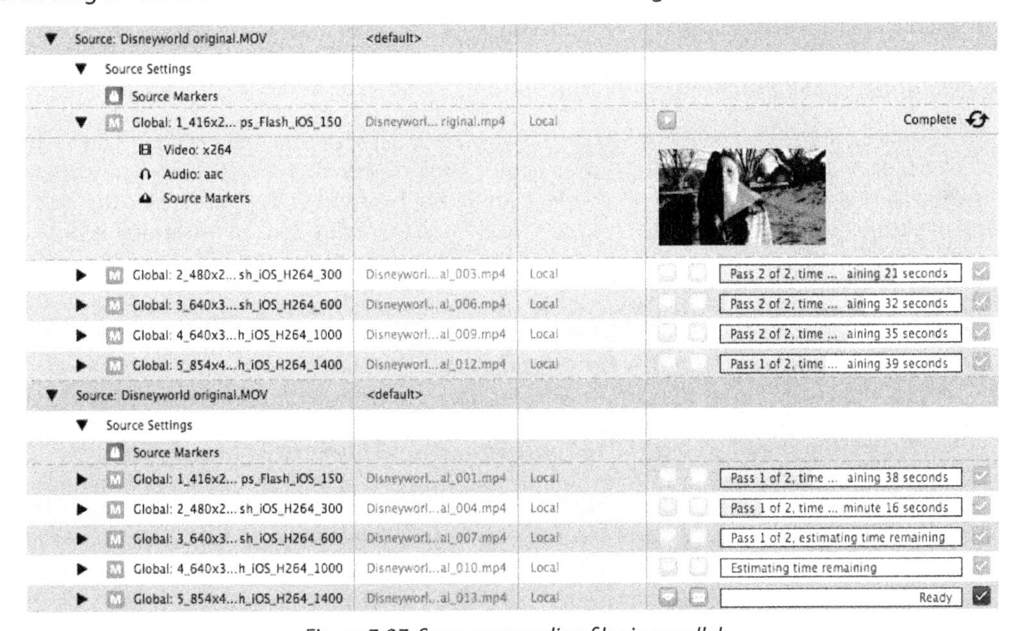

Figure 7-27. Squeeze encoding files in parallel.

You control how many files Squeeze will attempt to encode via the Compression System Load slider shown in Figure 7-28. When set to Heavy, Squeeze will consume virtually all system resources, loading files and starting to encode as necessary to do so. With a Light setting, Squeeze will load and attempt to compress fewer files to preserve system CPU for other tasks.

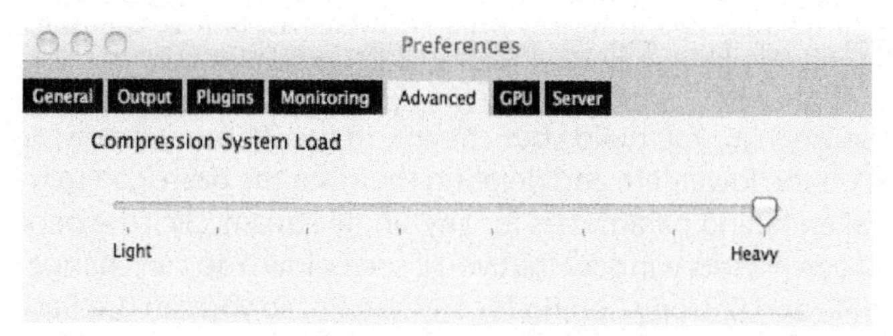

Figure 7-28. Adjusting the system resources Squeeze will consume.

This mechanism is a bit of a blunt instrument, although it's generally worked well in my tests and use. The one time it failed was when I was encoding longer files, from 10 to 60 minutes in length, with Heavy selected. This caused my 8-core Mac and 12-core HP Z800 to crash pretty regularly. So, I backed down to the middle of the range, and both systems encoded with no further problems.

Note that Squeeze 8 debuted a feature called "divide and conquer," which, in the words of Sorenson COO Eric Quanstrom, "means breaking files into smaller chunks, sending them to all the server's available cores, encoding them, and then passing them back to be rejoined with no loss in quality." Note that this mode only works on .MP4, .MKV, .MOV and WebM files, and only on files longer than 30 seconds in duration. This feature was designed to accelerate single-file encoding, and it works quite well.

> **Tip:** *If you're encoding several single long files, you may get better overall performance by loading multiple instances of Squeeze, loading a single file into each, and encoding separately. You can load multiple instances in Windows just by double-clicking the icon or using the Start menu multiple times. On the Mac, you need to follow a procedure specified in my article "Squeeze update: Opening multiple instances on the Mac," which you can find at* **bit.ly/squeeze_mac**.

Other notable features. Squeeze offers many notable features. These include:

- ***Watch folder functionality for shared use.*** Check out the tutorial at **bit.ly/Squeeze_watch** to learn how to set up and use watch folders.

- ***The ability to deliver via FTP to YouTube and to S3 buckets.*** Accessed via the publishing options in Figure 7-29. Double-click the preset to input account credentials.

- ***Access Squeeze from Adobe Media Encoder (Windows only).*** In the Windows version of AME only, choose the Squeeze Format in the Format list box, and then the preset in the Squeeze preset box below. In Final Cut Pro 7, you can access Squeeze presets via the File > Export > Using QuickTime Conversion option. Then choose Squeeze Export Pro as the Format on the lower left of the Save window, which opens the Squeeze Export Settings dialog. In that window, choose the output format and desired preset. Note that there is no way to access Squeeze from Final Cut Pro X.

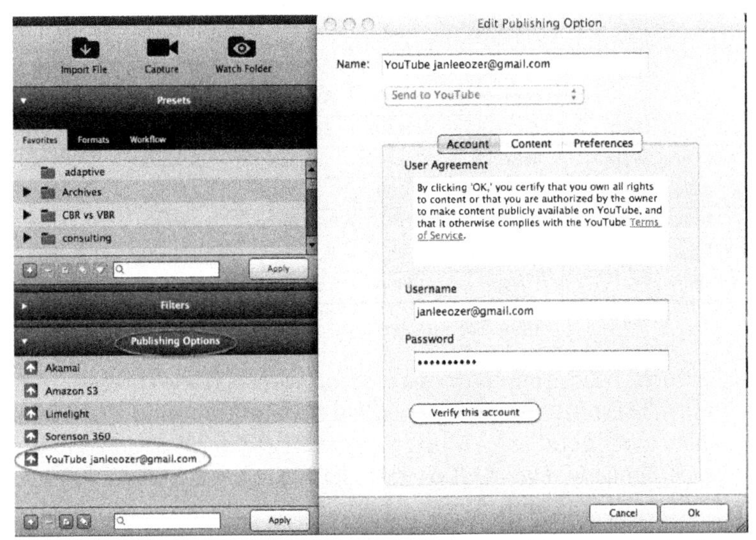

Figure 7-29. Squeeze can publish directly to YouTube, FTP sites and S3 buckets.

To be clear, in AME and FCP 7, you can only choose a preset from these apps; you can't access Squeeze's encoding controls, but this still eliminates the time and disk space involved with creating an intermediate file to then input into Squeeze.

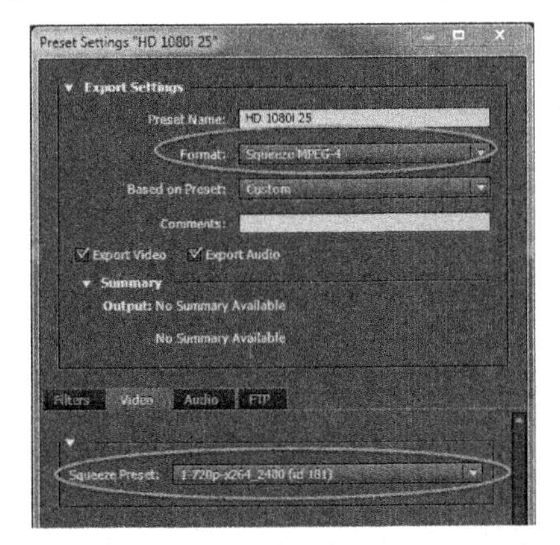

Figure 7-30. You can access Squeeze presets from the Adobe Media Encoder (Windows only).

General encoding controls. Very extensive.

H.264 controls. Very extensive, as you'll see below.

Preset	Profile	Rez	Video		Audio	
			Survey Average	Squeeze Preset	Survey Average	Squeeze Preset
Legacy_iPod_768Kbps_180p	Baseline	320x180	532 kbps	550kbps	95 kbps	128 kbps
Legacy_iPod_768Kbps_240p	Baseline	320x240	532 kbps	550kbps	95 kbps	128 kbps
Midrange_iPod_2500Kbps_360p	Baseline	640x360	1.14 Mbps	1.4 Mbps	113 kbps	128 kbps
Midrange_iPod_2500Kbps_480p	Baseline	640x480	1.14 Mbps	1.4 Mbps	113 kbps	128 kbps
iPad-iPhone4_7Mbps_720p	Main	720p	2.9 Mbps	3.5 Mbps	169 kbps	160 kbps
IPad-Apple_TV_1080p	High	1080p	5.08 Mbps	5 Mbps	158 kbps	160 kbps

Table 7-5. Squeeze's iDevice settings.

iTunes presets. See Table 7-5. Sorenson does a nice job with its iTunes presets, using the correct Profile with 16:9 and 4:3 SD presets and reasonable data rates throughout. In the mid-range presets, if your quality looks good at 1.4 Mbps, you may try to dial that down to the survey average of 1.14 Mbps. Similarly, if your 720p videos look good at 3.5 Mbps, you may try to lower that to the survey average of 2.9 Mbps.

Getting acquainted. If you're totally unfamiliar with Squeeze, there are many video tutorials that you can view, many produced by yours truly. For example, you can check out "Sorenson Squeeze 8: A video Tutorial," at **bit.ly/Squeeze_tutorial**, where I show the new features in Squeeze 8 and compare x264 quality with MainConcept quality in Squeeze (x264 wins). There's also a Sorenson-produced video about presets that you can see at **bit.ly/Squeeze_presets**, and I'm told there are 15 more coming from Sorenson. You should probably watch one or both of these videos to get acquainted with the program before reading any further.

Encoding with Squeeze. Essentially, producing H.264 with Squeeze is the tale of two codecs, MainConcept and x264. Although they share the same basics, their advanced encoding parameters are as different as children raised by different parents—which, of course, they kinda are. So I'll cover them individually, starting with the x264 codec.

Squeeze and the x264 Codec

One of the attractions of the x264 codec is that it offers the widest variety of configuration options of any H.264 codec, and Sorenson makes all configuration options available in its encoding interface. These can be incredibly valuable for some producers, but the amount of time required to learn how and when to use these controls is mind-boggling. Fortunately, by using the x264 presets in Squeeze, you can get most of the benefit of these controls with a minimal learning curve.

> **Tip:** *Squeeze 9 presets may have different names than their Squeeze 8 counterparts, so you may have to find a different preset to follow along.*

If you're working with Squeeze and want to follow along, click the Formats tab, open the MPEG-4 folder and double-click the 2000Kbps_720p preset, which uses the x264 codec to produce 720p video at 2000 kbps. Double-click the preset, and then, if necessary, click the Advanced button to reveal all x264 encoding parameters. Staggering, isn't it? No worries, though, since Squeeze has reduced the complexity of operation by controlling most encoding parameters via the preset options shown in Figure 7-31.

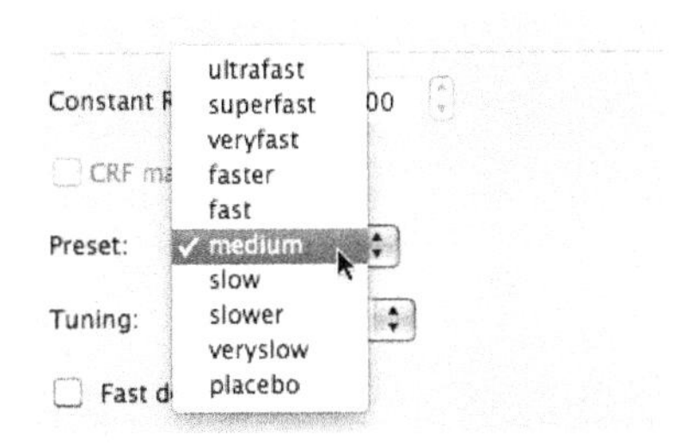

Figure 7-31. These x264 presets control most relevant encoding parameters.

Not surprisingly, these are the same presets seen in the section above titled "x264 Presets and Tuning," where I describe how to use the x264Encoder within Compressor. Most producers who use the x264 codec use these presets, and you'll see them again below when discussing Telestream Episode.

Within Squeeze, choosing a preset adjusts most codec-related parameters to the default settings for that preset. If you choose a preset and observe the settings, you'll see some of them change—particularly if you switch from one extreme, like ultrafast, to the other, like placebo. We'll use this to our advantage as we craft a strategy for working with x264.

Here's the high-level view. Once you choose your x264 preset and tuning option, it's time to customize other parameters in the setting that aren't affected by the preset and tuning option selections. These parameters fall into multiple categories, including:

- ***General encoding options unrelated to the codec.*** These are options like data rate, bit rate control and key frame interval.
- ***H.264 options not controlled by the preset.*** The only one in this category is Profile.
- ***H.264 options controlled by the preset that you want to override.*** These may include options like B-frame and Reference frame values, and entropy encoding.

Basically, it's a complicated mess that makes me want to dive into a pint of Ben and Jerry's Chocolate Therapy ice cream. But presuming at least one reader bought this book for detailed guidance on these settings, I'll push on. Actually, some of it is kind of interesting—in a glass-half-full kind of way.

With this as background, here's the workflow I would use to customize a Squeeze x264 setting:

1. Start by choosing the desired Preset and tuning. More on where to start in a moment.

2. Now let's attack the general encoding options. Based upon information you learned in previous chapters, you would need to set:

- ***Method (bit rate control)***
- ***Frame rate***
- ***Data rate***
- ***Frame size***
- ***Key frame interval and enable/disable key frame upon scene change.***

3. Now let's attack the H.264 options not set by the x264 preset. With x264, that's just the profile, so select the desired profile according to what you learned in Chapters 2, 3 and 4.

4. Now let's customize H.264 parameters controlled by the setting you want to override. There are two schemas used to override the options selected by the preset/tuning parameter. For some options, such as the Number of B-frames (Figure 7-32), you can just check the checkbox and choose the desired option. With others, such as CABAC (also Figure 7-32), click the desired option (on or off) to take that parameter off auto control. Note that when auto is selected, the option in capital letters (ON in Figure 7-32) is the option used for encoding. When you manually choose a parameter, it turns green. Note also that the CABAC control is not located directly below the number of B-frames in the Squeeze Setting as shown in the figure; I Photoshopped the two together for presentation purposes.

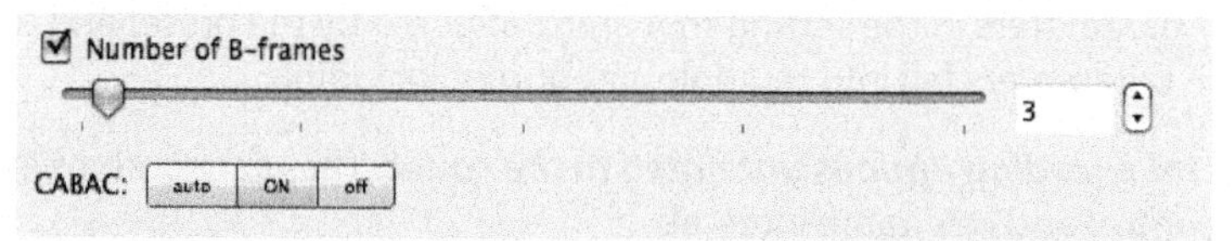

Figure 7-32. Two kinds of auto controls in the x264 configuration screen.

Here are the settings you should customize:

- ***Entropy coding.*** If encoding using the Main or High profile, set to On.
- ***Number of B-frames.*** If encoding using the Main or High profile, click the checkbox and set the number to 3.

5. If a parameter doesn't auto-adjust and you don't know what it does, leave it alone—particularly when encoding for devices, where an errant configuration option can render the file incompatible.

Tip: *If you choose a preset other than Medium, check the number of B-frames and make sure it's acceptable. For example, if you use Placebo, Squeeze inserts 16 B-frames, which is too high.*

Now, which preset should you start with? To figure this out, I encoded a 1-minute DV file to 500 kbps on a 2.93 GHz 8-core PowerMac using the presets shown on the top of Table 7-6.

Preset	Ultra Fast	Very Fast	Fast	Medium	Slow	Very Slow
Encoding Time (sec)	44	58	84	90	138	400
Increase in Time From Fastest Setting	0%	32%	91%	105%	214%	809%

Table 7-6. x264 encoding times at various x264 presets.

I compared the output quality from these tests and saw very little quality difference except that the Ultra Fast-encoded file was obviously degraded (Figure 7-33). Why so little difference? Not all advanced encoding techniques yield significant quality differences with all source materials.

Are my results representative? Hard to say, but I would guess so—at least for general-purpose video footage. Throw computer-generated animations into the mix, and you may see greater variations in quality. However, most x264 pundits recommend starting with the Medium preset, which roughly doubles encoding time over the noticeably degraded Ultra Fast preset. I recommend the same, as does Sorenson, since that's the default used for all its x264 encodes.

Figure 7-33. Ultra Fast was noticeably degraded, but all other presets looked virtually identical. For a better look, download the PDF of all figures in the book at **bit.ly/Ozer_multi.**

Again, I saw almost no difference in quality between even the Very Fast clip, which encoded in 58 seconds, and the Very Slow clip, which took 400 seconds, or 6:40 (min:sec). If time isn't an issue, there's no harm in using the higher-quality preset, but if you run a high-volume production shop and encoding time is money, you probably want to standardize on Medium or Slow, but not go beyond.

Tuning your x264 Encodes

In addition to the presets, Squeeze lets you tune your encode for different sources (film, animation, grain, still image) or optimize for a particular result (post-encode signal-to-noise or structured similarity quality tests) via the Tuning control. There are more extensive definitions of their functions from the Zencoder website just above Figure 7-19. Within Squeeze, these selections work like presets, so as you toggle through the different Tuning settings, you'll see adjustments flowing though the various configuration options below.

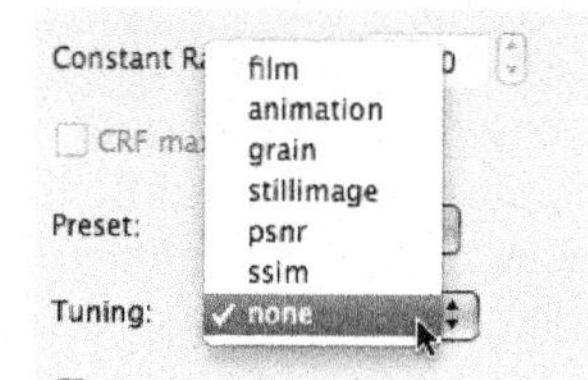

Figure 7-34. Tuning controls for x264.

These controls are applied ***after*** the options dictated by your preset selection, so choose the preset first, then Tuning, and then customize your controls as desired. In general, if your video falls in one of the first four categories, you should apply the fine-tuning option. Otherwise, I generally use None, although some notable compressionists, like Amazon's Ben Waggoner, recommend the Film tuning option. If you have representative footage you can test, you should test with Film and None and see if there's a difference with your footage. I ran these comparisons with my standard test files and saw no difference. For more on tuning, check the section titled "x264 Presets and Tuning" above for a list of resources.

We're ready to move onto the MainConcept codec. Be careful when you close a Squeeze preset. If you choose OK, you'll save any changes over the canned preset that Squeeze provides. Instead, click Cancel and you won't. If you'd like to save settings to a preset yet preserve the canned preset (which I recommend), make a new preset first and experiment with that one. To create a new preset based on an existing preset, right-click the preset and choose Copy Preset. Then you can make any desired changes in the new preset and click OK to save them.

If you want to read about working with the MainConcept codec, leave the preset you've been working on open.

Squeeze and the MainConcept Codec

If you don't have a preset open, in the Formats tab, click MPEG-4, then double-click the 2000Kbps_720p preset to open it. As the name suggests, this is an H.264 preset that encodes at 2000 kbps at 720p resolution. However, it uses the x264 codec, so we need to switch over to MainConcept.

> **Tip:** *Squeeze 9 presets may have different names than their Squeeze 8 counterparts, so you may have to find a different preset to follow along.*

On the top left of the Squeeze Presets dialog, click the codec list box and choose the MainConcept H.264 codec, as shown in Figure 7-35. Be careful when you close this dialog because if you click OK, you'll save changes and overwrite the template settings. As mentioned, if you'd like to save settings to a preset yet preserve the old, right-click the preset and choose Copy Preset. Then you can make the desired changes and click OK to save them.

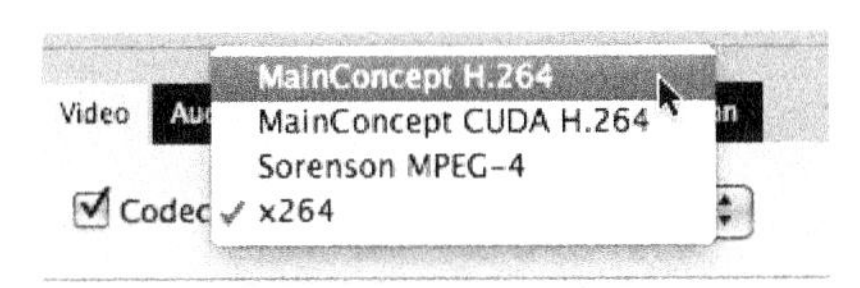

Figure 7-35. Choosing the MainConcept codec.

OK, so we've switched to the MainConcept H.264 (and not CUDA) codec. As with the x264 codec, there are an absolute ton of configurations. Unless you have a serious interest in reading up on and testing (and testing, and testing) these parameters, here's how I would approach configuring MainConcept encodes:

Figure 7-36. The MainConcept Performance slider.

Start in the upper left, and drag the performance slider to the left and right. Observe how most of the parameters in the grayed configuration options change as you adjust the performance value, which ranges from 0 to 16. For example, take a look at Entropy Coding on the lower right. When the performance slider is at 0, which dials in a lower quality but faster encode, the encoder uses CAVLC. At 16, it uses CABAC.

Basically, as you drag the slider from one position to another, Squeeze is choosing the configuration options and values that deliver the selected blend of quality and encoding time. So I recommend the following approach:

1. Start by choosing the desired value in the Performance slider. More on where to start in a moment.

2. Now let's attack the general encoding options. Based upon information you learned in previous chapters, you would need to set:

- ***Method (bit rate control)***
- ***Frame rate***
- ***Data rate***
- ***Frame size***
- ***Key frame interval and enable/disable key frame upon scene change***

3. Now let's attack the H.264 options not set by the x264 preset.

- ***Profile.*** Choose the desired profile.
- ***Entropy coding.*** If encoding using the Main or High profile, set to On.
- ***Number of B-frames.*** If encoding using the Main or High profile, set to 3.
- ***HRD Conformance.*** This will be enabled by default for CBR encoding. Uncheck to deselect for VBR encoding (if you hover your pointer over the HRD Conformance control, you'll see the tool tip recommending this).

4. Now let's customize H.264 parameters controlled by the setting that you want to override. With some options, such as the Hadamard Transformation (Figure 7-37), this means checking the checkbox and choosing the desired option. With others, like the Deblocking Filter (also Figure 7-37), it means choosing the desired option and taking that parameter off auto control. When auto is selected, the option in capital letters (ON in Figure 7-37) is the option used for encoding. When you manually choose a parameter, it turns green.

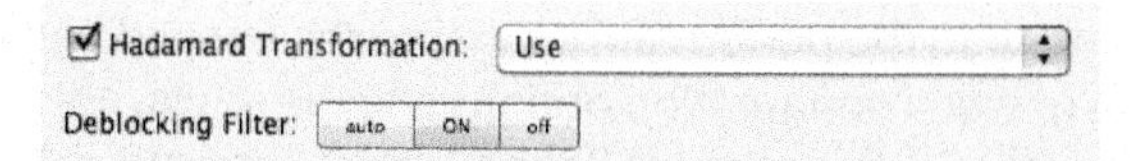

Figure 7-37. Two kinds of auto controls in the MainConcept configuration screen.

Here are the settings you should customize:

- ***Hadamard Transformation.*** Set to Use.
- ***Deblocking Filter.*** Set to On.

5. If a parameter doesn't auto-adjust and you don't know what it does, leave it alone—particularly when encoding for devices, where an errant configuration option can render the file incompatible.

Now, where should you start on the Performance slider? To figure this out, I encoded a 1-minute DV file to 500 kbps on a 2.93 GHz 8-core PowerMac using the slider values shown on the top of Table 7-7.

Preset	1	10	13	16
Encoding Time (sec)	41	42	49	69
Increase in Time From Fastest Setting	0%	2%	20%	68%

Table 7-7. MainConcept encoding times at various settings on the Performance slider.

As you can see, the difference in encoding time between the slowest and fastest setting was only 28 seconds, or about 68%. The difference between the settings of 1 and 10 was only 1 second, making it a total no-brainer to start at 10, particularly given the increase in quality that (hopefully) you can see between the 1 and 10 setting in Figure 7-38.

In contrast, I saw very little quality difference in the videos produced with the settings of 10 and 16. Still, for only a 68% increase in encoding time, unless I'm in a huge hurry, I will set the performance slider to 16 and encode away. Perhaps for draft work, I might use a setting of 10, but you get the idea. Of course, your mileage may vary with different source footage and encoding parameters, but you can easily duplicate this analysis in your shop if you want to determine the best starting point for your MainConcept encodes.

At the slider value of 16, most parameters are perfect. To fine-tune, I would set the Profile to High, the Use B-Pictures value to 3 and the Constrain Maximum Data rate value to 200%, and that's about it. Then I'd save the preset under a different name for later reuse.

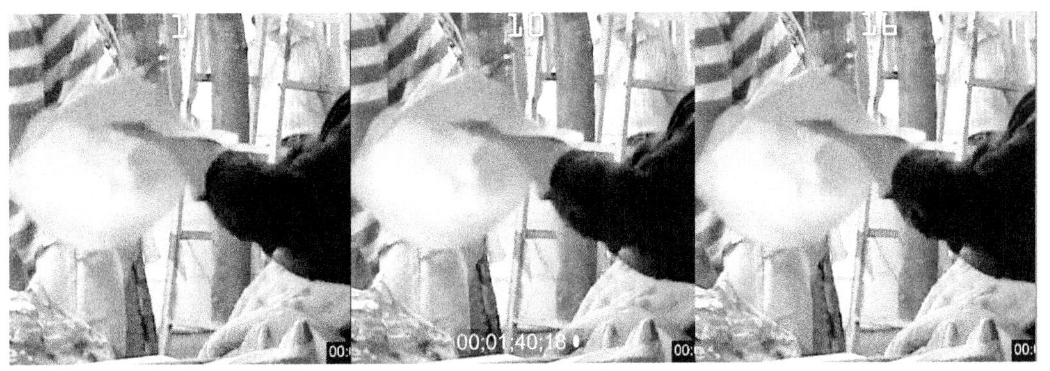

Figure 7-38. Big difference between a setting of 1 and 10; not so much between 10 and 16.

A couple of other relevant points when producing with the MainConcept codec:

- ***Tool tips.*** Sorenson added tool tips to Squeeze 8, so if you don't know what a parameter is, you can hover you mouse over it and get a description.

- ***MainConceptReference.pdf.*** This is the help file that comes with the MainConcept reference encoder most licensees use to test out the software before buying. It's a great reference for most functions. MainConcept doesn't offer a free download from its own site (as near as I can tell) but if you Google "MainConceptReference.pdf," you should be able to find a downloadable copy from somewhere else.

Remember, when you close the preset you've been working with, click Cancel, not OK, so you don't save your experimental adjustments and overwrite the preset.

Tip: *When using the MainConcept H.264 codec, you can choose between 2-pass VBR and multi-pass, which is also VBR. In tests I performed for* Video Compression for Flash, HTML5 and iDevices, *multi-pass took about twice as long to encode and I saw no visible difference between the encodes (see page 149). For this reason, I use 2-pass exclusively.*

MainConcept CUDA H.264 Codec

Briefly, CUDA is an acronym for Compute Unified Device Architecture, a computing architecture developed by NVIDIA, which manufactures the graphics processing chips (or GPUs, for graphics processing units) used in most workstation and gaming cards. In essence, CUDA is a parallel computing engine that application developers can use to perform certain types of instructions, including encoding H.264 video. In plainer English, you can make the GPU do the work rather than your computer's CPU, which is typically off doing lots of other things.

Figure 7-39. In some sequences, GPU encoding produces lower quality than CPU encoding. For a better look, download the PDF of all figures in the book at **bit.ly/Ozer_multi**.

If you choose the MainConcept CUDA H.264 codec, Squeeze will use the GPU (if an NVIDIA GPU is available) to encode the video, which should accelerate encoding. In fact, in my tests on several different multiple-CPU Mac and Windows workstations, GPU encoding cut rendering time on average by more than 50%. However, as shown in Figure 7-39, in some sequences—particularly those with both high detail and high motion—quality can suffer.

If you're encoding talking-head footage only, GPU acceleration (used interchangeably with CUDA acceleration) is worth a try. However, with high-motion footage or mixed high- and low-motion footage, you're better off using CPU encoding unless you can check each scene for quality issues. Since I encode primarily mixed- or high-motion footage, I always disable GPU encoding in the Preferences dialog (Figure 7-40) and never select it for encoding.

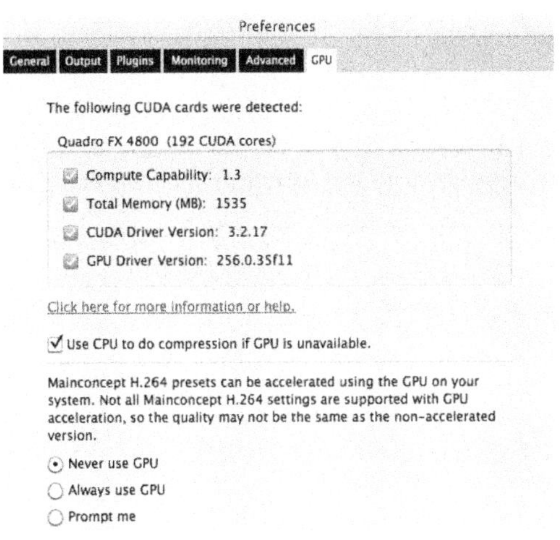

Figure 7-40. CUDA-related preferences.

Overall, it's best to consider CUDA encoding a work in progress. Obviously, both MainConcept and Sorenson are aware of the quality issue, and MainConcept is working to ensure that later versions of CUDA encoding produces equivalent quality to CPU encoding in all types of sequences. Until they get there, though, I would use it cautiously. That's all for Squeeze; let's move on to Telestream Episode.

Telestream Episode

Telestream Episode is another highly regarded, cross-platform desktop encoding tool with three versions: Episode ($495), Episode Pro ($995) and Episode Engine ($5,994). Note that x264 encoding costs $99 for all three versions, which I recommend because it delivers better quality than the MainConcept H.264 codec that ships with the product. In addition, the Pro Audio Option, which includes support for Dolby Digital AC3, and HE AAC and AAC-LC encoding, is $395 for all three versions. Without this option, you can still encode into AAC audio format; you'll just be using the QuickTime AAC encoder, not the Dolby encoder.

Note that maintenance and support is not included with Episode and Episode Pro, but that Engine includes one year of support and maintenance. The $999 in the Engine column is due one year after the purchase. All three products share the same interface; the other differences relate to format support and performance as detailed in Table 7-9.

	Episode	Episode Pro	Episode Engine
Cost	$495	$995	$5,994
X264 Option	$99	$99	$99
Pro Audio Option	$395	$395	$395
Maintenance and Support (Yearly)	$99	$199	$999

Table 7-8. Pricing for Episode versions and components.

Specifically, both Episode Pro and Episode Engine support high-end pro formats like .MXF, .GXF, .IMX, and MPEG-2/-4 Transport Streams (.TS). For more details on the formats supported by the various versions, Telestream offers a detailed PDF that you can view at **bit.ly/Episode_formats**.

	Episode	Episode Pro	Episode Engine
Support for High-End Formats	No	Yes	Yes
Encode in Parallel	No	2 files	Unlimited
Split and Stitch	No	No	Yes

Table 7-9. Key differences between Episode versions.

In terms of performance, the base version of Episode can only encode a single file at a time, while Episode Pro can encode two at a time. Episode Engine can encode an unlimited number of files simultaneously, limited only by the number of CPUs in your system. You can choose the number yourself or let Episode Engine recommend the best number. For example, as you can see in Figure 7-41, Episode Engine recommended 5 simultaneous encodes on my eight-core Mac Pro.

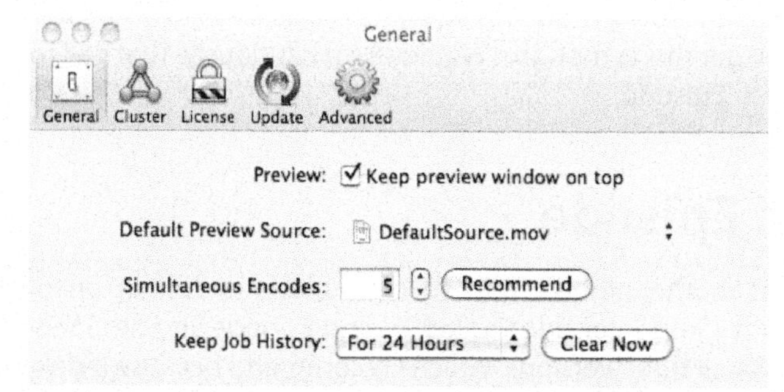

Figure 7-41. Choosing the number of simultaneous encodes in Episode Engine.

Split and stitch is a technology that speeds the encoding of long files on a multi-core computer by splitting the file into pieces, allocating different pieces to different cores, and stitching the resulting file back together. As Table 7-9 shows, it's only available on Episode Engine.

Note that you can easily upgrade from one version to the other by changing the license code in the Preferences panel; that's the cute license lock you see in Figure 7-41. You can also run a single version of any Episode version on multiple systems by deactivating the license on one system while activating it on the other.

H.264 codecs. As mentioned, Episode can use both the MainConcept and x264 codecs, which you can toggle between in the Video Codec field in the Inspector panel (Figure 7-42). As a reminder, x264 will only appear if you license the x264 Option from Telestream.

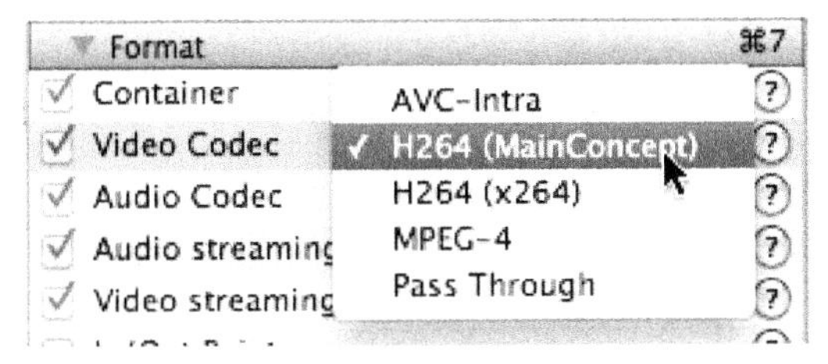

Figure 7-42. Episode's H.264 codecs.

Adaptive-bit-rate strategy. Episode can produce chunked and metadata files for HLS and Smooth Streaming, but only via the command-line interface enabled with Episode 6.2. You can read about these workflows in the "Episode 6.3 Advanced Users Guide" at bit.ly/Episode_CLI. I have not worked with this interface.

Encoding efficiency. Single-computer efficiency relates to which version that you purchase. With split and stitch technology and unlimited parallel file encoding, Episode Engine is very efficient. At the other end of the spectrum, Episode isn't that efficient on multi-core computers, since it encodes serially and without split and stitch. Like Compressor, however, Episode users on a single LAN can send jobs to one other via clustering. There's a nice tutorial of this feature in a tutorial titled "Episode 6 Cluster Browser" at bit.ly/Episode_demos.

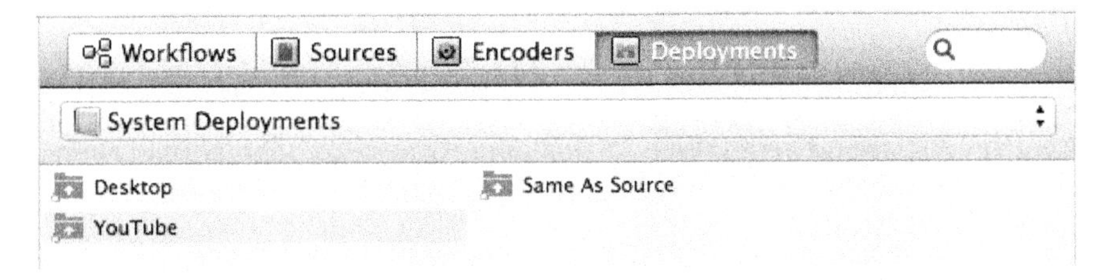

Figure 7-43. Episode can publish directly to YouTube, FTP sites and S3 buckets.

Other notable features. Here are two other notable Episode features:

- ***Watch folder functionality for shared use.*** Check out a tutorial on YouTube at bit.ly/Episode_watch to learn how to set up and use watch folders.
- ***Upload to YouTube.*** YouTube is a standard option available in the Deployments bin.

General encoding controls. Very extensive.

H.264 controls. Very extensive, as you'll see below.

iTunes presets. See Table 7-10. Telestream offers a wide variety of presets for iDevices—usually offering 4:3, 16:9 and letterboxed 16:9 presets at most relevant target resolutions, although the highest-resolution preset is only 720p. Telestream used the Main profile for 640-resolution encodes, which may produce files that won't play on older iPods/iPhones; I would use the Baseline profile. Otherwise, data rates were good in the 320 and 640 categories, although the 2.4 Mbps used for the iPad HD preset was very low compared with our survey. While Telestream typically used CBR for most iDevice encodes, some iPhone presets were VBR, which I would change to CBR.

	Profile	Rez	Video		Audio	
Preset			Survey Average	Episode Preset	Survey Average	Episode Preset
320x240 Group	Baseline	320x240	532 kbps	600 kbps	95 kbps	128 kbps
640x480 Group	Main	640x480/ 640x360	1.14 Mbps	1200 kbps	113 kbps	128 kbps
iPad HD	Main	720p	5.08 Mbps	2.4 Mbps	158 kbps	128 kbps

Table 7-10. Episode's iDevice settings.

Getting acquainted. If you're totally unfamiliar with Episode, there are several video tutorials available, including a helpful introductory video by yours truly available at bit.ly/Episode_ozer. Telestream also has several tutorials you can view at bit.ly/Episode_demos.

Encoding with Episode. Episode offers two H.264 codecs, MainConcept and x264, and their interfaces are totally different. I'll cover both, starting with x264.

Episode and the x264 Codec

Remember, x264 is a $99 option for Episode, so if you haven't purchased it, you don't have it. If x264 presets aren't appearing in your Episode program, that's obviously why. All the presets we'll work with in this section are the Web > Streaming > H.264 Encoders section, with presets starting with H264 using the MainConcept codec (H264 MP4 1Mbit) and presets starting with x264 using the x264 codec (x264 MP4 1Mbit).

As with most encoders, it's easier to edit a preset than to create one from scratch, so that's the approach I'll take here. So I'll start by duplicating the x264 MP4 2Mbit 16x9 preset by right-clicking the preset and choosing Duplicate Encoder. Episode puts the copy in the User Encoders section and should take you to that section, where you should find a preset named x264 MP4 2Mbit 16x9 2.

Click that preset to open it in the Inspector window. When working with x264, I recommend starting by choosing an x264 preset and tuning, then configuring other options that aren't controlled by the preset. To see the x264-specific encoding controls, click the Video Codec line in the top field in the Inspector window (see Figure 7-45).

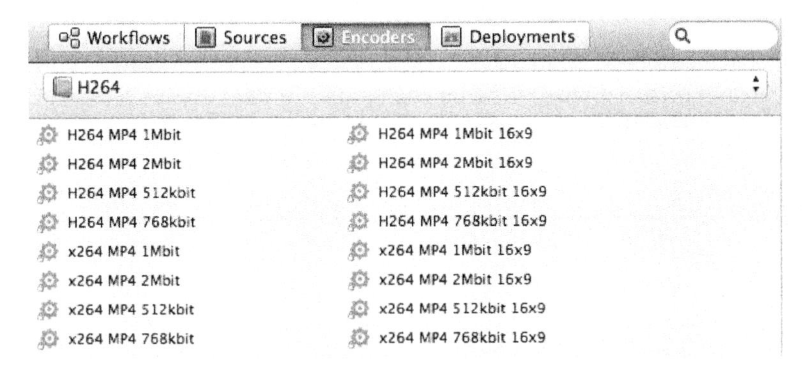

Figure 7-44. Episode's H.264 streaming presets.

As mentioned, the developers of x264 created the 10 presets shown in Figure 7-45 to enable x264 users to simply and easily balance encoding time against quality. You'll see the same presets whether you access x264 via plug-ins like x264Encoder, free programs like HandBrake, or other professional encoding tools that offer x264 encoding like Sorenson Squeeze. As with these tools, when you choose a preset, the x264Encoder applies a certain set of parameters that you can customize. If you're interested, you can find a complete breakdown of the encoding parameters used in each preset at **bit.ly/x264_explained** and some performance (but not quality) comparisons at **bit.ly/x264_speed**.

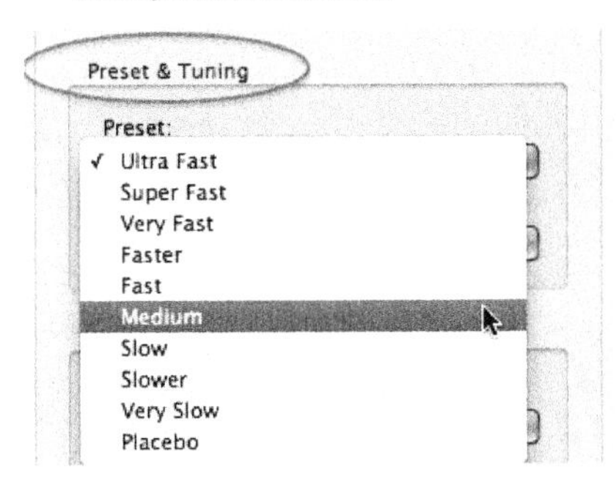

Figure 7-45. Episode's x264 presets.

Within Episode, choosing a preset adjusts most codec-related parameters to the default settings for that preset. If you choose a preset and observe the settings, you'll see some of them change—particularly if you switch from one extreme, like Ultra Fast, to the other, like Placebo. We'll use this to our advantage as we craft a strategy for working with x264.

Now, which preset should you start with? To figure this out, I encoded a 1-minute DV file to 500 kbps on a 2.93 GHz 8-core PowerMac using the presets shown on the top of Table 7-11. It's a different file than the one I encoded with Sorenson Squeeze, so please don't make any performance comparisons.

Preset	Ultra Fast	Fast	Medium	Slower	Very Slow	Placebo
Encoding Time (Sec)	14	24	26	56	121	370
Increase in Time From Fastest Setting	NA	68%	84%	292%	746%	2487%

Table 7-11. x264 encoding times at various x264 presets.

Looking at the output quality from these tests, it's clear that Ultra Fast is a no-go, since the footage is degraded throughout. Beyond that, the difference in encoding time between Fast and Medium was so small that it didn't make sense to consider using the former. So I focused my comparison on the difference between Medium, Slower and Very Slow to see if the dramatic jumps in encoding time delivered noticeably better quality.

Well, it depends upon your definition of noticeable. In the hardest-to-encode scene in the test file, comparing frame by frame, I did see a bit more detail in the Slower file than the Medium, which you can (attempt) to see in Figure 7-46.

Figure 7-46. Very minor differences between Medium and Slower.

Is this sufficient to warrant more than doubling the encoding time? That depends upon your circumstances. If you're encoding very high-motion or visually complex videos to very low data rates and aren't constrained by encoding time, I would use Slower. If you're encoding to pretty high data rates (bits per pixel of 0.15 and above), you probably wouldn't notice the difference.

Are my results representative? Hard to say, but I would guess so—at least for general-purpose video footage. Throw computer-generated animations into the mix, and you may see greater variations in quality. However, most x264 pundits recommend starting with the Medium preset, and that's the default value in Episode.

Choosing a Tuning Setting

In addition to the presets, Episode lets you tune your encode for different sources (film, animation, grain, still image) or optimize for a particular result (post-encode signal-to-noise or structured similarity quality tests) via the Tuning control shown in Figure 7-47.

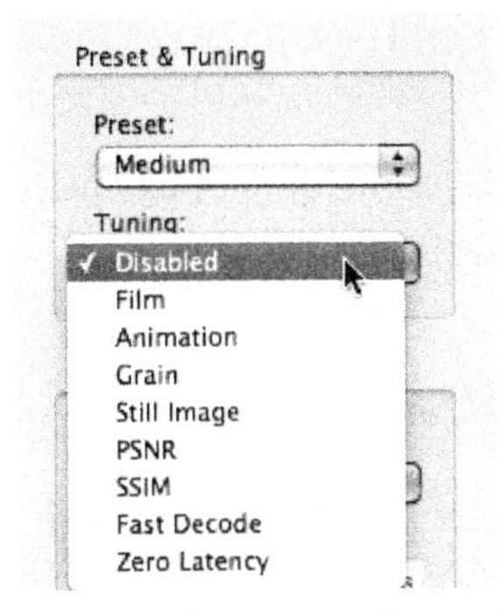

Figure 7-47. x264 preset tuning options.

Here are more detailed definitions of the options from the website of cloud encoding vendor Zencoder (**bit.ly/x264_tuning**).

- ***Film.*** Optimized for most non-animated video content (not only feature films).
- ***Animation.*** Optimized for animation. Note that most 3D animation behaves more like film and not like hand-drawn animation, so only use this for hand-drawn animation (anime, classic Disney, etc.)
- ***Grain.*** Optimized for film with high levels of grain.
- ***PSNR.*** Uses peak signal-to-noise ratio to optimize video quality.
- ***SSIM.*** Uses structural similarity to optimize video quality.
- ***Fast Decode.*** Reduces encoding complexity to allow for easier decoding.
- ***Zero Latency.*** x264 will keep an internal buffer of frames to improve quality; this setting removes that buffer, but reduces quality

Zencoder doesn't define Still Image, but it's worth trying if you're encoding slide shows or similar content. In general, if your video falls in one of the first four categories, you should at least try the fine-tuning option and see if it makes a difference. Otherwise, I generally use None, although some notable compressionists, like Amazon's Ben Waggoner, recommend the Film tuning option. If you have some representative footage you can test, you should test with Film and Disabled (Episode's default) and see if there's a difference with your footage. I ran these comparisons with my standard test files and saw no difference.

For those seeking additional information on the x264 encoding options, here's a list of x264 resources—although if you Google "x264," you can find thousands more.

- ***The x264 MeWiki.*** A comprehensive guide to x264 encoding options (mewiki.project357.com/wiki/X264_Settings)

- ***Diary Of An x264 Developer.*** The blog of Jason Garrett-Glaser, the primary developer of x264 (x264dev.multimedia.cx). Very topical and useful.

- ***Doom9 Forum.*** Doom9 has some of the most active forums on x264. If you have a particular question or issue, Google that issue plus "Doom9," and you'll probably find a discussion on point.

Customizing your x264 Preset

Once you choose your x264 preset and tuning option, it's time to customize other parameters in the setting. These parameters fall into multiple categories, including:

- ***General encoding options unrelated to the codec.*** These are options like data rate and key frame interval.

- ***H.264 options not controlled by the preset.*** Profile and Levels.

- ***H.264 options controlled by the preset you want to override.*** These include options like B-frame and Reference frame values, and entropy encoding.

With this as background, here's the workflow I would use to customize an Episode x264 setting:

1. Start by choosing the desired Preset and tuning.

2. Now let's attack the general encoding options. Based upon information you learned in previous chapters, you would need to set:

- ***Bandwidth Settings.*** This is a bit complex, but you can evolve it into these simple rules:

    - ***Rate control method.*** Always choose Average Bit Rate. Constant Quality, the other option, is only good for archiving and similar functions.

    - ***For CBR.*** Choose CBR in the HRD Compliance list box and enter the target data rate in the Average Bit Rate field, and the same value in the VBV Size field.

    - ***For VBR.*** Choose VBR in the HRD Compliance list box and enter the target data rate in the Average Bit Rate field, and the same value in the VBV Size field. Enter the maximum data rate in the Max Bit Rate field. For example, in Figure 7-48, the VBV Max Bit Rate of 3,520 is 200% that of the average bit rate, enabling constrained VBR with a maximum data rate of 3,520 kbps.

- ***Key frame interval and enable/disable key frame upon scene change***. Set as normal, differentiating between single-file and adaptive-file encodes.

- ***Buffering***. I recommend leaving the Look Ahead settings at their default.

- ***2-pass encoding.*** Typically enable unless you're in a real hurry.

- ***Frame rate.*** Set as normal.

- ***Frame size.*** Set as normal.

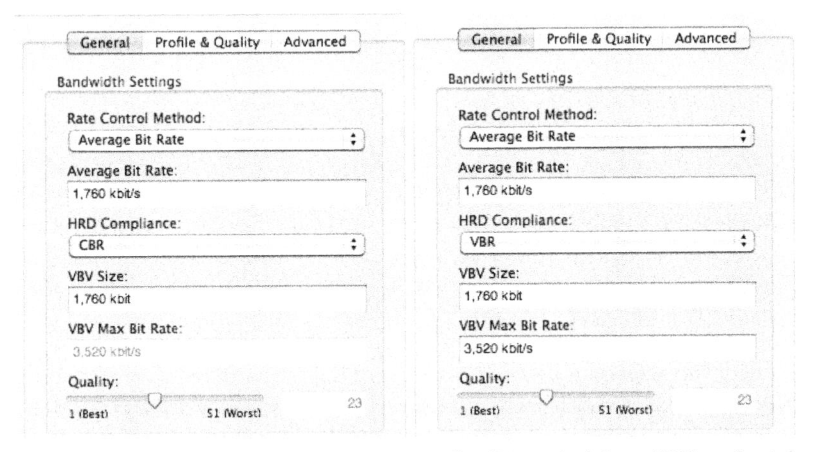

Figure 7-48. x264 preset tuning options. Settings for CBR on the left and VBR on the right.

3. Now let's attack the H.264 options not set by the x264 preset.

- ***Profile.*** Set as normal.

- ***Level.*** I recommend leaving the default value of Auto (recommended).

- ***B-Frames as Reference Frames.*** Enable this.

4. Now let's customize H.264 parameters controlled by the setting you may want to override, which have Auto (From Preset) as their values.

- ***Max Number of Reference Frames.*** Set to 5.

- ***Number of B-frames.*** If encoding using the Main or High profile, set to 3.

- ***Entropy Coding.*** If encoding using the Main or High profile, set to CABAC.

5. If a parameter doesn't auto-adjust and you don't know what it does, leave it alone—particularly when encoding for devices, where an errant configuration option can render the file incompatible. This includes all the settings in the Compatibility box in the Advanced Settings.

Those are the video settings you need to configure. Click Audio Codec in the Format box to reveal the Bit Rate setting, and set that at your target.

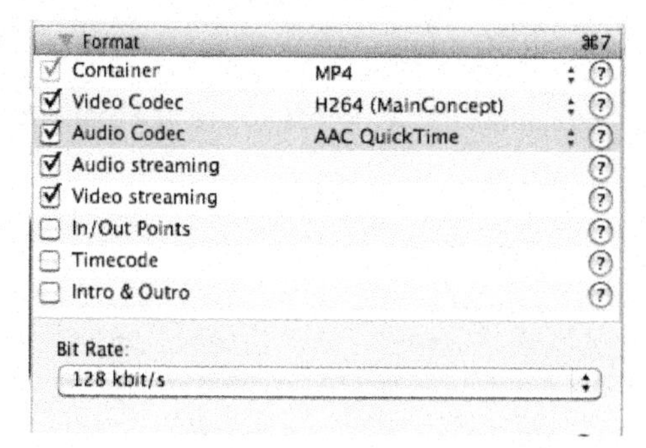

Figure 7-49. Setting the audio data rate.

Leave the Audio streaming and Video streaming options at their default. At this point, customize the name of the preset in the Task name box at the top of the Inspector, and don't forget to save it. Now let's take a quick look at encoding with the MainConcept codec in Episode.

Episode and the MainConcept Codec

After our work with the x264 codec, MainConcept is a breeze, with relatively few options. I'll trust that you know you have to set options like frame rate and resolution, and jump into those in the video codec section screen by screen. If you'd like to follow along, I made a copy of the H264 MP4 2Mbit 16x9 preset and I'm customizing the copy.

Figure 7-50 shows the General settings. Most of these you should know by now; here are the highlights and idiosyncrasies:

- ***Bandwidth control.*** There is no constrained VBR option because, according to Telestream, attempting to constrain the data rate didn't work. So use VBR for single file encodes and CBR for adaptive streaming.
- ***Key frames***. Natural and Forced with interval determined by application; I use 300 for single-file streaming and 90 for adaptive (10 seconds and 3 seconds).
- ***Adaptive B-frames.*** Enable this.
- ***Number of B-frames.*** Set to 3 frames when encoding into the Main or High profiles.
- ***Number of Reference Frames.*** Set to 5 frames.

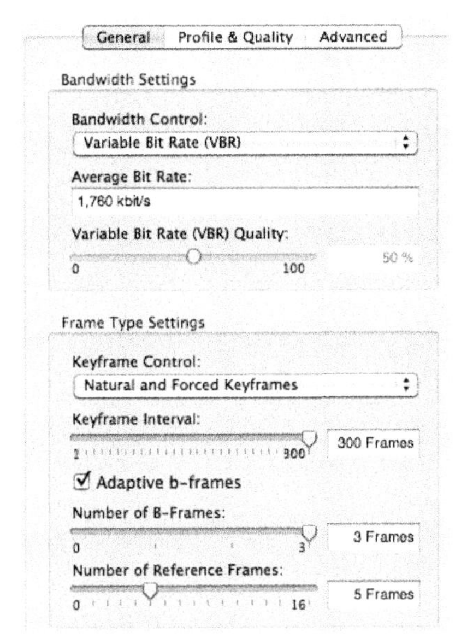

Figure 7-50. Episode's MainConcept controls in the General tab.

The left side of Figure 7-51 shows the Profile & Quality settings. Here are the highlights and idiosyncrasies:

- ***Profile settings.*** Choose the desired profile. If it's Main or High, choose CABAC.
- ***All other settings***. Use those shown in the figure.

The right side of Figure 7-51 shows the Advanced settings. Here are the highlights and idiosyncrasies:

- ***Number of Slices.*** Set this to 1 Slice for maximum quality.
- ***All other settings***. Use those shown in the figure.

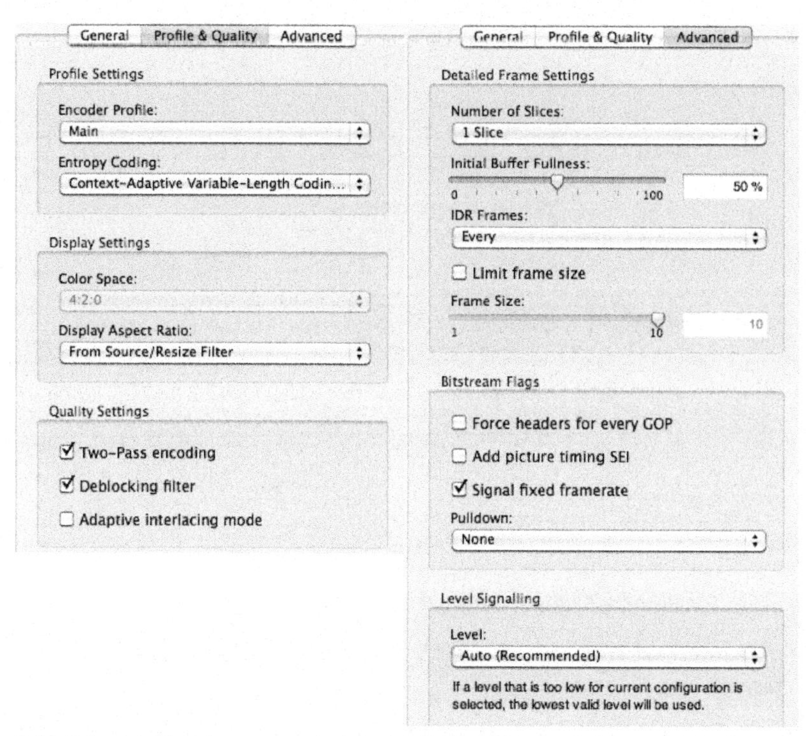

Figure 7-51. Episode's MainConcept controls in the Profile & Quality (left) and Advanced (right) tabs.

This concludes my review of desktop encoders; now let's take a much quicker look at several enterprise-level encoders, starting with Harmonic ProMedia Xpress. The point for these isn't an exhaustive review and instruction, but a quick flyover of the basic workflows and H.264 encoding options.

Harmonic ProMedia Xpress

The Harmonic Workflow System is an architecture for retrieving and encoding audio and video files into multiple single-file and adaptive formats, checking their quality, and delivering the encoded results to local or remote destinations. The primary encoding engine has been ProMedia Carbon, a Swiss Army Knife program with very extensive input and output options, excellent output quality and many other fine features. Unfortunately, encoding speed was not one of them.

To remedy this, Harmonic recently released ProMedia Xpress, a highly focused product designed to fit neatly into the encoding workflows utilized by many broadcasters. Specifically, in its first iteration, the product can only input MPEG-2 Transport Streams using either the MPEG-2 or

H.264 codec, and can only output H.264 encoded MPEG-2 Transport Streams. With additional licenses, the system can convert these outputs into chunked video files and metadata for HTTP Live Streaming (HLS), HTTP Dynamic Streaming (HDS) and Smooth Streaming.

While future Xpress versions will support additional file inputs like .AVI, .MOV and .MFX input, including both ProRes and Avid DNxHD, output support will continue to be limited. For example, although DASH is scheduled for the next software release, the development path does not include single or multiple .MP4 file output; if you're an RTMP streamer, either use ProMedia Carbon or find another solution.

You can purchase the software-only version for $9,000, or a rack-mounted turnkey system called the ProMedia 5200 Application server, which is the unit I tested, for $25,000. The license for each additional adaptive streaming module is $2,500, with an additional license of $2,500 for encryption for each adaptive format.

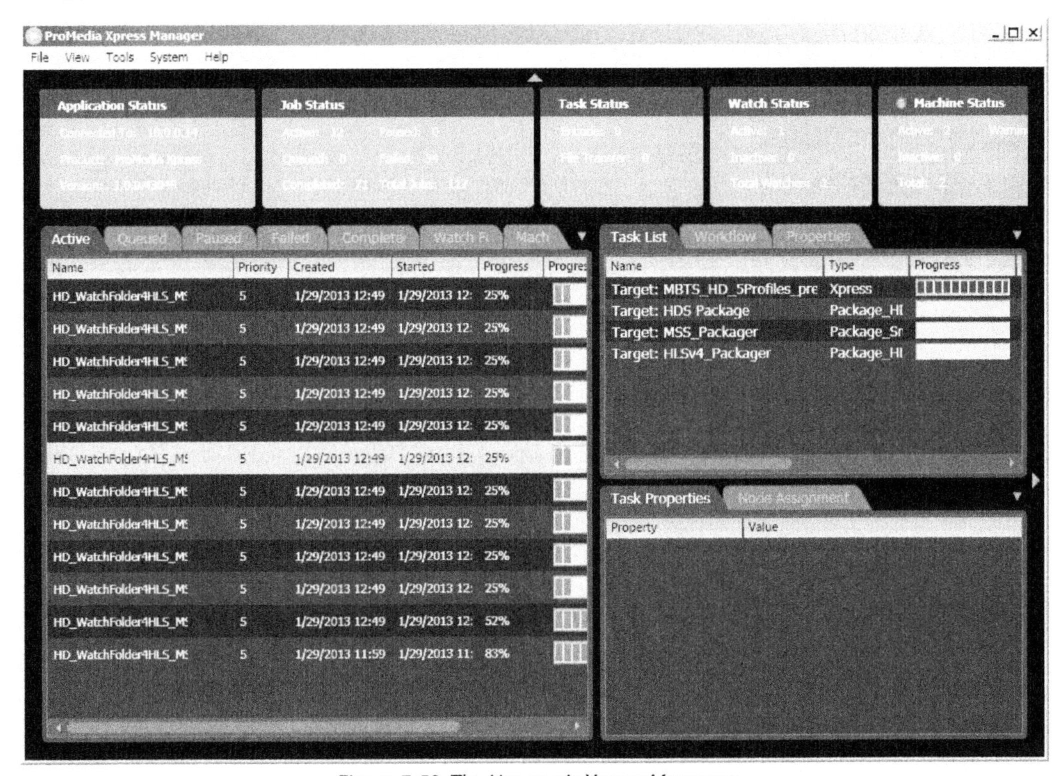

Figure 7-52. The Harmonic Xpress Manager.

The 5200 is a beast of a system, with two computers in a 1 RU rack-mount form factor: one serving as the controller, the other serving as the encoding node. Both are equipped with two 6-core 3.33 GHz Intel Xeon X5680 processors with 12 GB of RAM, and run Windows Server 2008

RT Standard, SP1. The system does not use GPU acceleration, and both units are equipped with a Matrox G200eW graphics card.

The special sauce Harmonic developed to speed encoding is called MicroGrid parallel computing, which, according to its marketing materials, "splits a large H.264 transcoding job into thousands of tiny ones, each of which is completed concurrently, removing the bottlenecks associated with traditional transcoding architectures." Xpress can operate stand-alone, or it can serve as an encoding engine within the Harmonic Workflow System, operating via a GUI or API. When you drive the system directly, you use software the ProMedia Xpress Manager and its components, which run exactly like the Harmonic Workflow System software.

Figure 7-52 shows the Xpress Manager. As you can see, tabs atop the interface provide details regarding active jobs, queued jobs, pending QC jobs and other functions. By way of background, all encodings are driven by a workflow—whether started manually, via watch folder or via API.

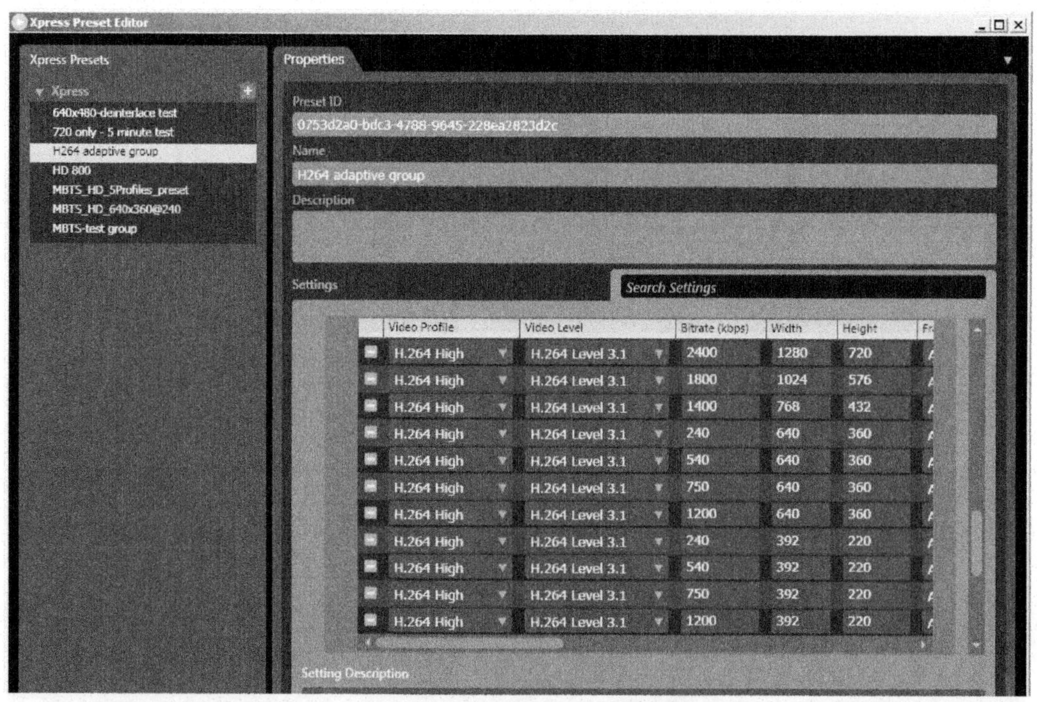

Figure 7-53. The Preset Editor. Note that I'm editing the H264 adaptive group that I deploy in Figure 7-55.

You build a workflow in the Workflow editor, which uses encoding presets created in the Preset editor and Packaging presets built in the Package preset editor. Figure 7-53 shows the Preset Editor, where you configure your adaptive streams and set your H.264 encoding options. Configuration options are very limited, including H.264 configuration options. For example, all encoding is single-pass, with CBR and average-bit-rate control options, but not true VBR. This enforces the most conservative theories of adaptive streaming, but may be too inflexible for some producers who seek the additional quality that VBR, or even 2-pass CBR, often delivers.

H.264 configuration options are limited to Profile and B-frame interval, without control over options like entropy encoding or search functions. In truth, for most users, fine control over H.264 encoding options is unnecessary, but if you know what you're doing, you may find this lack of control frustrating.

The Package Editor performs a similar function, allowing users to configure HLS, HDS and Smooth Streaming options like chunk size and closed captions. The Xpress packagers can integrate captions from multiple sources—including CEA-608 and 708, DVB-Subtitle, and Teletext—and convert the input for use in all three packaged formats, although some custom development may be necessary for complete integration.

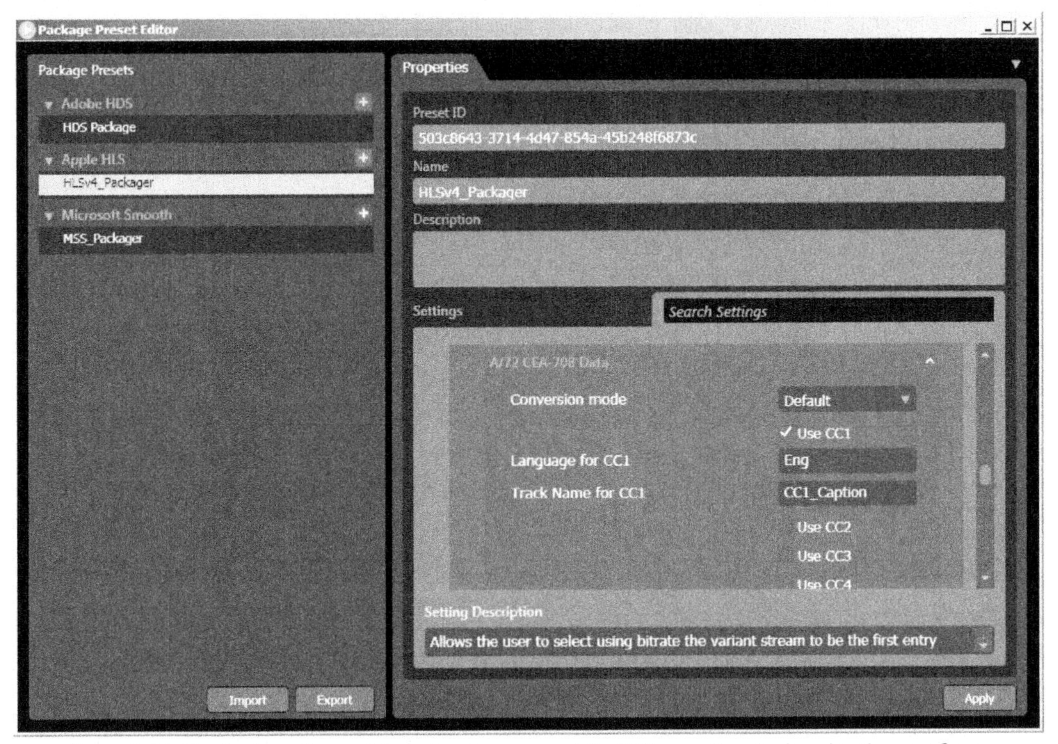

Figure 7-54. The Package Preset Editor. Creating a preset for HLS encoding deployed in the next figure.

Once you have your presets and packages configured, you build them into a workflow in the Workflow Editor, shown in Figure 7-55. There are three major classes of components—Pre-Transform tasks, Transform tasks and Job End tasks—with success and failure components available at each level. You see this in the tree-like structure shown in the middle of Figure

7-55. Pre-Transform tasks include quality control or file configuration checks that require the Harmonic Quality Control System (not included), or notification tasks.

Transform tasks are the primary encoding functions that are driven by presets. In Figure 7-55, you see that I'm using the preset created in Figure 7-53. Upon a successful encode of the source file to this preset, Xpress deploys the HLS and HDS packages created in the Package Preset Editor shown in Figure 7-54. The packagers input the MPEG-2 transport streams created during the initial encode, and create the unique container formats and metadata files necessary for each adaptive format. Since no additional transcoding takes place, these operations are very fast.

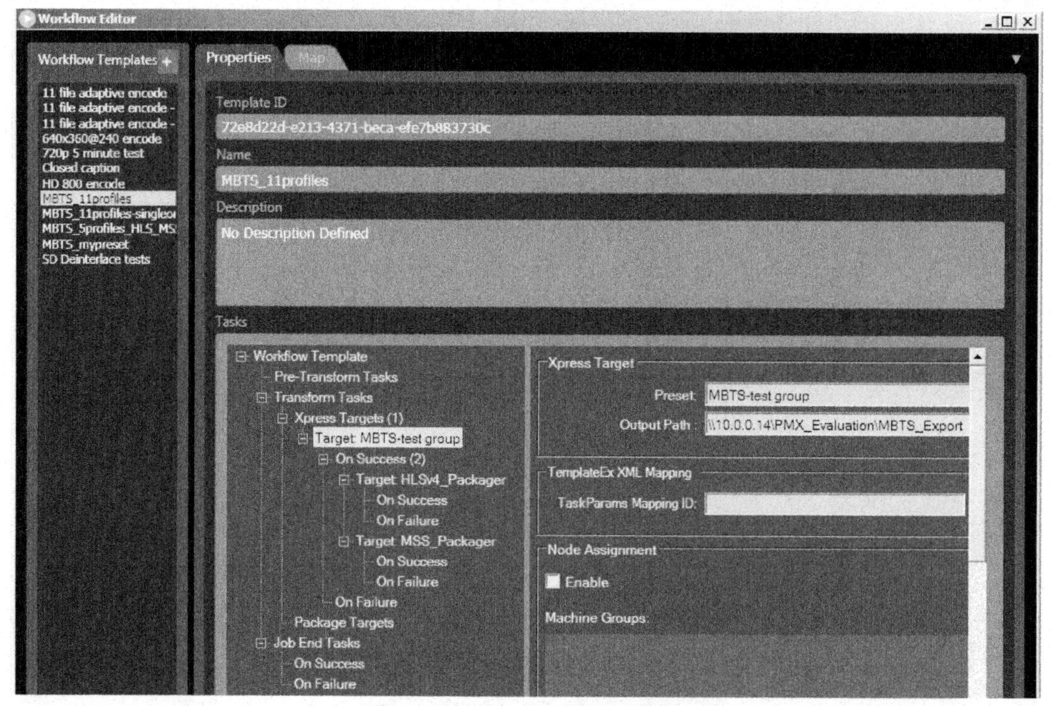

Figure 7-55. The Workflow editor.

All targets and packagers have output paths. At job end, you can trigger another set of tasks that deliver these files to CIF or FTP locations, invoke additional quality control stages, or create reports or notifications.

Once you create the workflow, you can trigger it via a watch folder that can poll UNC or CIFS drives or retrieve via FTP, by direct file input via another tool called the File Queuer, or via the API. Irrespective of the triggering technique, you can set an encoding priority on a scale from 1 to 10, with 10 the highest. While you can't save a File Queuer action, you can easily re-queue a file in the Xpress Manager by right-clicking the log entry and choosing Re-Queue. This is a convenient way to re-encode repetitive jobs, or tasks that fail.

Telestream Vantage

Vantage is Telestream's workflow system encoder. I recently had a chance to work with the new Multiscreen encoding engine running on the Lightspeed server, so I'll use that platform to demonstrate how Vantage operates.

You drive the Multiscreen encoder through the Vantage workflow software, where it shows up as a codec building block in the top toolbar. Architecturally, Vantage works around the concept of a workflow, which you create using graphical elements in the Workflow designer shown in Figure 7-56.

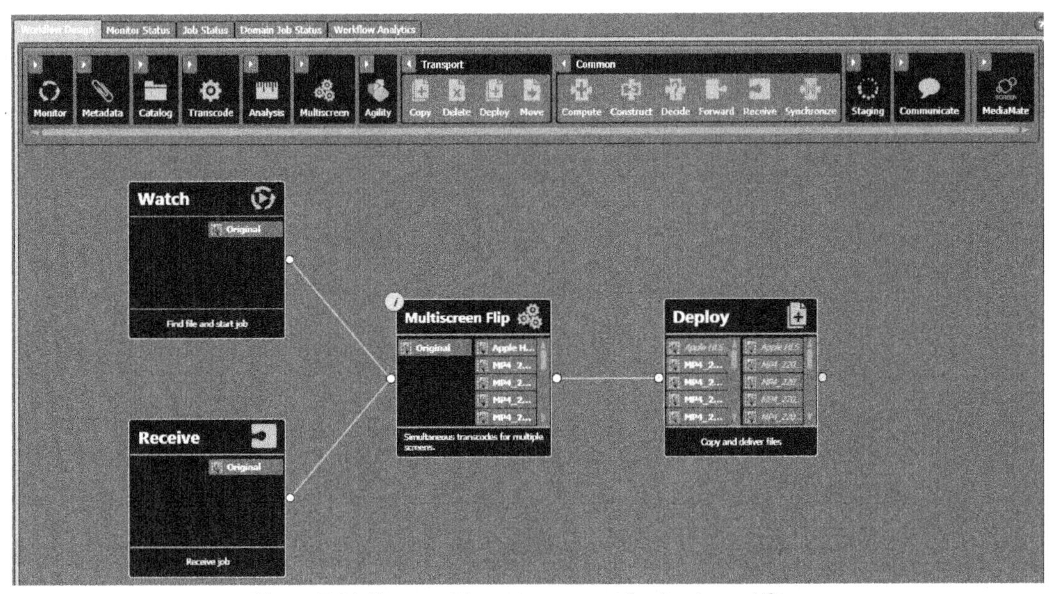

Figure 7-56. The graphics components of a simple workflow.

If you look closely, you'll note that icons on the top right of each element identifies the function that the element will perform. For example, in Figure 7-56, The Watch element is scanning a watch folder for incoming files to submit to the Multiscreen Flip, while the Receive function lets you manually load a file into the workflow. The Multiscreen Flip is the element that deploys the Multispeed encoder running on the Lightspeed encoder. After encoding is complete, the files are sent to the Deploy function for delivery.

You build a workflow by dragging a workflow building block from the icons on the top and connecting the elements in the Workflow Designer. Once inserted into a workflow, each element offers extensive configuration options that finely detail the task performed. For example, in Figure 7-57, you see the three high-level configuration options available in the Multiscreen Flip module: Inputs, Transcoders and Outputs.

Figure 7-57. The three high-level configuration options in the Multiscreen Flip module.

The input function defines and names the elements of the incoming file—Video 1 and Audio 1 in this case. The Transcodes function grabs the discrete elements of the incoming file and encodes them into single audio and video files, essentially elementary streams. For example, in Figure 7-58, you see that Audio 1 is sent to an AAC transcoder where it's converted into two files: one at 128 kbps, one at 96 kbps.

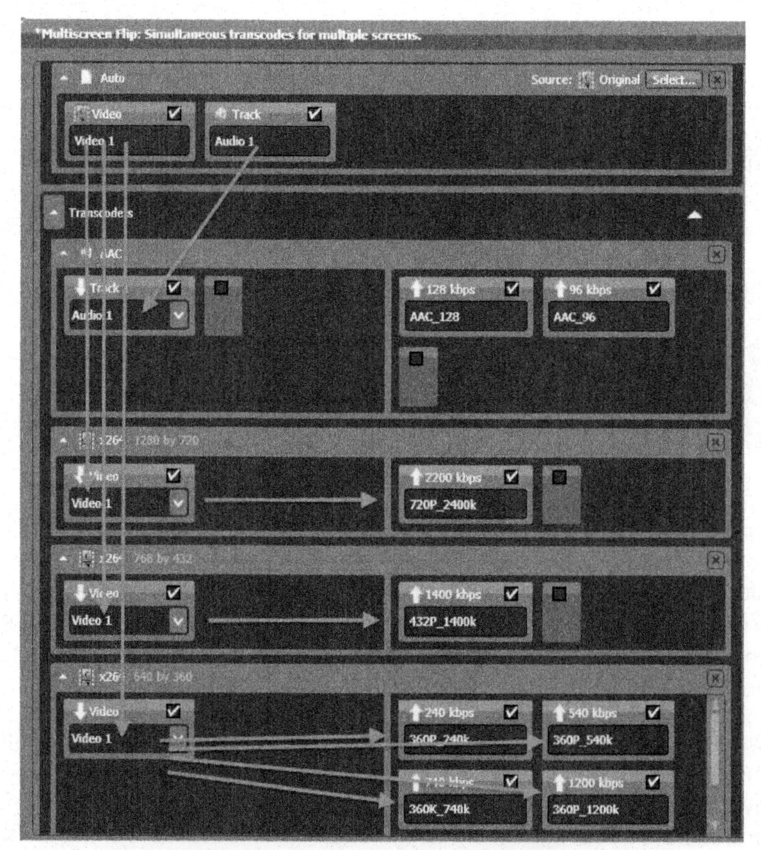

Figure 7-58. Converting the incoming audio and video into elementary streams using different encoding parameters.

Similarly, the Video 1 file is sent to multiple x264 converters—some encoded into a single resolution, some encoded into multiple versions at multiple resolutions. For example, in addition to unique steams at 720p and 768x432, this adaptive group encodes four streams around the 640x360 configuration: at 240, 540, 740 and 1200 kbps.

Since I'm starting with a 720p file, getting to 640x360 involves scaling, and for some source files, deinterlacing and other filtering as well. Rather than processing each file in the adaptive group individually, Vantage processes all files produced at the same resolution at one time, performing all scaling, deinterlacing and other filtering once, and then encoding the output four separate times. Most other encoders would encode each file separately, performing the scaling, deinterlacing and other filtering each time, which is less efficient.

Encoding Controls

Vantage uses the x264 codec with familiar presets and tuning profiles and a few key adjustments available in the GUI (Figure 7-59). If you really want to tweak the controls, you can drop in a command line argument and access all x264 parameters.

Figure 7-59. Vantage's encoding controls.

You apply all encoding parameters except data rate to each transcoder instance, essentially each resolution produced by the encoder. For example, in Figure 7-58, you would set all encoding parameters except the data rate in the each of the three x264 boxes. Then you would set the data rate for each iteration at that resolution within the box separately.

Packaging the Elementary Streams

Once the elementary streams are created, Vantage can incorporate them into multiple output files, as shown in Figure 7-60. There you see the same elementary screens packaged into two different output variants for HDS and HLS. Building these variants is simplified by drop-down lists that expose the elementary streams encoded in the previous step. Just click the Plus sign to add a variant, type in the name (or copy/paste/edit the name) and choose the desired elementary streams in the drop-down list.

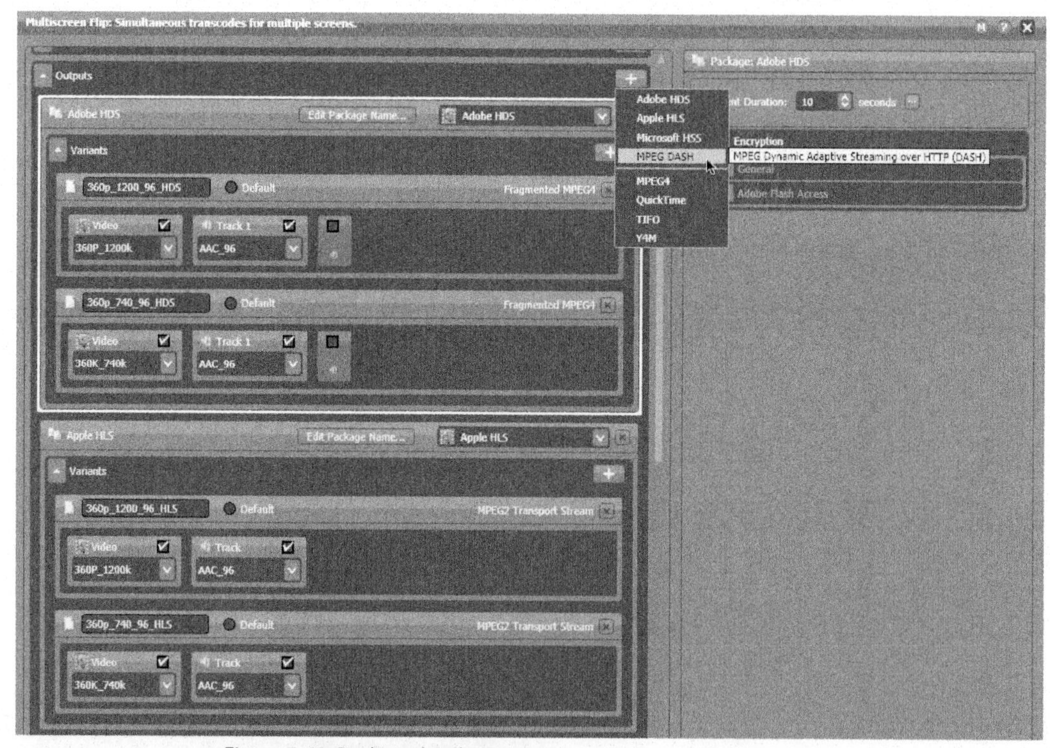

Figure 7-60. Packing the elementary streams into multiple outputs.

Less obvious in the screen shot is the drop-down box on the upper right, which shows that Vantage can also package the streams in DASH and Smooth Streaming formats as well as MPEG-4, QuickTime, .Y4M and .TIFO (Telestream Intermediate Format) files. To the right of the list are the fields for inserting encryption identifiers and credentials for the HDS files.

This workflow is extremely efficient when outputting into multiple output formats. Assume, for example, that I was outputting my adaptive group of 11 files into three outputs: .MP4 files for local and RTMP Dynamic Streaming, and packaged output for HLS and HDS. With my 11 test files, Vantage scaled five times—one for each resolution in the group—and encoded 11 times to produce the elementary streams, which it packaged into the respective output targets without re-encoding. In comparison, some competitive products would have to scale and encode 33 times to produce the same outputs.

To add closed captions to the streams, you enable a metadata processing filter that detects and parses out the captions and inserts them into compatible streams. Using this feature, I was able to insert closed captions into the HLS and HDS streams.

Elemental Server

Elemental Server is the prototypical high-volume encoding engine. The following description is adapted from my review of the Elemental Encoder that you can read at **bit.ly/Ozer_elemental**. In line with the other enterprise encoder descriptions, I'll focus only on the workflows and H.264 encoding controls. As a company, Elemental has been great about allowing me access to its encoder for competitive testing and don't mind if I discuss my findings, but the company likes to keep its UI private, and I respect that. So, I'll only show a screen or two.

Briefly, Elemental Server is a $26,000 hardware encoder that runs Linux using off-the-shelf components, including graphics processing units (GPUs) from NVIDIA. Elemental's special sauce is that the company has created its own GPU-accelerated H.264 codec that combines outstanding performance and competitive quality. Although there are several other GPU-accelerated encoders available, none offer the same level of quality and performance. Some are speedy but quality is substandard, and others offer comparable quality but performance is much slower.

You operate Server via three components: presets, profiles, and jobs. Presets contain an individual set of encoding parameters, plus any preprocessing or postprocessing options. Profiles contain multiple presets, making it easy to create an adaptive group that produces multiple output files. When you create a job, you typically insert a source file into the profile and press the magic Go button to start encoding.

H.264 Encoding Parameters

Elemental's encoding philosophy is to limit the number of configuration options available to their users as much as possible. For example, with H.264, you can specify critical options such as profile, level, entropy encoding setting, and the number of B-frames and reference frames, but you get no control over esoterica such as search shape and pyramid B-frames. Given the consistently high quality of the H.264 output, this approach makes sense, but if you need comprehensive access to H.264 encoding parameters to customize for a unique scene or output requirement, it's not available.

Like most other enterprise encoders, Elemental Server combines basic encoding with adaptive packaging, so profiles can incorporate .MP4 file output and multiple format ABR output, which you can see in Figure 7-61. You define the basic output parameters in the file group, then the output parameters for each file in each ABR group in a separate tab.

Figure 7-61. The Elemental Server can output multiple ABR groups with a single encoding pass.

Elemental has optimized its supported codecs to run efficiently on both GPUs and CPUs. In discussions with Elemental personnel, however, I learned that GPUs are more efficient with higher-resolution, higher-data-rate files, while CPUs crunch through smaller files faster. To optimize overall encoding performance, Server provides several controls to allocate encoding chores between GPUs and CPUs.

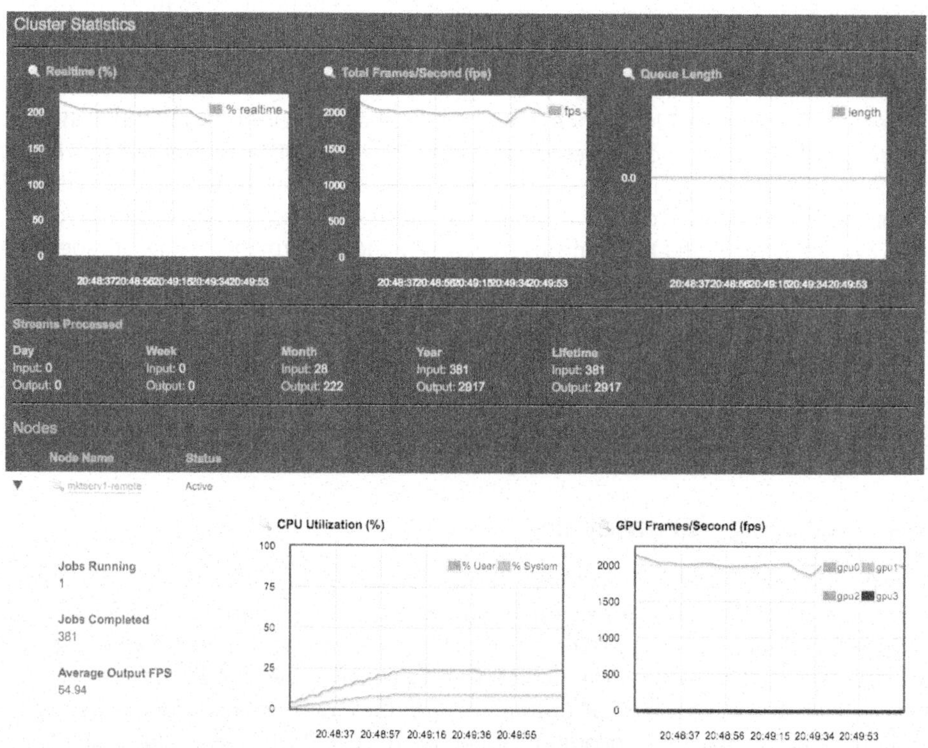

Figure 7-62. Load-balancing statistics provided by the Elemental Server.

To assist these load-balancing efforts, Elemental Server produces real-time encoding statistics as shown in Figure 7-62. On the bottom left of the figure, you see that CPU utilization is low, while the frames per second encoded by the GPU is high. If you were producing multiple files in an adaptive batch, you could adjust certain configuration controls to allocate more files to the CPU.

If you're looking for a high-volume encoder, Elemental Server should be first on your list.

ProMedia Carbon

ProMedia Carbon is a prototypical Swiss Army Knife transcoder with outstanding format support, but aging performance, which is why Harmonic is adding the ProMedia Xpress encoder to the family. Carbon costs $5,000 as a stand-alone product, and can be used as a component of the ProMedia Workflow Systems, or WFS. Carbon lacks the ability to output in any adaptive streaming formats, although that functionality is available via other components of the Workflow System.

By way of background, Rhozet was a spin-off from Canopus Corporation that was acquired by Harmonic Corporation in 2007. The company is in the process of rebranding the products, so the Carbon interface that you'll see still says Rhozet rather than ProMedia. Hey, it's only been six years—cut them some slack, OK?

When you install the product, Harmonic loads multiple programs, including the Carbon encoder and the Carbon Admin panel. We'll work through both interfaces in this short description.

Figure 7-63 shows Carbon's main interface. The tabs on the left—Source, Target and Convert—drive the workflow. Click Source to input one or multiple files. As mentioned, Carbon can accept a broad range of input files, including all the usual suspects (.AVI, .MOV, MPEG-2/-4, .WMV, etc.) and media containers like .HDV, .MXF, .GXF, .LXF and QuickTime. The program also supports standardized broadcast streams like ATSC, DVB and CableLabs; high-end cameras like Sony XDCAM and Panasonic P2; high-end video servers like Leitch VR and Nexio, Grass Valley Profile and K2, Omneon Spectrum, and Quantel sQ; and files from Avid, Final Cut Pro, Premiere Pro and Grass Valley Edius. Like I said, the prototypical Swiss Army Knife transcoder.

Once loaded, you can apply a range of filters to any source clip by double-clicking the clip and choosing a filter. Key filters include 601 correction, which expands the color space of your TV content for the web; 601 to 709 color correction for SD to HD conversions; adaptive deinterlacing; gamma correction; Line 21 extraction for closed-captioned text; cropping; and bitmap keying.

Once you've loaded your source files and applied any filters, click Target to select target output parameters. Carbon supports both individual presets, which contain one set of output parameters, and profiles, which contain multiple presets.

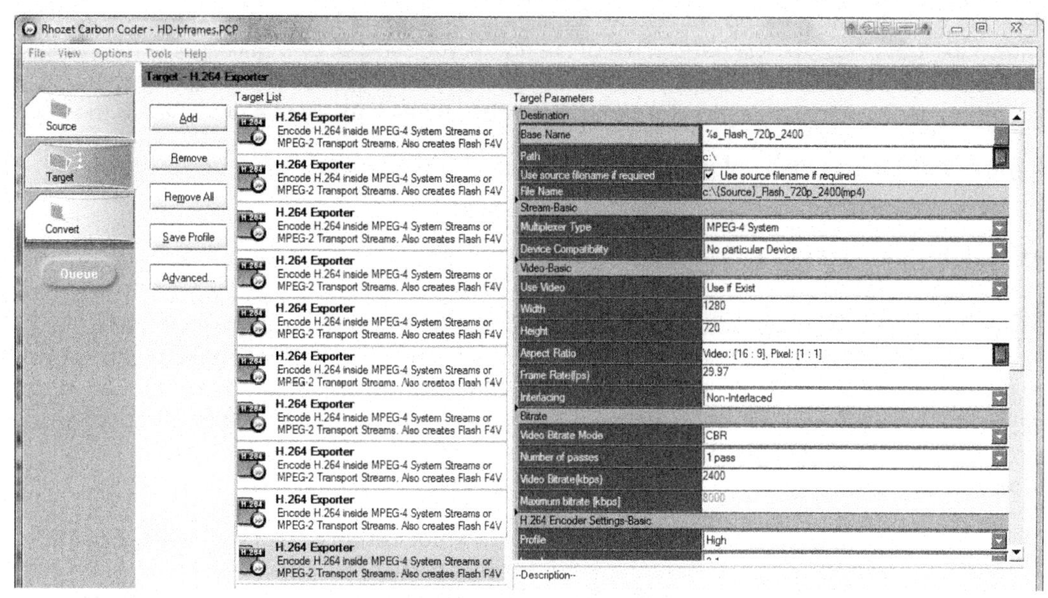

Figure 7-63. Carbon, the big picture.

For example, I applied the 10 encoding Presets shown in Figure 7-63 by choosing one profile. Once you've customized your presets and profiles, you can access them in the main program or via watch folders, adding a highly useful layer of shared encoding capabilities.

Carbon uses the MainConcept H.264 codec, and its interface offers an exceptional range of H.264 adjustments, a portion of which are shown in Figure 7-64. Fortunately, Rhozet also provides a very comprehensive guide to these options in a PDF file you can download at **rhozet.com/rhozet_H264guide.pdf**.

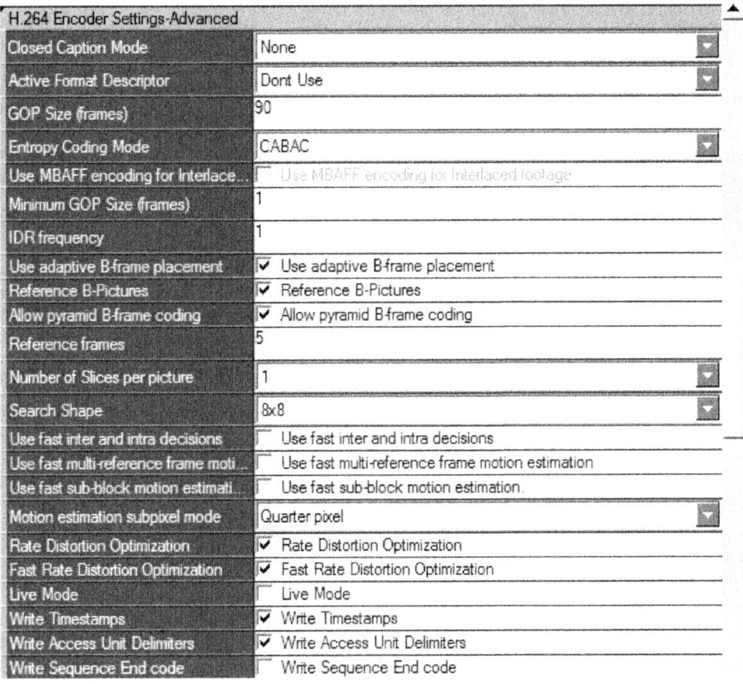

Figure 7-64. Carbon provides a very extensive range of H.264 configuration options.

You can render using two techniques. Click Convert, and Carbon will encode the files serially with Carbon's main interface. Or, you can click Queue to encode the files in the updated Carbon Administrator, which opens the Job Queuing window shown in Figure 7-65.

Job Queuing ? X
Job
Job Name : Job (02-25-13,16:45:29)
Job Description : Job submitted Monday, February 25, 2013, 16:45
Job Priority : 5
One Job for each Target:
Virtual Source Name: Band0016
Target File Name: Rename if exists
Execution Machine: Local
Render in Network Grid: Segments : 4
WFS Controller IP Address:
WFS Controller Port:
Queue Cancel Help

Figure 7-65. The Job Queuing window.

On a multi-core system, click the One Job for each Target button in the middle of the window, which enables Carbon to more efficiently assign processor cores to the encoding job. If your version of Carbon is part of a rendering farm, you would click the Render in Network Grid checkbox on the bottom to send the files to rendering farm.

Then click Queue to send the encoding jobs to the encoding engine. To watch your progress, open Carbon Administrator, the other program installed when you set up Carbon, and shown in Figure 7-66. Operationally, you'd always want to check the Kernel Settings to make sure you have as many Transcoding Slots as you have cores in your system.

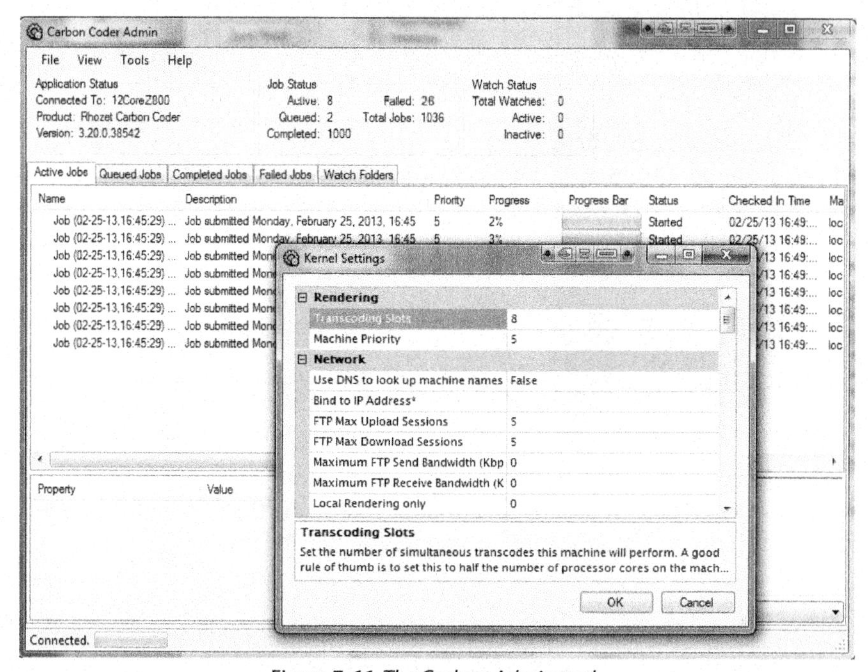

Figure 7-66. The Carbon Admin tool.

Despite using the MainConcept codec, Carbon has always produced excellent quality. As mentioned at the top, however, performance—which was once top-notch—has suffered compared with new systems. It's great as a Swiss Army Knife transcoder, but it's definitely not a speed burner.

Conclusion

Now you know how to use your encoding tool; let's have a quick look at encoding for iTunes.

Chapter 8: Producing for iTunes

This chapter focuses on producing for delivery via iTunes. While you don't need an iDevice to use iTunes, to a degree, most producers seem to customize their videos for specific classes of iOS devices. For example, many producers still encode to the 320x240 format, which is the only video that plays on the first few generations of video-capable iPods.

For this reason, I'll start by reviewing the classes of iDevices you first visited in Chapter 4. Then, I'll review the results of some recent research that reveals which groups of iDevices most producers target and the configuration they were using to do so. I'll conclude with a look at the optimal encoding parameters to use when encoding for iTunes delivery. In this chapter, you will learn:

- The playback requirements for different groups of iDevices
- The device targeting strategies used by prominent iTunes publisher and the encoding parameters they use
- The optimal encoding parameters to use when encoding for free distribution on iTunes
- The parameters used by producers selling music videos and SD and HD TV episodes on iTunes.

Let's begin.

Producing for iTunes Delivery

Table 7-1 details the device and playback specifications of all Apple iDevices. These are critical, because in most instances, iTunes won't load files that exceed a device's playback capabilities.

	Original iPod (to-5g)	iPod nano/ classic/iPod touch/iPhone to V4	iPhone 4/ iPod touch 4/ iPad 1	iPod Touch 5	iPhone 4S/5 iPad 2/New iPad
Video Codec	H.264	H.264	H.264	H.264	H.264
Profile/Level	Baseline to Level 1.3	Baseline to Level 3.0	Main to Level 3.1	Main to Level 4.1	High to Level 4.1
Max Video Data Rate	768 kbps	2.5 Mbps	14 Mbps	50 Mbps	62.5 Mbps
Max Video Resolution	320x240	640x480	720p	1080p	1080p
Frame Rate	30 fps	30 fps	30 fps	30 fps	30 fps
Audio Codec	AAC-LC	AAC-LC	AAC-LC	AAC-LC	AAC-LC
Max Audio Data Rate	160 kbps	160 kbps	160 kbps	160 kbps	160 kbps
Audio Parameters	48 kHz, stereo	48 kHz, stereo	48 kHz, stereo	48 kHz, stereo	48 kHz, stereo

Table 7-1. Specifications for various iDevices.

For example, if you try to load a 640x480 file onto a first-generation iPod, iTunes will present an error message that looks something like Figure 7-1.

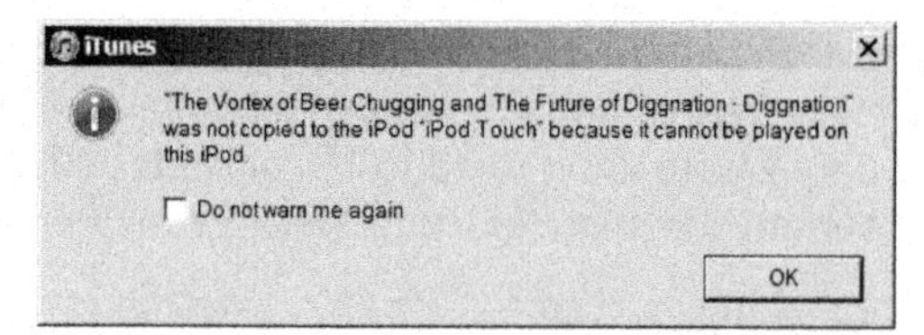

Figure 7-1. This is what happens when you try to sync a file that exceeds the iDevice's playback capabilities.

Let me share the same comments about these specs that I mentioned in Chapter 4. Specifically, Apple's specs don't always conform with reality, but they're usually conservative. For example, while the spec sheet says the original iPad can only play H.264 video encoded using the Main profile, I've tested 720 video encoded with the High profile and it plays fine. Although my iPhone 4S shouldn't play video with audio encoded using HE AACv2, it plays without problem, and while Apple's maximum audio data rate is 160 kbps, virtually all music videos shipped on iTunes use 256 kbps.

The spec sheets are confusing because they present separate specs for "Audio Playback" and "TV and Video." For example, for the iPhone 4, audio paces include "AAC (8 to 320 kbps), Protected AAC (from iTunes Store), HE AAC" and others. However, in the TV and video section, the spec states "H.264 video up to 720p, 30 frames per second, Main Profile Level 3.1 with AAC-LC audio up to 160 kbps, 48kHz, stereo audio." From this, I infer that the device can play stand-alone audio files up to 320 kbps, but only 160 kbps when delivering audio with the video. However, I've tested multiple devices, and they all can play at least AAC audio at 256 kbps when delivered with video.

Still, when you're targeting specific devices, or classes of devices, you should be conservative and conform your streams to the specs shown in Table 7-1. Scanning the table, you see that there are five levels of playback capabilities:

- 320x240 Baseline profile for pre-5G iPods
- 640x480 Baseline profile for later iPods and iPhone and iPod touch devices before 4G
- 720p Main profile for iPhone 4, 4G iPod touch and iPad 1
- 1080p Main profile for the iPod touch 5
- 1080p High profile for the iPhone 4s/5 and iPad 2 and New iPad.

Obviously, producing at 320x240 will deliver 100% compatibility but a poor-quality experience on newer devices, particularly current iPads and iPhones.

This raises the following four questions:

- Should you abandon older iPods that only support 320x240?
- Should you try to support all post-5G iPod devices with a single 640x480 maximum file?
- Should you distribute multiple files?
- Should you distribute 1080p files to the new iPad?

To answer these questions, I downloaded 52 files from 34 different producers, downloading the most popular podcasts as designated by iTunes, several from three-letter networks and many from popular technology sites like CNET.

Once I found an episode from a specific producer, I searched iTunes to determine if other episodes, at other resolutions, were available. For the purposes of this survey, all downloads were free; I'll discuss paid TV show episodes and music videos below. Let's address our four questions in order.

Should You Abandon Older iPods?

Here's what I found:

- Of the freely downloadable files, six of 52 files, roughly 12%, were 320x240 or smaller, and one was a 640x480 file encoded using the MPEG-4 codec that will also play on the oldest classes of iPods. These producers included Oprah, Face the Nation, Tekzilla, Car Tech, Bill Moyers Digital Archive and Reason TV.
- This percentage actually increased from the first time I ran this test, where only five of 98 files were 320x240 or smaller.

So, if you're seeking to reach the broadest possible audience, you should continue to encode at 320x240. On the other hand, if you're selling highbrow products to upper-class prospects, it's probably safe to assume that they have a more recent playback device.

Should You Support Post-5G Devices with a Single Stream?

Twenty-eight of the 52 files were 640x480 or 640x360, by far the largest single configuration group. For 23 of these producers, this was the only size offered. Single-stream producers included a wide range of websites, including *The New York Times*, *Sesame Street*, CNBC, NOVA, the White House and Reason TV. Overall, so long as you're producing at 640x480 or 640x360—as opposed to a larger resolution, which won't play on older iPods, iPod touches and iPhones—offering a single stream is a solid strategy

Should You Distribute Multiple Files?

There are three scenarios where you should consider distributing multiple files. First, if you're a tech site—like CNET, Revision 3 or This Week in Tech (TWIT)—you should consider distributing multiple files just to keep up with the Joneses. All tech sites do it; you probably need to do it also.

The next scenario where you should consider multiple streams is if you're offering one stream at 720p or larger. In the survey, five of 10 producers who offered a 720p stream also offered a lower-resolution stream, typically in the 640x360 range. Unless you're adapting a high-end strategy that deliberately excludes small-screen viewers, I would offer a lower-resolution stream (probably 640x360) if I offered a 720p stream.

The final scenario where producing multiple files makes sense is when you have content that lends itself to big-screen viewing. For example, if you're producing software screencasts, 720p is a great resolution for viewing on computers and iPads. Ditto for visually impressive footage like NASA's space shots. On the other hand, if you're distributing simple talking-head footage of an interview or presentation, 720p seems like a waste of disk space.

One interesting example of this strategy is the Obama White House. Routine matters like press briefings are distributed at 640x360 resolution, while more glamorous functions, like traveling with the First Lady, are distributed at 720p.

Should You Distribute 1080p to the new iPad?

None of the free podcasts I downloaded were configured at 1080p. On the other hand, all HD TV episodes I downloaded were larger than 720p—a dramatic shift over the last 12 months. I personally feel that 1080p is a total waste in most instances, but then I don't own a new iPad.

Mistakes to Avoid

What were the most common mistakes I noticed from my survey results? Several come to mind.

- ***Wrong profile.*** Several sites produced their lower-resolution videos using the Main or High profile, which won't play on the middle generations of iDevices. For example, NOVA was encoded at 640x360 using the Main profile, which excludes the second class of devices, while Old Jews Telling Jokes was encoded at 640x360 using the High profile, which does the same.

- ***Using resolutions that won't play on lower-resolution devices and aren't optimal on higher-resolution devices.*** Several producers used resolutions like 424x240 (MSNBC) or 512x288 (Washington Week). These resolutions won't play on the oldest iPods and will appear grainy on the newest high-resolution iDevices. A resolution like 640x360 wouldn't exclude any viewers compared with the lower resolutions used by MSNBC and Washington Week, but would look better on new devices.

What About B-frames?

I was preparing for a seminar on Producing for iDevices last October in London, when I happened upon this quote: "Apple recommends that you not use B-frames when targeting iPhone and iPod devices." The source was a page about serving iDevices via Microsoft's IIS Media Services (bit.ly/iisidevice). Unfortunately, there was no link to an Apple source for this information, so I couldn't verify it.

I did notice, however, that in the one Apple podcast I downloaded that used the Main or High profile, Apple did use B-frames. Of other videos produced using the Main or High profile, 20 of 22 also used B-frames, and they loaded and played fine on all tested devices. From this I'd have to conclude that when encoding with the Main or High profile, B-frames are fine.

44.1 or 48 kHz Audio?

I tend to produce at 44.1 kHz most of the time because it's the best frequency for Flash output. On the other hand, when producing solely for iDevices, what's the best frequency? In the survey of free podcasts, 63% were produced at 44.1 kHz, 34% at 48 kHz and 3% at 32 kHz.

All the for-fee music videos I downloaded were produced at 44.1 kHz, which seemed strange, but may relate back to CD-ROM audio being produced at that sampling rate. Interestingly, all HD versions of TV shows were produced at 48 kHz, while 10 of 11 of their SD counterparts were produced at 44.1 kHz. Makes no sense to me, but there it is.

If producing for delivery to Flash and iDevices, I would use 44.1 kHz. If producing solely for iDevices, I would use the sample rate of the source audio, which usually would be 48 kHz, but could be 44.1 kHz. If I had to convert the sample rate for encoding for iDevices—for example, if mastering at 96 kHz—I would use 48 kHz.

Let's synthesize this information into recommended encoding parameters for free podcasts.

Recommended Encoding Parameters

Essentially, when producing for iTunes, you should choose one or more of three file sizes: 320x240 (or 320x180 for 16:9) for ultimate compatibility; 640x480 (640x360 for 16:9) for compatibility with low-resolution iDevices, excluding the earliest units; and 720p for the iPad 1 and 2 and iPhone/iPod touch 4G and 4S. Over the next few sections, I'll share the average data rate used by producers each category plus the relevant device maximums, which should give you a pretty good feel for how to configure videos bound for iTunes distribution.

In all categories, I'll focus on the resolution and data rate, and identify the appropriate H.264 profile. You should fill in the rest of the blanks using recommendations from Chapter 3.

Encoding 320x240/320x180 Podcasts

You should produce at these resolutions using the Baseline profile when targeting the oldest video-capable iPods, or to produce a single file that plays on all iDevices. The first column shows the results of the survey, while the second shows the maximum for this class of device.

	Survey	Device Maximum
Video Codec	H.264 codec, Baseline profile	H.264 codec, Baseline profile
Data Rate Average	532 kbps	768 kbps
Frame Rate	Match source	30
Audio Codec	AAC-LC	AAC-LC
Data Rate	95 kbps/stereo	128 kbps/stereo

Table 7-2. Producing for 240p and 180p resolutions.

At 532 kbps, a 320x240 file has a bits-per-pixel value of around 0.231, which should translate to pristine quality irrespective of the source footage. When encoding your videos for this class, I would encode using 2-pass VBR, with a target data rate of 532 kbps and a maximum of 730 kbps.

Encoding 640x480/640x360 Podcasts

Data rates in this group averaged 1.139 Mbps, which translates to a bits-per-pixel value of 0.124 for 640x480 footage, and 0.165 for 640x360 footage. When encoding for this group, I would use the Baseline provide and 2-pass VBR encoding, with an average data rate of 1.139 Mbps and a maximum of around 2.4 Mbps. If quality isn't sufficient, I would boost the target until it is, but I would keep the maximum rate at around 2.4 Mbps.

	Survey	Device Maximum
Video Codec	H.264 codec, Baseline profile	H.264 codec, Baseline profile
Data Rate Average	1.139 Mbps	2.5 Mbps
Frame Rate	Match source	30
Audio Codec	AAC Low	AAC Low
Data Rate	113 kbps/stereo	128 kbps/stereo

Table 7-3. Producing for 480p and 360p resolutions.

Encoding 720p Podcasts

With an average data rate of 2.908, the bits-per-pixel value in this category is 0.105, which should be fine at this resolution since codecs are more efficient at higher resolutions. There's not a lot of meat on the bone, however, so if you need to bump the data rate a bit to achieve suitable quality, go right ahead. Certainly the device maximum can support it.

To translate this prose into action, I would encode using the Main profile and 2-Pass VBR with a target of around 3 Mbps and maximum of around 6 Mbps, or 200% of the target. If the quality weren't sufficient, I would go no higher than 4 Mbps, 8 Mbps maximum.

	Survey	Device Maximum
Video Codec	H.264 codec, 5 of 10 are High profile	H.264 codec, Main profile
Data Rate Average	2.908 Mbps	14 Mbps
Frame Rate	Match source	30
Audio Codec	AAC Low	AAC Low
Data Rate	169 kbps/stereo	160 kbps/stereo

Table 7-4. Producing for 720p resolution.

When producing 720p videos using a canned preset from an encoding tool, be sure to check the data rate, since many encoding tools use very high data rates in this class. For example, Compressor's 720p template uses a data rate of 10 Mbps, resulting in a porcine bits-per-pixel value of 0.362. Back when HD TV episodes were encoded at 720p, the average Hollywood data rate was around 4.1 Mbps, so I would resist the urge to go higher than this.

This concludes our look at free podcasts; now let's examine how producers encode paid content like music videos and network TV. Even if you're not encoding the types of videos, the configurations used for these encodes reveal some interesting techniques and the data rates necessary to sustain Hollywood-acceptable quality for podcasts delivered via iTunes.

Encoding Music Videos

I downloaded nine music videos to determine the parameters used in these instances, and I found a great degree of uniformity as shown in Table 7-5.

Music Videos	Width	Height	Data Rate	FPS	Bits per Pixel	Profile	Audio Data Rate	Channels
Cher Lloyd	640	480	1,548	23.976	0.21	Baseline	256	Stereo
Christina Aguilera	640	480	1,571	23.976	0.213	Baseline	256	Stereo
Jennifer Lopez	640	480	1,537	23.976	0.209	Baseline	256	Stereo
Justin Bieber	640	320	1,516	23.976	0.309	Baseline	256	Stereo
Katy Perry	640	480	1,425	23.976	0.193	Baseline	256	Stereo
Nicki Minaj	640	352	1,543	23.976	0.286	Baseline	256	Stereo
One Direction	640	480	1,588	23.976	0.216	Baseline	256	Stereo
PSY	640	352	1,529	23.976	0.283	Baseline	256	Stereo
Taylor Swift	640	352	1,501	23.976	0.278	Baseline	256	Stereo
Averages	**640**	**420**	**1,529**		**0.244**		**256**	

Table 7-5. Encoding parameters for music videos.

The key points here were:

- All videos were 640x480 or 640x360-ish, and all encoded using the Baseline profile. Producers ignored the oldest iPods but didn't customize for the higher-resolution devices—at least not yet.

- All were produced at 23.976 fps.

- The target audio for all songs was 256 kbps stereo, although sometimes the effective rate was lower. I doubt listeners could discern between 128 and 256 kbps on their cute earbuds, but clearly the music industry believes that 256 kbps delivers superior quality. As mentioned, it's interesting that all iDevices have a specified maximum audio data rate of 160 kbps for audio included with video, although these files played with no problem on all iPods I tested.

Given the frenetic pace of most music videos, you can also view the 0.244 average bits per pixel as the absolute maximum needed for adequate or better quality at this resolution.

One technique used in this category was to encode at a resolution of 640x480, but an aspect ratio of 16:9, which told the display program to play the video at a resolution of 854x480. Since iTunes checks resolution, not aspect ratio, to determine if an encoded file is compatible with a particular device, these files will load on older devices, but the 854x480 resolution looks better on iPads and iPhones with higher-resolution displays, as well as computers.

You can see this in Figure 7-2, from the "Wide Awake" Katy Perry music video. On the lower left, MediaInfo informs us that the pixel resolution is 640x480, with a display aspect ratio of 16:9. When played in QuickTime, the Movie Inspector tells us that the display resolution is 853x480. This file loaded on my iPod Nano, which has a maximum playback resolution of 640x480.

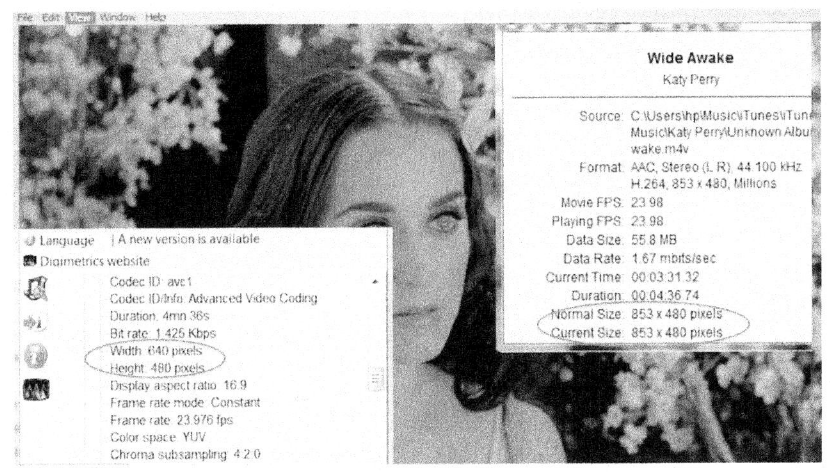

Figure 7-2. Encoding at 640x480 resolution but displaying at 853x480.

You can achieve this effect with Compressor's iPod/iPhone (Anamorphic) device preset, shown on the right in Figure 7-3, using a pixel aspect ratio of 0.75. You should be able to produce this same effect in any program that lets you specify the aspect ratio of the encoded video.

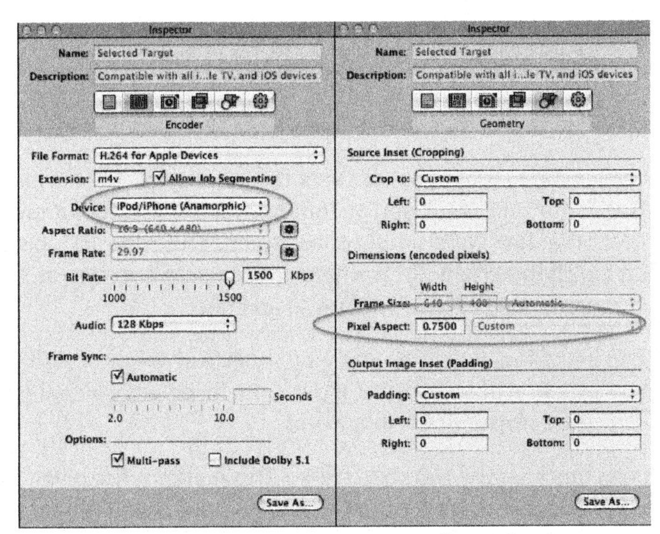

Figure 7-3. Here's how you produce that effect in Compressor.

Encoding TV Episodes

I downloaded 14 HD episodes, and you can see their configurations in Table 7-6. Note that in the past, whenever I purchased an HD episode, iTunes downloaded an SD episode as well, probably to have content to load if the purchaser of an HD episode didn't have an HD-capable device. In this survey, only three of the 14 files downloaded an SD version, which I mention only as an interesting data point.

	Width	Height	Codec	Data Rate	FPS	Bits per Pixel	Profile	Audio Data Rate	Channels
Revolution	1440	1080	H.264	5,062	29.97	0.109	High	160	Stereo
Bones	1912	1072	H.264	5,157	23.796	0.106	High	160	Stereo
How I Met Your Mother	1912	1072	H.264	5,103	23.796	0.105	High	160	Stereo
New Girl	1912	1072	H.264	5,240	23.796	0.107	High	160	Stereo
Glee	1916	1076	H.264	5,008	23.796	0.102	High	160	Stereo
Breaking Bad	1920	1080	H.264	5,023	23.796	0.102	High	160	Stereo
Captain	1920	1080	H.264	5,078	23.796	0.103	High	160	Stereo
Castle	1920	1080	H.264	5,241	23.796	0.106	High	160	Stereo
Elementary	1920	1080	H.264	5,131	23.796	0.104	High	160	Stereo
Emily Owens M.D.	1920	1080	H.264	5,064	23.796	0.103	High	128	Stereo
The Good Wife	1920	1080	H.264	4,755	23.796	0.096	High	160	Stereo
Revenge	1920	1080	H.264	5,236	23.796	0.106	High	160	Stereo
Vegas	1920	1080	H.264	5,102	23.796	0.103	High	160	Stereo
The West Wing	1920	1080	H.264	5,023	23.976	0.101	High	160	Stereo
Averages	**1,884**	**1,078**		**5,087**		**0.104**		**158**	

Table 7-6. Configuration options used with HD TV episodes.

Looking at the downloaded shows themselves, I saw a great deal of uniformity. For example:

- All files were encoded for 1080p (or thereabouts) display. The *Revolution* file that was encoded at 1440x1080 displayed at 1080p when screen resolution was sufficient to support it. Most other files were simply encoded at or near the 1080p display target. Note that this was a big change from 2011, where all files were encoded at 720p or for display at 720p. You probably can call this "The new iPad effect."

- The range in encoding data rates was very small, with a low of 4,755 kbps and a high of 5,236 kbps, a variation of about 10%. If data rate fixing were a crime, there might be evidence of anti-trust violations in Hollywood.

- All audio was produced at 160 kbps stereo. This makes sense given the production value devoted to audio in these programs; unless you're doing the same with your productions, you can stick with 128 kbps stereo, or even 64 kbps mono if the audio content is primarily speech.

• All producers encoded using the High profile, reflecting the playback capabilities of the New iPad and all desktop computers.

What about SD shows? I present data for eight SD programs in Table 7-7.

TV	Width	Height	Codec	Data Rate	FPS	Bits per Pixel	Profile	Audio Data Rate	Channels
Emily Owens M.D.	640	480	H.264	1,515	23.796	0.207	Baseline	128	Stereo
Nashville	640	480	H.264	1,507	23.976	0.205	Baseline	168	Stereo
30 Rock	640	480	H.264	1,510	23.946	0.205	Baseline	106	Stereo
Chicago Fire	640	480	H.264	1,552	23.976	0.211	Baseline	128	Stereo
How I Met Your Mother	640	480	H.264	1,568	23.976	0.213	Baseline	107	Stereo
Magic City	640	480	H.264	1,595	23.976	0.217	Baseline	128	Stereo
My Shopping Addiction	640	480	H.264	1,551	23.976	0.211	Baseline	128	Stereo
New Girl	640	480	H.264	1,551	23.976	0.211	Baseline	128	Stereo
Averages	**640**	**480**		**1,544**	**24**	**0.210**		**128**	**128**

Table 7-7. Configuration options used with SD TV episodes.

Note that while all SD episodes were encoded at 640x480 pixel resolution, most producers used the Katy Perry music video technique discussed in Figure 7-2 to produce a video with an actual pixel aspect resolution of 640x480 but a display resolution of 854x480. This allowed the video to load onto devices in the second category that support a maximum H.264 resolution of 640x480, but look much bigger than a 640x360 video on devices with larger-resolution displays.

Otherwise, the data rates averaged 1544 kbps, for a bits-per-pixel value of 0.21, compared with a bits-per-pixel value of 0.244 for music videos, which usually contain many more quick cuts and high motion. This was roughly double the bits-per-pixel value used to encode HD episodes—living proof of the veracity of the Power of .75 rule discussed in Chapter 2.

Conclusion

So that's encoding for iTunes, a nice discrete little chapter. In all previous chapters, we've explored either technology fundamentals, or how to encode your videos into H.264 format. Now it's time to start focusing on how to distribute video to your target viewers, which we explore in Chapter 9.

Chapter 9: Distributing Your Video

> ***Note:*** You can download a PDF file with all figures from this book at **bit.ly/Ozer_multi**. Since this chapter contains lots of images, now would be a good time to do so.

By this point, you know all you need to know about encoding your video. Now it's time to distribute it to the world. You have three basic approaches.

First, you can encode the video files yourself, create the necessary player and all the links, and upload the files to your own website or a web server leased from a third-party provider like Amazon. As long as viewing numbers stay fairly modest, this approach should work from a technology standpoint, although it may not be the best way to get maximum eyeballs for your videos.

The second alternative is to host your videos on a free, user-generated content (UGC) website like Vimeo or YouTube. These UGC sites relieve you of the encoding and player-creation chores and assume the task of hosting and distributing the videos for you. You can still embed the video on your own website, but by offering your video on a UGC site, you also expand the number of potential viewers, which can help from a marketing perspective. However, there are some negatives to consider, as well as some benefits that are only possible via the third alternative.

That alternative is to use a fee-based service to host and distribute your videos for you. Multiple online video platform (OVP) vendors offer hosting, encoding, customizable players and detailed statistics to help you maximize the effectiveness of your videos. They help distribute your videos to other sites to acquire more viewers. Even a few short months ago, these types of features would have cost hundreds of dollars per month. Today, however, depending upon the amount of video you distribute, you can sign on for less than $20 a month, with one service offering unlimited video views for less than $50 a month.

This chapter will review the costs and the benefits of all three alternatives. While there is no one-size-fits-all solution, any business seeking to truly leverage the value of its video should be familiar with the benefits of the second and third alternatives. Many businesses will find that an amalgam of these two is the best option of all.

Before jumping into the three options, let's look at the roles related to video distribution.

Understanding the Roles

Table 9-1 delineates the functions related to successful video distribution and how they're divided based upon the three distribution options. I've added a fourth—using a cloud-based streaming server—because it's gaining in popularity.

	Encode	Server	Feature Support	Platform Support	Player Creation	Get Eyeballs	Deliver
Do It Yourself	You	You	You	You	You	You	You
User Generated Content (UGC)	UGC	UGC	UGC	UGC	UGC	UGC	UGC
Online Video Platform (OVP)	OVP	OVP	OVP	OVP	OVP	You	OVP
Cloud Server	You	Cloud	You	You	You	You	Cloud

Table 9-1. The roles related to video distribution.

The encoding role is what you've learned in this book; whether you want to do it on a daily basis is up to you. The server role relates to installing and maintaining the streaming server. While a streaming server isn't essential for simple streaming, it is for advanced features like digital rights management (DRM), comprehensive analytics and some adaptive streaming formats.

Feature support relates to these ancillary features. As we'll discuss, when choosing an OVP, you'll compare dozens of features like content management capabilities, various syndication and monetization models, and support for adaptive streaming. If you use a third-party service, it will develop and implement these features; if you DIY, you DIY. Ditto for platform support. If you're rolling your own distribution and want to support iOS and Android devices, you have to build it yourself. Use a UGC or OVP site, and it does the heavy lifting. Sound far-fetched? Maybe, but by using YouTube to host some of its videos, IBM was one of the first corporate sites to post videos that played on an iPad.

Speaking of that, player creation is a huge feature supplied by OVPs and UGC sites. Creating a simple player is a breeze, but a simple player may not be sufficient. As I discuss in "Is your video player as good as your content?" (**bit.ly/gezmZg**), to maximize the impact of the video, a fully featured player needs to:

- Foster viewer engagement via features like number of viewers, ratings and comments

- Link to social networking sites
- Enable sharing options like linking, embedding and emailing
- Contain links to other content that viewers might want to watch
- Allow viewers to easily find the content they want to view.

For example, Figure 9-1 shows Kohler's video player, which is provided by OVP Brightcove. What's fun about this particular player is that I grabbed it on my iPad, and it looks and performs identically to the Flash Player I viewed on my computer, except that it includes iPad-specific control features, like the ability to use touch-based gesture commands to scroll through the library of videos on the right. Bet that wasn't cheap to implement.

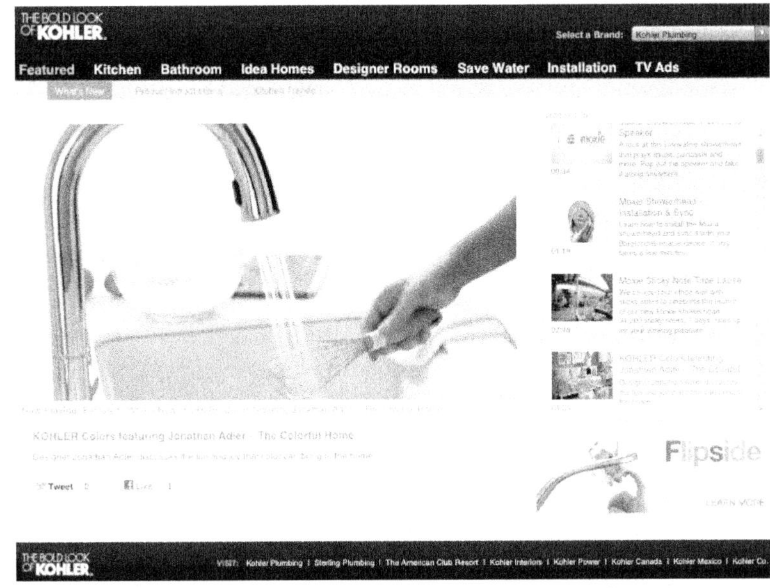

Figure 9-1. The Kohler video player, courtesy OVP Brightcove.

Beneath the player are links to Facebook and Twitter, and if you hover your mouse over the desktop player, you'll expose links for embedding and emailing. On the right is a playlist containing related videos that might prove enticing to the viewer, which makes the site much more likely to retain the viewer's interest, or "sticky." On the upper left are categories (What's New, Product Introductions, Kitchen Trends) that expose even more content. Many OVPs offer these features, but to develop them yourself might take days or even weeks of programming time.

Finishing up with Table 9-1, syndicate means getting additional eyeballs for your video. Some UGC sites, like YouTube, do this inherently, as you doubtless already know. While OVPs don't offer the same benefits, as we'll discuss, some OVPs can automatically syndicate your videos to sites like YouTube or Metacafe to garner additional attention.

The final role is delivery. If you host the videos on your own website, you can deliver effectively to a limited number of users, but your server and Internet connection are likely not optimized for high-volume video delivery. In contrast, most UGC and OVP sites have contracts with content delivery networks (CDNs) to ensure high-quality delivery of high-bit-rate streams.

Distributing Your Own Videos

As we've seen, if you distribute your own videos, you perform all roles from encoding to delivery. Certainly you can spread the load by licensing access to a server in the cloud, which saves you the server installation step and should take care of delivery as well. Or you could hire a CDN to distribute your videos. But you would still have to configure and maintain the server, develop all features, encode the videos, develop support for additional distribution platforms like iOS and Android, create the player, and chase the eyeballs.

In my experience, for most companies, this approach only makes sense when you have a need that can't be satisfied by a UGC or OVP site. For example, one consulting client sells music-training videos on a subscription basis and couldn't find an OVP that could satisfy the requirements of its revenue model. So it licensed a Flash Media Server from Amazon Web Services, programmed the required features into the server, and created its own player. For most general-purpose users, however, a UGC site or OVP likely makes more sense.

If you distribute your own videos, the most critical decision is the streaming server you buy or license. Remember, you don't absolutely need a streaming server for simple progressive delivery, but you will for advanced features like transmuxing from one format to another, DRM and other features. If you do buy or license a server, obviously you should choose one that supports all of your current and short-term future platforms.

For these applications, there are four basic choices: the Adobe Media Server line of products (**bit.ly/AMS_family**), Microsoft's IIS Media Services (**iis.net/media**), RealNetworks' Helix Server line of products (**realnetworks.com/helix**), and Wowza Media Server (**wowza.com**). There's also an open-source streaming server called Red5 that uses some secure protocols to deliver to Flash, but it doesn't currently convert streams for delivery to iOS or support any adaptive streaming technology. This would eliminate it from contention for most streaming producers.

There is no one-size-fits-all server. By this point, all the mentioned servers can input one set of streams and transmux them into (at least) HTTP Live Streaming (HLS) output. Beyond this, to choose the best server for your needs, you need to consider a number of factors, including:

- ***Streaming features.*** One important feature is live DVR, which lets viewers perform their own instant replays and then catch up to the live stream. Make sure this feature is available in all supported formats. Also important is cross-platform DRM support if you need to protect your content.

- ***Output formats.*** Most streaming servers support at least one flavor of Flash adaptive streaming (RTMP or HTTP), plus Apple's HTTP Live Streaming, but support for Silverlight and/or QuickTime is more rare.

- ***Input formats.*** Most streaming servers can accept Flash RTMP input, but few support broadcast inputs like MPEG-2 transport streams. Be sure to verify that the server you choose can handle the input format you'll be providing.

- ***Supported protocols.*** If you'll be serving internal as well as external viewers, Adobe Media Server can distribute via multicast and peer-to-peer, which are the most efficient protocols for delivering on corporate LANs and WANs. Not many other servers offer these alternatives.

- ***Supported applications.*** If you need interactive features like access to webcams or microphones, get this on the table early, because not all servers support these.

- ***Supported platforms.*** All servers run on Windows, but support for Linux, Unix and Mac is rare. If you're interested in streaming from the cloud, check the availability of cloud versions.

- ***Pricing.*** Beyond perpetual license costs, check the availability of daily or monthly rental, which can save big dollars for those who produce infrequent live events.

If you're looking for additional comparative input, Wowza Media has a great features table that compares Wowza Media Server, Adobe Media Server and Microsoft's IIS Server that you can access at bit.ly/Wowza_FT. Not surprisingly, it points out all the unique features of the Wowza streaming server and none of the deficits (like no multicast or peer-to-peer), so do your homework. Still, it's a great place to start. In addition, you might check out my article "A Closer Look at Streaming Servers" at bit.ly/Ozer_servers, which provides an overview of the four servers identified above.

Adding Video to Your Website

If you're distributing video from a website without a streaming server, the process is simple, although it does require some technical know-how. For example, Step 1 is uploading the file to your website, which is typically accomplished with an FTP program like freeware FileZilla, shown in Figure 9-2.

Once the video file is uploaded, you can create a link on a web page to play the video file. You can also email a link to a viewer, who can play the video file by either clicking the link in the email or copying and pasting the link into a browser. For example, if you go to doceo.com/Disney_World.mp4, you'll play the video file I uploaded in Figure 9-2.

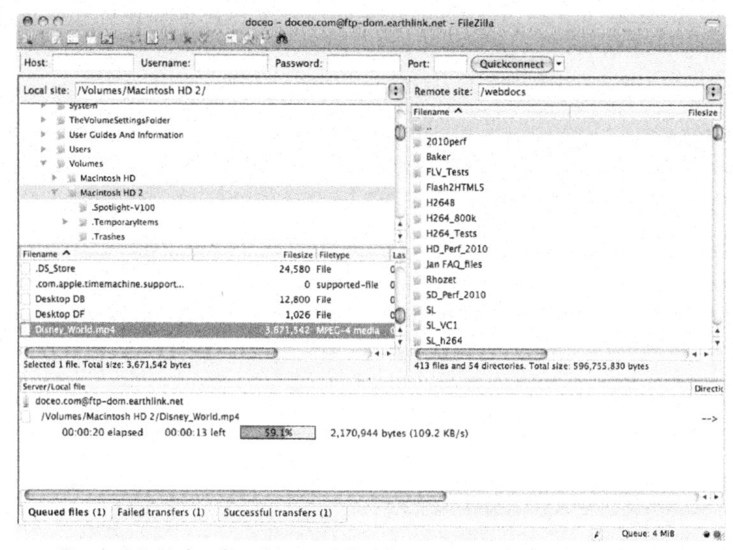

Figure 9-2. Uploading Disney_World.mp4 to my doceo.com *website.*

If you're going to email a link, things will go easier if you don't leave any spaces in the name of the file you upload, since that tends to drive some browsers crazy—that's why there's an underscore between Disney and World in the file name. Also to avoid sending the wrong URL for the link, I first enter the link in my browser, make sure it pulls up the right file, and then copy it into my email. Note that things like capital letters matter, so it's easy to type in the wrong link.

Linking and Embedding

If you want the video to play within a web page, you have to embed the video into the web page. The technique and language for doing this varies according to the technology you're using.

With HTML5, for example, the syntax is simple. Here's the embed code for a WebM file I used on my own website at **doceo.com/HD_WEBM.html**: Note that you'll have to use HTML5-compatible versions of Firefox, Chrome, or Opera to play the video, since IE and Safari don't play WebM files.

```
<video src="http://www.doceo.com/HD_800.webm"
       type='video/webm; codecs="vorbis,vp8"'
           controls
           autoplay>
```

Figure 9-3. HTML5 embed codes are very simple.

You can find a good explanation of the controls available for HTML5 video at **w3schools.com/html/html5_video.asp**. However, note that simply adding the tag to an HTML page on your site may not be sufficient—you may also have to tell your web server how to deal with this new type of video file, which is typically called a MIME type. For example, to get any file to play via HTML5 on my website, I had to add the following MIME types to the .htaccess file on the doceo.com website.

```
AddType video/ogg .ogv
AddType video/ogg .ogg
AddType audio/ogg .oga
AddType video/mp4 .mp4
AddType video/webm .webm
```

Figure 9-4. MIME types on the doceo.com *website.*

If you need help with these MIME issues, particularly with an Apache web server, click over to diveintohtml5.org/video.html and search for "MIME type" on that page.

Since not all of the installed base of browsers are HTML5-compatible, you'll probably want your embed codes to direct older browsers to fall back to Flash or other plug-in based playback technology. This is a bit far afield for me, but Google "HTML5 video tag" and "Flash fallback," and you'll find some great resources.

Embedding Flash Video

Embedding Flash videos is complicated because there are multiple options for displaying Flash video, and all use complex embed codes. I don't try to hand-code Flash embed codes any longer; I usually create the web pages in a tool like Adobe Dreamweaver or Adobe Flash Catalyst. You can see a tutorial on using Adobe Flash Catalyst at bit.ly/ozercatalyst.

The only frustration is that these tools typically won't input .MP4 or even .F4V files (in the case of Dreamweaver), even if the video is encoded in an H.264 format that the Flash Player can play. The simple fix is to manually change the extension to .FLV in Windows Explorer or Finder so the program will recognize the file. Flash Player will still be able to play the file, so changing the extension shouldn't cause any problems.

Now let's look at using a UGC site to distribute your videos.

Distributing via UGC Sites

When you work with a UGC site like YouTube, first you set up your free account and then you upload your video to the site. UGC sites will display the videos on their own pages within the site and provide an embed code you can use to display the video on your own web pages.

Uploading is straightforward, so I'll trust you to work through that one yourself, although we'll talk about encoding for upload at the end of the chapter. Embedding is also simple. First you find the embed code; with YouTube you click Share, then Embed to open the embed codes. Then you choose your options, then click inside the embed code box and copy the embed code to your clipboard.

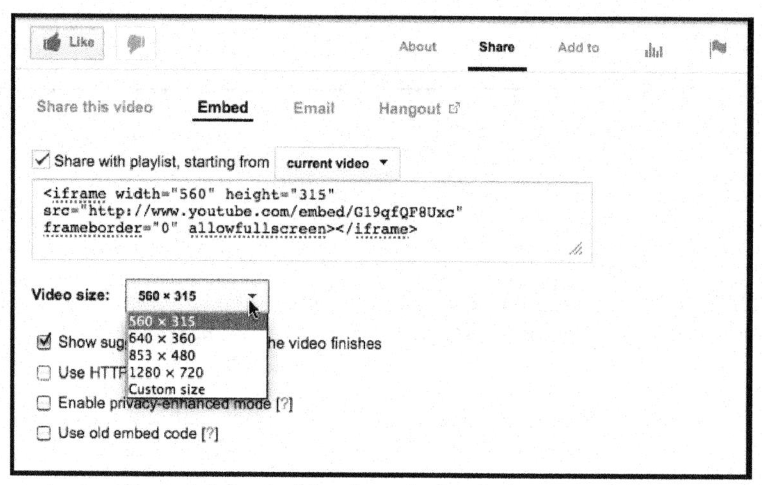

Figure 9-5. Customizing the embed codes from YouTube.

Typically, the most important option is size, because you'll want the video to fit within the horizontal boundaries of your web page. My blog is 700 pixels wide, which is why I would use either the 560x315 or 640x360 option for this video.

The only tricky part is pasting the embed code into your web page. Specifically, remember that you're pasting HTML embed codes into your site, which you must do directly in the HTML editor provided by your authoring program or content management system. For example, I use the Interspire content management system for *StreamingLearningCenter*, and it has an HTML editor that I access via the circled icon on the upper right of Figure 9-6. I don't paste the embed codes into the normal text entry area; I paste them into the HTML Source Editor you see in Figure 9-6.

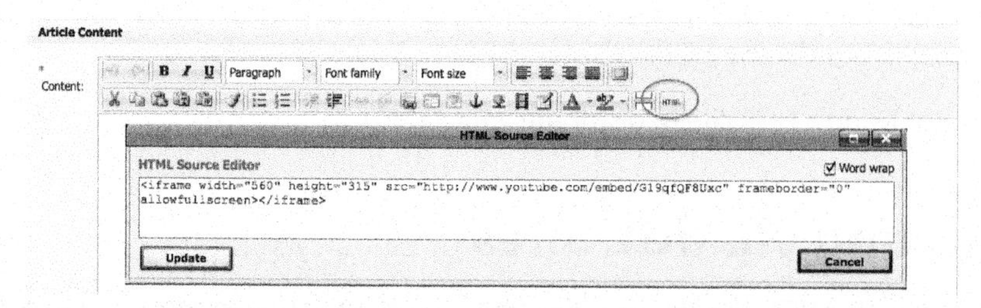

Figure 9-6. Accessing the HTML editor in my Interspire content management system.

It's a bit easier to see in the WordPress blog shown in Figure 9-7, which has two design areas: one for entering text and HTML codes, and the visual design area, which gives you a WYSIWYG (What You See is What You Get) view. Basically, if you ever see the HTML codes themselves in your website text, you've probably copied them into the wrong editor.

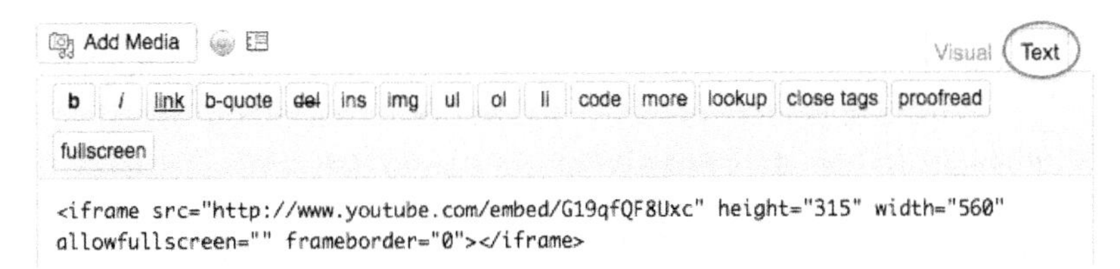

Figure 9-7. Accessing the HTML editor in WordPress.

Note that this embed procedure is the same whether you're pulling embed codes from a UGC site or an OVP. Either way, you get the embed code and paste it directly into the HTML editor on your website. If you'd like to see a tutorial on uploading to and embedding pages from a UGC site, click over to **bit.ly/uploadembed**.

Choosing a UGC Site

Now that you know how to use a UGC site, let's focus on which one to choose. There are two aspects: Which sites do you want to use to reach potential viewers, and which do you want to use to host the videos that you embed in your own site?

Chasing Eyeballs

If it's attention you want, it's hard to go wrong with YouTube. If you scan YouTube's channels, you'll note some pretty interesting names with some impressive view counts. E-Trade has about 26,000 subscribers with more than 64 million video views; the White House has more than 274,789 subscribers with more than 116 million video views; IBM has more than 15,000 subscribers and about 6.8 million video views; Intel has more than 43,000 subscribers and more than 48 million views; Nike Football (which is really soccer) has more than 421,000 subscribers and more than 160 million video views, which is topped by Taylor Swift's 2.5 million subscribers and 1.2 billion video views.

Clearly, exposure on YouTube isn't for everyone. But for many products and services, it's an inexpensive route to a vast audience you probably couldn't reach by posting videos on your own website.

The other thing about YouTube is that it isn't particular. That is, so long as you're not violating someone's copyright or uploading pornography or other "bad stuff," you're free to sell and market your products or services as you wish. This isn't the case with many other sites, particularly Vimeo, which says:

> You may not upload videos containing ads that are displayed before, during, or after the video unless given prior written permission from an authorized member of the Vimeo staff. Videos with advertisements in them will be removed.

When chasing eyeballs, my theory is that more is always better. In this regard, you should know about a website called TubeMogul, which offers a free service that distributes your videos to multiple sites from a single upload. Target sites include UGC sites like YouTube, DailyMotion, Metacafe and Vimeo; social media sites like Facebook and MySpace; and OVP sites like Brightcove and Viddler. Operation is simple: You upload your video, add metadata, choose the target sites and enter the required login information (Figure 9-8). TubeMogul does the rest.

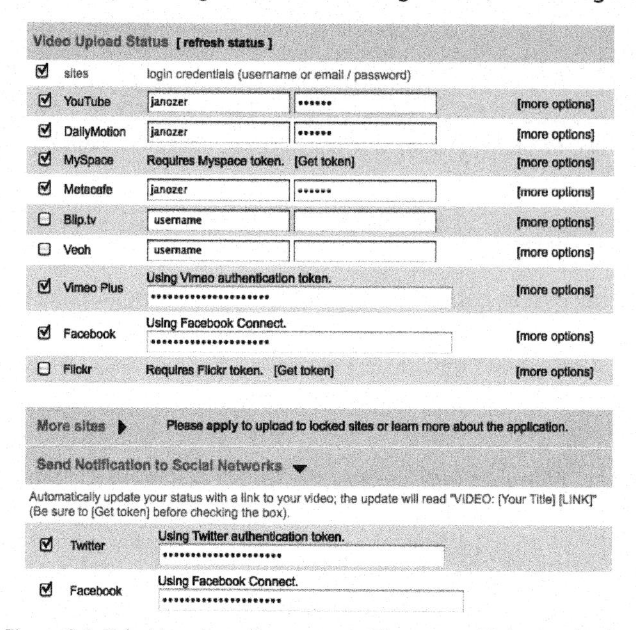

Figure 9-8. TubeMogul syndicates your videos to multiple sites for free.

If you want your videos to end up in multiple places, TubeMogul can be a huge time saver.

Choosing a UGC Host for Your Videos

Eyeballs are one thing, but which UGC site should you use to host the videos you embed in your website? That's a different set of concerns.

As mentioned above, whatever site you choose, make sure you're not violating its terms of service. I use Vimeo to serve some of the videos on my website, but only encoding-related tutorials or other demonstration videos. Once I started adding advertisements to these videos that would clearly violate Vimeo's terms of service, I started using YouTube and OVP Sorenson 360.

In early 2011, I reviewed five UGC sites plus WordPress' VideoPress service to determine their suitability for hosting videos embedded in another site (see "Choosing a UGC Site," **bit.ly/chooseugc**). I did a lot of work uploading videos that were embedded into the OnlineVideo.net site, so you can see the comparative quality—at least the quality at that time.

One of the comments asked, "Is there a conclusion or a summary?" I responded, "If and when I have to switch from Vimeo, I'll either go to YouTube or VideoPress," the latter being the OVP site from the WordPress folks. I haven't done a lot of testing on UGC sites other than YouTube since then, but in other projects, I've been surprised at the number of companies that have YouTube channels and use YouTube as their OVP. For example, in an article titled "HTML5 Is Taking Over the Web, Right? Not so Fast..." (**bit.ly/Ozer_HTML5**), I reviewed how the top 20 worldwide brands were using HTML5.

The short answer was that they weren't—at least as of November 2011, when I wrote the article (and they're still not). However, 19 of 20 sites had YouTube channels. Of these sites, six of them used YouTube videos on their own sites or to enable playback on iPads. In addition, I've noticed YouTube videos embedded into the sites of a number of corporations who certainly could afford an OVP, including IBM and Deloitte. This makes perfect sense, because not only does YouTube deliver maximum eyeballs, it also shoulders the load of ensuring playback on a range of devices, from smartphones to over-the-top devices. All for free.

One concern raised by some readers of the aforementioned article is YouTube's terms of service, as shown below:

> For clarity, you retain all of your ownership rights in your Content. However, by submitting Content to YouTube, you hereby grant YouTube a worldwide, non-exclusive, royalty-free, sublicenseable and transferable license to use, reproduce, distribute, prepare derivative works of, display, and perform the Content in connection with the Service and YouTube's (and its successors' and affiliates') business, including without limitation for promoting and redistributing part or all of the Service (and derivative works thereof) in any media formats and through any media channels.
>
> You also hereby grant each user of the Service a non-exclusive license to access your Content through the Service, and to use, reproduce, distribute, display and perform such Content as permitted through the functionality of the Service and under these Terms of Service.
>
> The above licenses granted by you in video Content you submit to the Service terminate within a commercially reasonable time after you remove or delete your videos from the Service. You understand and agree, however, that YouTube may retain, but not display, distribute, or perform, server copies of your videos that have been removed or deleted. The above licenses granted by you in user comments you submit are perpetual and irrevocable.

Pretty broad rights, right? Still, if it's acceptable to artists like Taylor Swift and IP-savvy companies like Intel, IBM, Deloitte, Coke and many others, you'd have to say that the benefits are more than worth the cost.

So, where does that leave us? In the testing I have done, YouTube continues to push the envelope with higher data rates and more versions than most competitors. So, I don't see any reason to change my conclusion: If you're going to use a UGC site for hosting your corporate videos, YouTube is the best choice—particularly if you're uploading your videos on the site to chase eyeballs anyway.

That's choosing a UGC site; now let's move on to choosing an Online Video Platform.

Online Video Platforms

Let's take a short look back at the history of streaming delivery to help explain the value proposition of the current crop of Online Video Platforms (OVPs). Back in the late 1990s, when streaming became mainstream, the biggest category of service providers was content delivery networks (CDNs). These companies would store your videos on their servers, distribute them on their networks, and ensure their smooth distribution to the ultimate viewer far and wide.

The CDNs and their services spawned a range of third-party service providers offering ancillary services such as player interactivity, viewer statistics, syndication of your video to other websites and hooks to advertising-related services—the latter two so you could make money from your video content. Over time, these services were merged into a single, consolidated product offering by companies such as Brightcove, Kaltura and Ooyala.

As you would suspect, these companies initially targeted media companies with high stream counts and the distribution budgets to match. There are only so many media companies around, though, so the OVPs' focus inevitably turned to corporate users, large and small.

As their focus changed, so did minimum pricing, which transitioned from low four-digit figures to pay-as-you-go. Sign-up procedures, which initially involved much paperwork and significant up-front financial commitments, have also become much more streamlined. One vendor, Ooyala, claims that you can download its Backlot software and start streaming in less than 15 minutes without speaking to an Ooyala employee or making any long-term financial commitment. As we'll see, pricing varies dramatically from vendor to vendor, so do your homework in that department early in the selection process.

However, although getting started with an OVP is simple, choosing one isn't. There are dozens of companies entering and exiting this market—some via consolidation, some via bankruptcy.

As you read about the introductory features discussed next, recognize that unless otherwise indicated, most are generic and not unique to the company identified. The purpose of this section is to identify the high-level features and benefits of using an OVP, not to tie features to specific providers.

Comparing OVPs

At a high level, you work with an OVP the same way you work with a UGC site: You upload the file to the service; it encodes and supplies a player and embed code. You embed the player into a web page and move on to the next item on your to-do list. For most companies, the basic feature set offered by most OVPs will do. That said, understanding the universe of features available is always beneficial to selecting the right service provider. Here's a structured analysis of the factors I would consider when choosing an OVP.

Note that I'll be showing you lots of screens from Brightcove, one of the largest OVPs. Primarily, this is because Brightcove is the OVP for *Streaming Media Magazine*, so it's the OVP that I have the most access to and most experience with. It's a fine OVP, but you shouldn't conclude that it's better than similar OVPs in its class, like Ooyala and Kaltura, simply because I show more screens.

Account Basics

By account basics, I mean details that involve media organization and the number of users who can access the system. For example, some consulting clients that I worked with had multiple divisions and wanted a completely unique data store of videos for each division, and a different look or feel for each division. With some OVPs, that meant multiple licenses, which typically costs more, while others could enable this via a single license or uber-license. If you have multiple stakeholders who may want separate facilities, get this issue on the table early.

Another big issue is the number of users that could access the system. Some OVPs limit the number of users, then charge as much as $1,800 for each additional user each contract year. Others enable unlimited users. If you'll be allowing multiple users to upload video to the service—say, in a UGC-like context—you need to understand who counts as a user and how much each costs.

Another basic relates to who owns the code and where it lives. Some producers prefer open-source alternatives, like that offered by Kaltura, because it allows them to customize their platform and provides protection if the OVP goes out of business. Some also want the OVP running inside their own firewalls, while most OVPs run only as an external-facing Software as a Service (SaaS).

Pricing

There are two high-level aspects to pricing: First, what are you paying for? Second, what can you exclude? The second part is easiest; if you have an existing relationship with a CDN, ask if you can send all videos to the CDN for delivery and pay it directly, rather than having the OVP deliver and paying the OVP. If you're already paying big dollars to your CDN for other content, this may drop the cost per gigabyte of your bandwidth down so that it's more affordable than paying the OVP.

Otherwise, before comparing pricing, you should create a chart like that shown in Table 9-2.

	Needs	What's Included	Overage
Base Fee		$15,000	
Number of Accounts	1	1	Separate accounts required
Number of Users	15	10	$0 - $1,800/ user
Size of Stored Content	1 TB	1TB	Size for overages?
Streams Viewed (or Videos Opened)	800,000	400,000	$1-$4/1,000
Bandwidth Charges	15 TB	10 TB	$0.16 - $0.90/GB
Other Services Like Account Setup	?	$1,000	Price?
Desired Customer Service Level			Price?

Table 9-2. Components of OVP pricing.

In the Needs column, you should identify your estimated needs for the first contract year, broken down into the following categories:

- ***Number of accounts.*** See Account Basics, above. Most simple accounts will need only one account, but if your OVP will serve multiple organizations that need a separate data stores—and particularly a unique look and feel—you may need two or more. This may have a huge impact on comparative pricing, so get this issue fleshed out early.

- ***Number of users.*** Typically this is the number of users who need log in access into the OVP. However, if you'll be enabling UGC-style uploading, get this on the table early and determine if each UGC uploader will counts as a user.

- ***Size of stored content.*** Estimate the total storage required for all iterations of your encoded video files. Typically, this isn't a large component of cost, but it's a question you'll get asked.

- ***Streams viewed.*** Some (but not all) OVPs charge by the number of streams viewed, which should be the same as video views. So you should know or be able to estimate this number before you go shopping.

- ***Bandwidth.*** All OVPs charge for bandwidth, so you'll need to be able to estimate your total consumed bandwidth for that first contract year. If you don't know, you should be able to make a rough estimate by multiplying the number of views by the average duration of your videos and the average data rate. Remember to divide your video data rate by eight to convert from bits to bytes, since data rates are in megabits or kilobits per second, while bandwidth charges are in gigabytes. Also factor in that most viewers will only watch less than half of your video.

I make an estimate of this calculation in Table 9-3 with the caveat that it's really easy to drop a zero or two and totally hose the calculations. Note that the number of videos isn't a component of the calculation, but is there to help estimate the number of streams. Right or wrong, I come up with an estimate of 1.125 TB, which is the number I would use for estimating bandwidth charges for my cost comparisons.

Videos	Average Duration	Streams	Average Bit Rate	Fall Off Estimate	Estimated Bandwidth
500	6 minutes	2,000,000	1,500,000	50%	1.125 TB

Table 9-3. Estimating your bandwidth consumption.

- ***Other services like account setup.*** Some OVPs have these, some don't, and it's always good to ask.

- ***Desired customer service levels.*** OVPs have different customer service levels; sometimes they are associated with certain account levels, sometimes not. If video is mission-critical to your organization, you should identify the response times and methods (phone, email) associated with the different service levels and their respective costs.

Note that this list isn't comprehensive; other OVPs adjust pricing for factors like the number of players created or files encoded. But the table should represent a good start on the information necessary to estimate your annual charges.

With many OVPs, after gathering this information, you can calculate your costs simply by looking at their website, but often some detail or another won't be available. For example, Figure 9-9 shows Brightcove's pricing. There is no cost per stream (there is with its higher-end packages), and the $499 service covers the 500 videos plus 3TB of total bandwidth, more than adequate for the estimated 1.125 TB.

Figure 9-9. Brightcove's pricing structure.

As you see in Figure 9-9, each pricing plan will come with usage limits, like the 250 GB monthly bandwidth for the $499 Brightcove account and the three users. Then you have to ascertain what overages cost. If you click the question mark over bandwidth on the Brightcove website, you'll see that it's $10 per month for every 10 videos over the 500-video limit, and $0.90 per GB for usage over 250 GB a month. The cost for extra users isn't disclosed here, nor is the service level associated with each pricing category. So, you'd have to call.

These issues aside, many lower-capacity users should be able to estimate their costs fairly closely simply by identifying their requirements and checking the price charts on most OVPs' websites. With larger, more complicated accounts, you'll have to speak with a rep, provide them the details and get more detailed cost estimates. As hinted at in Table 9-2, pricing is all over the map, so it definitely pays to shop around.

Once you know the pricing, it's time to identify the features that are key to your implementation. Each website and producer is different; here are the categories of features, in no particular order, that differentiate the various OVPs.

Upload-Related Features

Most producers upload very high-bandwidth files to their OVPs to preserve quality, which is time-consuming. At this point, most larger OVPs offer upload acceleration technology like that enabled by Aspera technology, which is said to speed uploading by as much as 20 times. If you'll be uploading lots of bulky files, ask about upload acceleration.

Also ask about available uploading techniques. All OVPs enable single- and usually multiple-file uploads via a web-based GUI. Some offer desktop watch folder integration so when an editor drops a file into a watch folder, it's automatically uploaded. If you're a high-volume operation, look for FTP upload or the ability to upload directly from your content management system (discussed in more detail below).

Also ask about application programming interface (API)-driven file and metadata ingest. If you'll be uploading 500 files or transferring 500 files from one OVP to another, the ability to automatically suck in metadata from an existing database and populate a new database can save dozens of hours of boring work with 100% accuracy. If you have to manually upload and enter metadata for each file, it's a big negative.

Encode/Transcode

By this point in the book, I'm guessing you're sold on adaptive streaming, and I am too. Be advised, however, that not all OVPs offer adaptive streaming, so if this is important to you, check early.

OVPs that enable adaptive streaming let you choose the encoding profiles that will be applied to each uploaded file. Brightcove's are shown in Figure 9-10, with stream 1 shown in editing mode. As with most OVPs, there aren't a lot of dials and levers that you can adjust; beyond the file basics (resolution, bit rate) and general encoding controls (bit rate type, 1- or 2-pass) the only H.264 configuration you can adjust is profile via the control circled in the figure. If you like to tinker, you're out of luck.

Actually, Brightcove users should be glad you can adjust any parameters as some OVPs limit your options to which streams to include or exclude, with no configuration options whatsoever.

This simplifies OVP operation in many ways, as players only need to adjust to a few known resolutions, and bit rates are always consistent. However, if you have odd window sizes on your website and need to customize streams accordingly, you'd better ask early how much you can tweak these transcode settings.

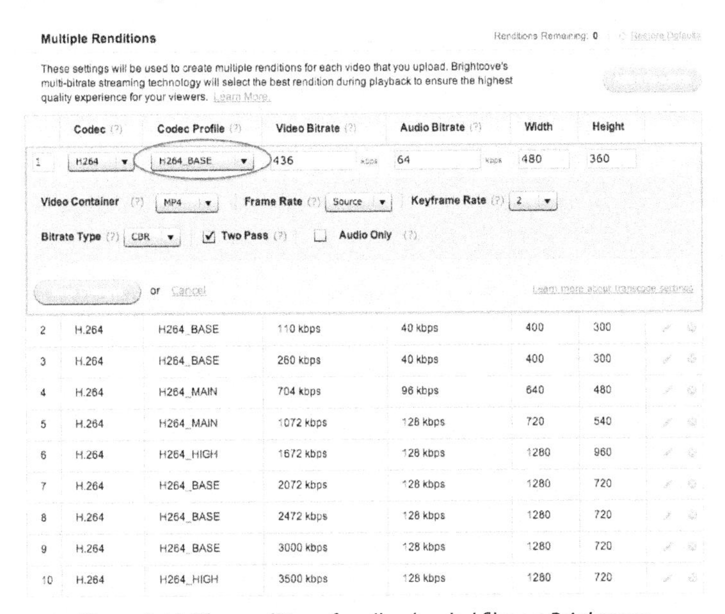

	Codec	Codec Profile	Video Bitrate	Audio Bitrate	Width	Height
1	H264	H264_BASE	436 kbps	64 kbps	480	360
2	H.264	H264_BASE	110 kbps	40 kbps	400	300
3	H.264	H264_BASE	260 kbps	40 kbps	400	300
4	H.264	H264_MAIN	704 kbps	96 kbps	640	480
5	H.264	H264_MAIN	1072 kbps	128 kbps	720	540
6	H.264	H264_HIGH	1672 kbps	128 kbps	1280	960
7	H.264	H264_BASE	2072 kbps	128 kbps	1280	720
8	H.264	H264_BASE	2472 kbps	128 kbps	1280	720
9	H.264	H264_BASE	3000 kbps	128 kbps	1280	720
10	H.264	H264_HIGH	3500 kbps	128 kbps	1280	720

Figure 9-10. The renditions for all uploaded files on Brightcove.

Supported Platforms

While encoding files to .MP4 format enables adaptive delivery to Flash Players and AIR applications via RTMP Dynamic Streaming, if you're relying upon browser-based iOS playback, you'll also need the chunking and metadata files required for HTTP Live Streaming (HLS). Most advanced OVPs offer this capability; many smaller ones do not. There's also a player side to this equation that I'll address below.

Content Types

Also important are the types of content supported by the OVP. For example, many producers of on-demand content are now starting to produce live events. For this reason, many larger OVPs are now starting to add live support to their product offerings. This makes things simpler for live event producers, because the interface should be familiar and because live events will automatically convert to on-demand events once they're completed. It's also more cost-efficient because all of a producer's bandwidth will be transmitted through a single vendor, rather than two. So, if you're currently producing or plan to produce live events, ask if your OVP supports live events.

Another example is PowerPoint support. As you probably know, many seminar and training-related presentations involve both PowerPoint and video, but attempting to combine the two into a single video window is awkward and often time-consuming. Some OVPs are now adding the ability to incorporate PowerPoint slides into on-demand and even live video, through either internal or third-party tools. If this capability is important to you, you'd better ask about this early and determine cost, because it can get pricey when third-party tools are involved.

Player Creation

All OVPs will let you build and customize a basic player, typically with most of the expected playback controls and links to social media sites. These players are almost always Flash-driven, with fallback to HTML5 for mobile devices and HTML5-compatible browsers without Flash installed. Obviously, mobile support is an important checklist item, preferably with support for HTTP Live Streaming to iOS and other compatible devices.

Beyond this, most advanced OVPs offer at least three kinds of players: a single player, a playlist player and a tabbed playlist player. As the names suggest, a single player would display a single video in a window. Referring to Figure 9-11, this would include just the video component showing yours truly and my *Streaming Media Producer* editor Stephen Nathans.

Figure 9-11. A tabbed playlist player created in Brightcove.

In contrast, a playlist player contains the video and the content shown on the right, which is a playlist of other videos associated in some way with the video being shown. Typically, playlists can be created manually or via a program, the latter using metadata to include any containing a specified tag or tags. As mentioned earlier, the benefit of a playlist is that it entices your viewer to watch more of your videos and spend more time on your site, thus increasing the site's stickiness.

A tabbed playlist player incorporates both playlists and the tabs you see on the upper right to make even more content accessible to the viewer. This isn't just beneficial for site stickiness; it's also a very convenient way to categorize videos so that they're presented simply and logically to viewers interested in videos addressing different content areas.

Regarding the player, most vendors offer the ability to create a branded player with all the normal playback controls and embedding and email options, if desired. Be sure to check the extent to which you can customize the player and video libraries presented within the page, and note how easy or how difficult these features are to customize. For example Figure 9-12 shows Kaltura's player editor for a similar tabbed playlist player. As you can see, I can include or exclude many features via simple checkboxes.

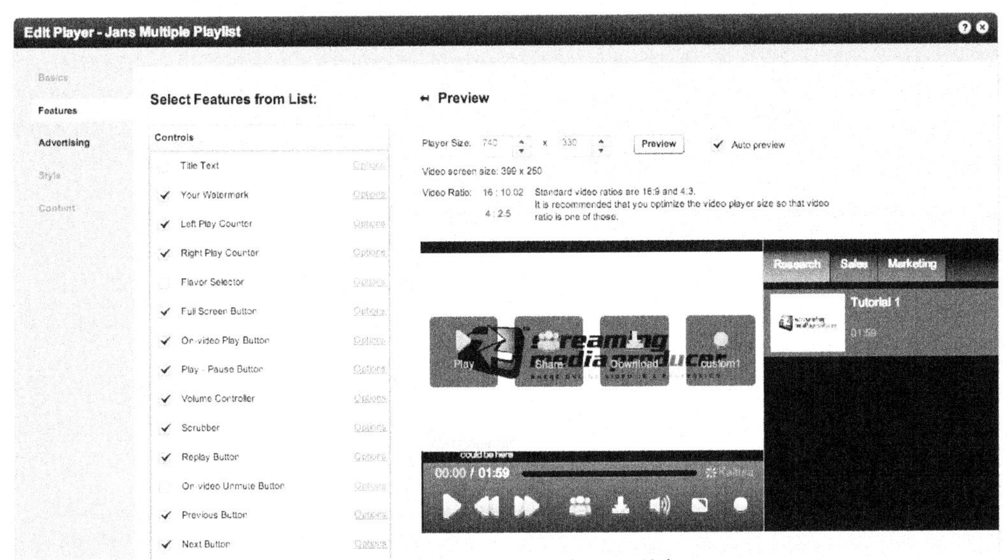

Figure 9-12. Customizing a player in Kaltura.

On the left, the style tab lets me adjust the colors of most style elements, while the advertising section lets me insert pre-roll advertisements from any supported advertising server. Most OVPs offer several layers of player customization: simple changes via controls like those shown in Figure 9-12, more advanced changes available via scripting, and/or extensive customization via coding or a markup language. When comparing OVPs, try to identify the level of customization you'll need and how simple it will be to accomplish that using the candidate OVP's customization schema.

Supported Playback Platforms

Also identify the playback platforms the OVP supports and the level of support. For example, what's particularly impressive about Brightcove's tabbed playlist player implementation is that operates identically on touch-driven devices, supporting gesture commands on iPads and similar devices. Most high-end OVPs also do this, but many smaller OVPs may not.

Given the importance of mobile viewing, it's also important to check for the availability of software development kits (SDKs) that enable OVP integration into iOS and Android apps, with Windows 8 on the near-term horizon. Again, most larger OVPs offer these, but many smaller do not. Broadcasters, educators and government agencies also need to ask about Section 508 compliance, which provides screen reader accessibility, keyboard control and closed captioning.

Content Management

All OVPs supply basic content management capabilities, typically including the ability to tag media for search and retrieval, select preview images or thumbnails for the videos, and create playlists either manually or via metadata tags. Surprisingly, however, many OVPs fall down in simple organization—for example, lacking the functionality to create folders or subfolders for your media bin. If you'll be uploading hundreds of videos, or videos from different departments or organizations, ask how you can organize them in the main content storage area.

Customization of metadata can be another critical issue, particularly if you're working with a structured database in your content management system. Some OVPs make it simple to modify metadata fields; others require scripting or coding to make this happen.

Another big issue relates to moderation and the related segmentation of functionality. Some organizations want certain users to upload content for approval by others before publishing. This requires the ability to segment functionality so that some users can only upload without access to player creation or other back-end functions. Most higher-end OVPs can assign different rights and capabilities to different users, but many smaller ones can't.

A similar capability is the review and approval workflow facilitated by Sorenson Media's OVP, Sorenson 360. This lets video producers upload a video and send a link to the client to review the video and make comments. If you frequently produce video for third-party approval, this can really speed up the process.

Beyond mere content management, some OVPs are adding features like trimming to remove frames from the beginning and end of videos, or even splitting and other editing functions. These can be very convenient when the alternative would be to edit locally and then re-upload.

Third-Party Integration

Many large customers need their OVP to integrate with their content management system (CMS). A tight integration between OVP and CMS can help in a number of ways, from synchronizing metadata to a unified permissions structure and integration of the OVP functions into the client's website, which can reduce implementation and training costs.

Most larger OVPs have modules for common CMSs like Drupal, WordPress or SharePoint; most smaller OVPs don't. If you're running a CMS and need to integrate the functionality of your CMS and OVP, determine whether such a module exists early in the process.

Support for Your Business Model

At some level, all producers need to monetize their videos—whether via third-party or in-house advertising, or some other technique. So be sure the OVP supports your intended models. If you'll be serving ads, make sure the OVP supports the IAB Digital Video-related advertising formats you plan to use (for more on IAB advertising formats, see bit.ly/advertformat). Also ask if the OVP will support the advertising networks you plan to work with.

If you're not serving third-party ads, ask about intro and outro videos you can use to market your own products and services. All OVPs can add intro and outro videos to a single video or player; what's key here is the ability to easily insert an advertisement—say, 15 seconds about your upcoming conference—in front of every video distributed by the OVP.

If you'll be selling your content, be sure the OVP supports your intended model, whether subscription, pay-per-view or other hybrid model. If you're producing content just to get the word out, make sure the OVP offers syndication to UGC and other sites.

Security

Many producers need to protect their content from access or copying via technologies like encryption or SWF verification on a cross-platform basis that encompasses desktop and mobile devices, which many smaller OVPs don't support. Taking this one step further, some producers need the ability to securely side load content to consumer devices like Blu-ray players or Smart TVs via technologies like Widevine or DivX. If you're selling content, the ability to access these popular consumer players can represent a significant revenue opportunities, but very few OVPs offer Widevine or DivX integration.

You may also be interested in geo-restricting the viewing of certain content, which is important when you're selling or syndicating content internationally. The most advanced systems allow you to select specific countries that can view your video—or, if it's simpler, to select countries that can't. You can see how Ooyala enables this in Figure 9-13. Not all OVPs offer this capability, and some require that you purchase expensive third-party modules to do so.

If you allow your content to be embedded, you may want to limit embedding to a few known domains or, again, exclude specific domains. Ooyala enables both, along with the ability to set time restrictions for playback or devices.

If you'll be integrating OVP operations with internal security structures, ask if the OVP can integrate with your existing log-in infrastructure, whether it's LDAP, SharePoint or SAML. This saves the hassle of creating a new database and enables very granular reporting detailing all the videos contributed or watched by each employee. Working off classes or roles defined in your existing log-in infrastructure should also simplify creating and applying role-based permissions for the various functions enabled by the product.

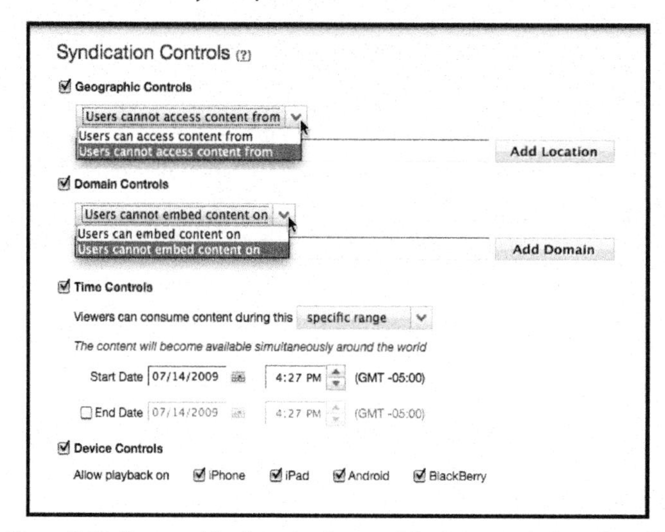

Figure 9-13. Geographic, domain, time and device controls from Ooyala.

Reporting and Analytics

Once you make your videos available, you'll want to know how and where your videos are played. At the very least, most OVPs detail video views and report download bandwidth and details like the country and domain of your viewers. Some offer the ability to download .CSV files so you can further analyze this data in Excel. If you're a current user of analytics programs from providers such as Google, Omniture or Visible Measures, look for the ability to integrate your data into these packages.

Beyond these basics, many OVPs also present true viewer analytics that allow you to identify patterns within the statistical data. One common and exceptionally useful report relates "drop-off" statistics that identify where viewers stop watching the video (Figure 9-14). The figure shows the performance of a video I produced titled "How to Create a Great Video Case Study," which is hosted on the Sorenson 360 OVP. The bad news is more than 50% of viewers dropped off before the first minute. The better news is that the drop-off rate was much slower after this point.

Information like this provides hard data to back up recommendations such as ditching the CEO's greeting or the 20-second introductory animated collage. If few viewers make it through to the end of your 4-minute product demo, you know that you need to get to the point more quickly or, perhaps, present that video later in the sales cycle. If you're displaying advertisements in your videos, you'll be able to tell where they have the most impact and where they simply drive viewers away.

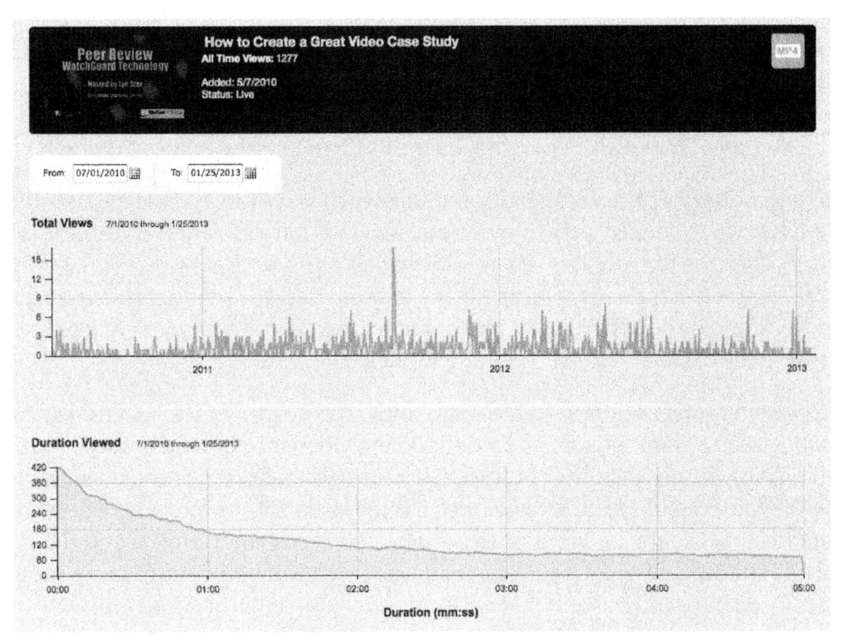

Figure 9-14. Drop-off statistics from Sorenson 360.

Different reports will appeal to different users and different stakeholders within a group of users. So talk to all the stakeholders before you start your search and make sure you've identified the relevant reports they want to see.

Professional Services

The bigger the installation, the more you'll likely need professional services, perhaps to help customize your database or player, perhaps to help import vast quantities of metadata and content from other sources. Ask about the availability of professional services and pricing.

There are literally hundreds of features touted by the various OVPs, but hopefully I've covered the majors. In terms of the decision-making process, following is the procedure used by Bruce Colwin, a serial entrepreneur who has selected multiple OVPs for his various projects.

Choosing an OVP: A Six-Step Process

Bruce Colwin is a serial entrepreneur who has used multiple OVPs for his different companies and has fine-tuned the selection process to six simple steps. He currently works with LegalMinds.tv, and this is the process he used to choose an OVP for this site. His first step was to visit **vidcompare.com**, a comparison service for decision makers seeking to choose an OVP. You can read a *Streaming Media* review of VidCompare at **bit.ly/vidcompare**.

Briefly, VidCompare lets you search for OVPs using categories such as delivery and payment structure and provides basic information such as a feature list, a pricing structure, and the date the company was founded. Even though he was familiar with the OVP market, Colwin found VidCompare useful because the comprehensive listing identified candidates he shouldn't miss. He estimated that he spent about six hours total on the VidCompare site before making his decision.

After identifying potential options on VidCompare (which was Step 1), Colwin used information on VidCompare and the vendor sites to weed out services that didn't fit his business or pricing model (Step 2). Then, he read the company information and any user reviews on VidCompare and polled his business connections to see if any of them had any relevant experience with the remaining candidates (Step 3). Next, he contacted the candidates to ask any remaining questions and to make sure he understood their pricing structures (Step 4).

Then, he requested trial accounts with the final candidates (Step 5). Over a period of a few weeks, he uploaded content, tagged it, embedded the video into browsers and played it back on multiple computers. During this process, he found that one potential vendor had problems with WordPress that could have been a huge issue with some of his potential clients. With another final candidate, he found the pre-sales service lacking, which soured him on the prospect of its after-sales service. After the trials, Colwin consolidated his findings regarding quality, usability, pricing and customer service, and then made his decision (Step 6).

Certainly, I would take all the steps that Bruce recommends, plus ask to talk to reference accounts that are currently using the service. While the OVP will obviously point you to happy customers, I doubt the references would lie to you when you ask about the service. In particular, I would ask how quickly the OVP fixes bugs and other problems, and its approach to new feature requests.

OK, that's all on choosing an OVP. Now let's focus our attention on encoding for uploading to a UGC or OVP site.

Encoding for Upload

We spent most of the early part of the book learning how to encode video to final form for distribution to your viewers. All UGC sites re-encode your video, so when you're uploading to a UGC site, you're uploading for re-encoding. Some OVPs will distribute the files you upload, but many re-encode, so you should ask about this before you start uploading files.

Recognize that when you're encoding for re-encoding, you're trying to provide the highest-quality starting point that can be reasonably uploaded to the UGC site/OVP. If you're getting tired of reading, I've created a video that outlines the factors you should consider when encoding for upload, which you can access at **bit.ly/encode4upload**. As you would guess, the content of this chapter will follow that video pretty closely.

Basic Considerations

Most UGC sites have a maximum upload limit, which can range from 100 MB to 2 GB. So you'll want to know that number first and make sure that your file doesn't exceed that limitation.

With SD files, I always scale and deinterlace before uploading. That is, if I shoot in SD—say, 4:3 aspect ratio—the resolution is 720x480 and the file can be interlaced or progressive. Before uploading, I would scale to 640x480 and deinterlace (if necessary) during the encoding process. For a 16:9 DV file, I would scale to either 640x360 or 848x480 and deinterlace during output.

	SD - 4:3	SD - 16:9	720p	1080p
Resolution	640x480	640x360/ 848x480	1280x720	1920x1080
Scan	Progressive	Progressive	Progressive	Progressive
Data rate (Target/Max)	5/10 Mbps	5/10 Mbps	10/15 Mbps	20/30 Mbps
Codec	H.264	H.264	H.264	H.264
Profile	High	High	High	High
Entropy Coding	CABAC	CABAC	CABAC	CABAC
Bit Rate Control	2-pass VBR	2-pass VBR	2-pass VBR	2-pass VBR
Key Frame Interval	5 seconds	5 seconds	5 seconds	5 seconds
B-Frame/Reference Frame	3/5	3/5	3/5	3/5
Audio Codec	AAC	AAC	AAC	AAC
Audio Channels	Native	Native	Native	Native
Audio Sample Rate	Native	Native	Native	Native
Audio data Rate	128/256 kbps	128/256 kbps	128/256 kbps	128/256 kbps

Table 9-4. Encoding for upload recommendations.

I always encode files for upload with the H.264 codec at high data rates, following the guidelines shown in Table 9-4. Typically, I encode using the same encoding parameters I use for maximum quality (High profile, CABAC, 3 B-frames, 5 reference frames) although at the high data rate recommended, these parameters probably have minimum impact on quality. In addition, I always use 2-pass VBR encoding limited to 200x, with a key frame interval of 5 seconds.

As you'll see below, it's often important for the moov atom to be located at the front of the file, which enables the encoder to start compressing immediately, rather than wait for the file to completely upload. So, when available, you should always encode with the Fast Start option enabled. I discuss the moov atom back in Chapter 4 in a section titled "Where Dat Moov Atom?"

For audio, I recommend uploading the number of native channels in the stream: one channel if mono, two if stereo. I encode to AAC format at 128 kbps for mono, 256 kbps for stereo. Most of my work is simple speech, though; if you're recording the London Symphony Orchestra (or some reasonable facsimile thereof) you might try encoding and uploading the audio component at higher rates.

Table 9-4 consolidates these recommendations into a more usable form. To be honest, although I've used these parameters extensively for UGC and my uploads to various OVPs, I haven't extensively researched the recommendations that may be posted by specific OVPs. My basic assumption is that at these data rates, minor adjustments to the H.264 encoding parameters make very little difference. If you're about to start a long relationship with an OVP, I would ask for its recommendations. If it has none, I would try a couple of short files using the recommendations in Table 9-4, and tweak them if necessary.

Let's conclude with a look at encoding for YouTube. This is derived from an article I wrote titled "Encoding for YouTube: How to Get the Best Results," which you can read at **bit.ly/YouTube_specs**.

Encoding for Upload to YouTube

I started my article with a look at the recommended specs YouTube provided. By way of background, when YouTube first came to prominence, the company seemed to enjoy working as an opaque black box, providing little direction regarding the best way to prepare your file for upload. Now YouTube provides very specific encoding parameters, which you can find at **bit.ly/YT_encode**.

Note that these are true recommendations in the sense that YouTube won't reject the files if you don't follow them. Rather, YouTube will attempt to encode pretty much any file you throw at it; you'll just get the most predictable results by following the recommendations.

At a high level, YouTube provides recommendations for two classes of users: standard and those with "professional-quality content" and "enterprise-quality Internet connections." After covering the basics, I'll share these specific resolution and data rate recommendations with you.

- ***Support file formats:*** Although YouTube accepts multiple formats including .MOV, .AVI, .WMV, and .FLV, the advanced specifications page recommends H.264/AAC in an .MP4 container format. YouTube recommends putting the moov atom at the start of the file, which for many encoders means activating the Fast Start option.

- ***Audio recommendations:*** YouTube recommends uploading stereo or 5.1 audio in either 48 or 96 kHz. As you'll see, many of the presets I reviewed use 44.1kHz, which I recommend changing to 48.

- ***H.264 parameters:*** YouTube recommends using the High profile with CABAC entropy coding, with variable-bit-rate encoding. Other recommendations include a B-frame interval of 2, which is curious given that YouTube produces its H.264 video with no B-frames. Recommendations also include a closed GOP with a GOP size of half the frame rate, which means 2 key frames a second, an interval that none of the encoders matched.

- ***Frame rate:*** Don't change the frame rate for uploading; if you shoot at 24p, you should upload at 24p.

- ***Frame composition:*** If you're working with interlaced source content, YouTube recommends deinterlacing before uploading. While you can upload at resolutions up to 4K, you shouldn't burn letterboxing or pillarboxing into your video; produce native 16:9 video at a 16:9 resolution and native 4:3 video at a native 4:3 resolution. YouTube also recommends against uploading files with a pixel aspect ratio other than 1.0, otherwise known as square pixel output.

Again, YouTube identifies two classes of users: standard and high-quality. Here are the suggested data rates for varying video resolutions:

Standard uploads:

Type	Video Bit Rate	Mono Audio Bit Rate	Stereo Audio Bit Rate	5.1 Audio Bit Rate
1080p	8,000 kbps	128 kbps	384 kbps	512 kbps
720p	5,000 kbps	128 kbps	384 kbps	512 kbps
480p	2,500 kbps	64 kbps	128 kbps	196 kbps
360p	1,000 kbps	64 kbps	128 kbps	196 kbps

Table 9-5. YouTube recommendations for standard-quality uploads.

High-quality uploads for creators with enterprise-quality Internet connections:

Type	Video Bit Rate	Mono Audio Bit Rate	Stereo Audio Bit Rate	5.1 Audio Bit Rate
1080p	50,000 kbps	128 kbps	384 kbps	512 kbps
720p	30,000 kbps	128 kbps	384 kbps	512 kbps
480p	15,000 kbps	128 kbps	384 kbps	512 kbps
360p	5,000 kbps	128 kbps	384 kbps	512 kbps

Table 9-6. YouTube recommendations for high-quality uploads.

After reviewing YouTube's recommendations, I turned my attention to the presets provided by common desktop encoders and encoded a standard test file using YouTube presets from Adobe Media Encoder, Apple Compressor, Sorenson Squeeze and Telestream Episode, which all used different of data rates. Then I uploaded all the files to YouTube, then downloaded the files produced by YouTube from those files, and compared the results.

At 1080p, Compressor's YouTube preset encoded at 20 Mbps, compared with Adobe Media Encoder and Telestream Episode at 8 Mbps, and Squeeze at 6 Mbps. Comparing the YouTube output from these inputs, I concluded in my article, "Remarkably, with the exception of one clip with extreme high motion and lots of fine detail, the YouTube output was virtually identical. … Again, video compression is a garbage in/garbage out medium, and more is always better when it comes to data rate. However, if upload time is a concern, as it is with me, I wouldn't go higher than 8 Mbps unless the footage was extremely difficult to encode."

At 720p, Apple Compressor's YouTube preset was at 10 Mbps, while all others were at 5 Mbps. After comparing the results, I concluded in my article, "I ran the same quality related tests with video uploaded from Compressor, Adobe Media Encoder and Squeeze with the same results; the 5 Mbps files looked identical to Apple's 10 Mbps files. The bottom line is while you're free to encode at any data rate, you're unlikely to see any benefit from rates beyond 5 Mbps unless you're working with exceptionally hard-to-encode video."

I made no observations about SD video except to note that YouTube hosed the aspect ratio of uploaded DV files and that it was better to upload square pixel output in the resolutions identified in Table 9-4.

Synthesis

With this as background, why are my recommendations in Table 9-4 so high? A couple of things happened. First, in response to the article, I got the following note from Colleen Henry, a Video Hacker from Google's Video Infrastructure Group.

> It's important to think of the files you upload to YouTube as golden masters, as they will be used as source material to generate video streams for years to come. Simply put, the better the quality of the file you upload to YouTube today, the better quality the viewer's experience will be throughout your video's life on YouTube.
>
> As displays increase in size, compression techniques become more efficient, playback devices become more sophisticated, and Internet connections improve, so will the quality YouTube will be able to provide to the viewers of your videos. This means, while you may reach a limit on perceived benefits from higher bit rates or more efficient encoding if you were to test it today, that does not mean you should stop there. You will see a huge benefit over the lifetime of your video being available on YouTube, as Internet speeds, hardware, and software evolve. Upload the best-quality video that you can create and squeeze through your Internet connection!
>
> Bonus tips:
>
> - Many encoders can spend more CPU time to create a much more efficient file. If you have a powerful computer but a slow Internet connection, look into using more complex and efficient encoding to save upload time.
> - You can noticeably improve the quality of your video on YouTube by using a sophisticated, scene-aware denoising filter prior to uploading.
> - Key frame interval doesn't really matter much at this moment in time, but please keep it under 5 for VOD.
> - The sample rate of your audio should match your source's sample rate in which it was produced.

- If you make sure to use a streaming format—like .MKV, .MP4 or .MOV—with the metadata at the front of the container, we will begin processing your video while you are uploading it, drastically reducing overall turnaround time. This will make things much faster, with no negative side effect. You can add the metadata atom to the front of your file with something like qtfaststart, or select it when you are creating the file in Squeeze, Episode, etc.

- It is ideal to use constant-quality encoding. This will let you create a high-quality, variable-bit-rate file at the speed of a single pass. It will maintain a consistent target quality throughout the file, rather than trying to allocate bits to hit an arbitrary bit rate, which can easily under-shoot or over-shoot and, with 2-pass, take extra long to create.

- You can put uncompressed PCM audio in an .MOV or .MKV container and deliver it to us if you like. However, make sure not to create multiple discrete mono streams when you do it.

Something else happened that made me take a closer look. Briefly, as a part of a consulting project, I was testing the output quality of "Encoder A," which wasn't (at that time) compatible with ProRes input files, the codec used for many of my test clips. "No problem," I thought. "I'll just encode my ProRes test files into high-bit-rate H.264 and encode from those on Encoder A." These were 720p files, and I rendered them at 30 Mbps because Adobe Media Encoder topped out at that value.

I then rendered files from the H.264 source in Encoder A and compared them to files produced by Encoders B, C and D ***from the ProRes source***. Then I was concerned that comparing Encoder A's output from even high-data-rate compressed H.264 input to the output from Encoders B, C and D from the ProRes source wasn't an apples-to-apples comparison. Easy enough to check, of course—just encode the H.264 source in Encoders B, C and D, and compare those files with the file Encoder A produced from the same compressed H.264 source.

Now that I had all these data points, I wondered how the files encoded from ProRes compared with those encoded from H.264 by the same encoder. The difference was so significant that I re-encoded my ProRes test files into H.264 format at 50 Mbps, this time using Sorenson Squeeze, to make my tests as realistic as possible.

To supplement these tests, I converted a 1080p test file to 50 Mbps and 20 Mbps iterations and uploaded them to YouTube. Then I downloaded and compared the quality of the files produced by YouTube from these sources. In the 1080p files that YouTube encoded at 5.8 Mbps, there was no noticeable difference. In the 640x360 files YouTube produced at 636 kbps, the files produced from the 50 Mbps source showed more detail.

To be fair, the difference would be unnoticeable to the casual user. Within the context of my consulting project, however, the developers of Encoder A weren't casual, so I had to redo all my tests using the 50 Mbps H.264 source clips, and yet again when the encoder that I was testing was able to encode ProRes files.

In a more general sense, the quality differentials between the footage produced from the low-data-rate H.264 files and the 50 Mbps files made me rethink the value proposition of using 5 to 10 Mbps H.264 as an intermediate format to save uploading time. As hard as we work to preserve quality throughout the production pipeline, this no longer looked like the most appropriate tradeoff. So if you want absolute top quality for your mission-critical videos, observe the guidelines in Table 9-4. If speed is more important than quality, as it often it is, you can drop your data rates down to the guidelines shown in Table 9-5 for YouTube and for other UGC and OVP sites.

Conclusion

Now you know how to distribute your on-demand videos via UGC and OVP sites. Next up is an introduction to producing live events.

Chapter 10: Introduction to Live Streaming

This is the first of four chapters on live streaming. In this chapter, I'll introduce you to the technologies involved with live streaming and outline some of the technology decisions you'll have to make. In Chapter 11, I'll detail your distribution options for live video; in Chapter 12, I'll describe encoding options; and in Chapter 13, I'll discuss a mélange of practical elements, like monitoring audio, shooting a live event and other tips.

In this chapter, I'll take a 50,000-foot view; in subsequent chapters I'll go into the nuts and bolts. Specifically, in this chapter you will learn:

- The six elements involved in a live production
- Various options for encoding your video
- Various options for transferring the video from encoder to server
- Various options for buying or leasing a streaming server
- The ideal features of a highly competent landing page
- Options for delivering video to your viewers.

Again, this chapter is the high-level view, and throughout the chapter, I'll point you to more detail in subsequent chapters.

If you're unfamiliar with the process of live streaming, you may want to start by watching a tutorial I produced in 2010 titled "Producing Live Webcasts with ViewCast's Niagara 2100." While few producers are using the Windows Media codec any longer, otherwise the setup and configuration steps are identical to units streaming H.264 today, and you'll get a useful overview of the process in about 10 minutes (**bit.ly/introtolive**).

Components of a Live Production

As an overview, in a live event, you shoot the video, encode the video into a file and send that to the streaming server. From there, the video is presented in a landing page and delivered to your viewers. Let's look at each component briefly, and then take a deeper look.

- ***Video signal.*** By video signal, of course, I mean the output of a webcam or camera or multiple cameras via a video mixer—basically, the video from the event that you want to broadcast.

- ***Live video encoder.*** Whether the video signal is from a single camera or webcam, or from a video mixer, you now have to compress the video into a stream or streams to send to the streaming server. Here you have myriad options that I'll outline below.

- ***Transmit video to streaming server.*** This is the biggest problem for many live producers since outbound bandwidth is often limited, particularly while on-location. For the most part, you'll use an Ethernet or Wi-Fi connection to get this done. 4G transmission is worth considering for some production.

- ***Streaming server.*** You can buy your own streaming server, lease a server, or hand off the server and player creation aspects of your live events to a live streaming service provider (LSSP).

- ***Landing page.*** This is the page where you direct the viewer to watch the live video.

- ***Delivery to the viewer.*** Options here including delivering yourself via your own Internet connection, or using a content delivery network (CDN), either directly or via a LSSP.

All live event producers need to understand their options for each element and manage each element for every production. Let's take a deeper look.

Video Signal

Again, at the simple end, live broadcasts can start with the output from a single webcam or camera. However, using multiple cameras adds polish to any live event, which is why many producers now mix multiple camera feeds on site and then stream the mixed feed live. You have multiple options for mixing: some driven by the software encoder you might use, some computer-based systems, and some dedicated hardware devices like the Roland VR-3 mixer shown in Figure 10-1.

Figure 10-1. Roland's VR-3 Mixer.

Whether you're using a single camera or mixer, you have to encode your video using the next live production component: a live encoder.

Live Encoder

Choosing a live encoding tool used to be simple: You typically would encode a single stream for delivery to your desktop viewers, and budget was the most important buying criteria. When buying today, of course, you've almost certainly expanded your target viewers to include both mobile and desktop clients, with adaptive streaming preferred over single-file delivery.

You have a host of new workflow options to consider, from live cloud or server-based transrating to server- or content delivery network (CDN)-based transmuxing. You also may have several new requirements, from digital rights management to closed captions to advertising insertion. In Chapter 12, I detail the questions you should ask before buying a live encoder and lay out your options in all relevant categories.

Transmission to Streaming Server

Most live producers transmit their streaming video to the streaming server via Wi-Fi or Ethernet—or, for larger organizations, over dedicated fiber or satellite trucks. As mentioned above, one new technology that's increasingly being used to distribute video from remote locations is 4G. As you probably know, 4G is the fourth generation of cell phone mobile communications standards with a theoretical peak performance of 1 Gbps from a fixed location (and 100 Mbps from a moving platform like a train or car). To connect via 4G, you need a 4G-compatible device (like a mobile phone) or an external 4G modem, which typically connect to notebook computers and portable encoders via USB connections.

There are two deployed 4G technologies—Mobile WiMAX and long-term evolution (LTE)—and performance will vary by service provider and location. In a May 2012 story, *PC World* tested AT&T, Sprint, T-Mobile and Verizon 4G in 13 cities, and found a significant variation in upload performance, with Sprint and T-Mobile at the low end at 0.97 and 1.32 Mbps, respectively, compared with AT&T at 4.91 Mbps and Verizon at 5.86 Mbps (**bit.ly/4G_speeds**).

These are average real-world numbers, and the AT&T and Verizon results are impressive. Remember, though, that the 4G connection simply connects the 4G modem to the tower. While each 4G connection is separate and unique, simultaneous 4G connections share the bandwidth from the tower to the central office. The more connections sharing the bandwidth to the central office, the lower the bandwidth for each. Fortunately, most mobile connections are downloading data, rather than uploading it, so contention for upload bandwidth should be minimal, but it does vary widely by location or even time of day.

While 4G is becoming more widely available, it's far from pervasive, so you shouldn't assume that coverage will be available at all potential broadcast locations. Fortunately, most 4G service providers have maps where you can plug in an address or zip code and check availability, like Verizon's map at **network4g.verizonwireless.com**.

Gee? No, 4G

When broadcasting via 4G, you have three basic hardware options:

- ***A 4G modem in the USB port of an encoder.*** This transmits a single 4G stream to the streaming server.
- ***Link aggregation and cellular bonding.*** The devices use multiple 4G connections to stream at higher data rates back to the server.
- ***An encoder/aggregator.*** These devices combine H.264 encoding and 4G transmissions.

The differences between the technologies are best explained via a simple table (Table 10-1). Basically, a modem in a USB port enables a single 4G transport over any supported carrier—this is the approach taken by the Livestream Broadcaster, shown in Figure 10-2, which has a USB port for a 4G modem. This can work extremely well, so long as you're the only user attempting to push high-bandwidth files via that same cell tower. If you're one of 40 video journalists pushing a live feed through the same tower from the Super Bowl, your results may suffer.

	4G Transport	Link Aggregation	Multi-Carrier Redundancy	Adaptive Encoding
4G Modem in USB Port	Yes	No	Yes	No
Link Aggregator	Yes	Yes	Yes	No
Encoder/Aggregator	Yes	Yes	Yes	Yes

Table 10-1. Features of various 4G-capable products.

You can read about my experience testing the Livestream Broadcaster and 4G in an article titled "Livestream Broadcaster Review Revisited: Testing 4G Performance" at **bit.ly/BC_redux**.

Figure 10-2. A Verizon 4G modem atop the Livestream Broadcaster on-camera encoder.

A link aggregator aggregates multiple 4G modems into a single higher-bandwidth connection and can deploy modems from different carriers (See Figure 10-3). That way, if Verizon is having a bad day, T-Mobile or AT&T can kick in. While this might not help in our Super Bowl scenario, in day-to-day use, link aggregation is more reliable than a single modem because it can access multiple carriers. Plus, you should be able to push a higher-bit-rate stream because of the shared bandwidth.

This shared bandwidth concept is shown in Figure 10-4, a view of Teradek's Sputnik server, which communicates with the Bond cellular aggregation unit. The green and blue lines on the bottom show the effective throughput via the only two 4G modems that found a signal: two Verizon modems on Port 3 and Port 5. These combined to produce a 4.308 Mbps signal that's represented by the yellow line atop the graph.

Figure 10-3. The Teradek Bond 4G aggregator sitting atop the Teradek Cube.

However, this shared bandwidth comes at a cost. Specifically, you'll need to deploy a server component to receive and reassemble the multiple streams into a usable format before sending it to the streaming server. In contrast, when communicating with a single 4G modem, there's nothing to reassemble, so you can send the signal directly to the streaming server.

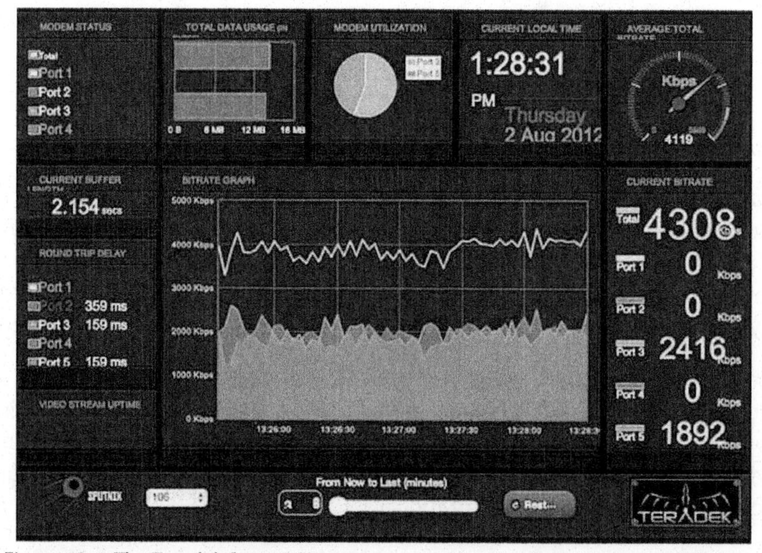

Figure 10-4. The Teradek Sputnick server reassembles the signals into a usable form.

Figure 10-4 is a screen shot from the software server that you deploy with the Teradek products, while Figure 10-5 shows the LGR or VMS hardware playout unit required for the AirStream Video TX cellular multiplexer.

The third option, an encoder/aggregator, combines encode and transport functions in a single system—whether in a single piece of hardware, like the AirStream Video TX shown in Figure 10-5, or two separate components, like the Teradek Bond and Cube shown in Figure 10-3. These systems offer the benefits of link aggregators—plus, since the encoder and 4G transmission unit can communicate, the ability to adapt encoding to changes in effective throughput.

For example, with the Teradek Bond/Cube system, if you initially configure your encode to produce a 4 Mbps stream and effective bandwidth drops, the Cube transmitter will tell the Bond encoder to throttle down the data rate. At your option, you can also drop the frame rate to preserve frame quality at the cost of smoothness of motion.

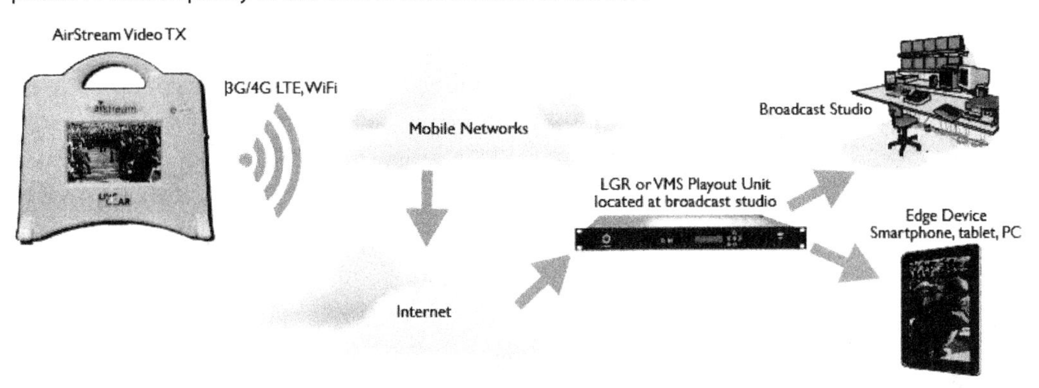

Figure 10-5. The AirStream Video TX multiplexer needs an LGR or VMS playout unit to reassemble the signals from the multiple modems.

Note that it's possible for stand-alone encoders to adjust their encoded bit stream in response to changes in throughput, although the responses are typically less sophisticated than encoders/aggregators working together. For example, Livestream Broadcaster monitors the outgoing queue, which fills when outbound bandwidths are inadequate to carry the stream. First it provides warning messages to the operator to consider dropping the quality level. If the buffer gets too high, it will drop video frames and, if necessary, audio segments. Note that it does this for all transport mechanisms, whether Ethernet, Wi-Fi or 4G.

However, because there's no standardized communication between Broadcaster and the communications device, it's tough to implement more sophisticated correction techniques, like changing the stream data rate or frame rate. That's why encoder/aggregators offer more advanced options. Overall, for occasional 4G use on a non-mission-critical basis, a single 4G modem is worth a try. For mission-critical applications, you'll need a link aggregator or preferably an encoder/aggregator. Here are factors to consider when shopping for a 4G-capable device:

- ***Total price.*** Remember to include the price of the encoder (if needed), plus the aggregator and server.

- ***Form factor.*** On-camera devices are nice from a portability perspective, although battery life is often an issue.

- ***Battery life.*** Speaking of battery life, consider that too. If you plan on broadcasting very long events, find a unit that can be externally powered or has hot-swappable batteries. For shorter events, I prefer units with internal rechargeable batteries.

- ***Number of modems and configuration.*** Most systems max out at five modems. Some are configured internally, like those on the AirStream Video TX, which is more convenient because you don't have to carry them around separately.

- ***The availability of error correction.*** Which helps ensure signal integrity and video quality.

This concludes my coverage of 4G transmission, although I'll cover on-camera and portable H.264 encoders in Chapter 11. For more information on 4G transmission, check out my review of the Teradek Bond and Cube units in an article titled "Streaming Over 4G with the Teradek Cube" at bit.ly/Bond_cube. For a tutorial on how to use the AirStream Video TX, go to bit.ly/Airstream_ozer.

Streaming Server

All live streaming requires a streaming server, and you have four basic options. You can host your own streaming server, either locally or in the cloud; you can choose a live streaming hosting provider like PowerStream or Streaming Media Hosting, which supplies access to a streaming server; you can sign on with a traditional online video platform (OVP) provider like Brightcove or Kaltura, which has expanded to offer live streaming; or you can engage a live streaming service provider (LSSP) like Livestream or Ustream. In this section, I'll discuss the pros and cons of all four approaches so you can make the best decision for your organization.

Know the Roles

When evaluating these alternatives, it's helpful to break the elements of a live streaming event into distinct "roles" that you can see in Table 10-2. The first role is server purchase, installation, hosting and administration. If you host your own server, you're in charge of all those elements, so you'll need technical resources on staff to handle these tasks. Self-hosting locally means providing a physical server and the associated costs, which you can offload for a fee by hosting in the cloud.

	Host Your Own	LSHP	OVP	LSSP
Server Purchase, Install, Hosting and Administration	You	Them	Them	Them
Production	You	You	You	Them
Player and Landing Page creation	You	You	Them	Them
Mobile, OTT and Other Platform Support	You	Them	Them	Them
Distribution	You	Them	Them	Them
Eyeballs	You	You	You	Them

Table 10-2. Roles relating to various streaming server options.

If you use a service provider, it will purchase, install, host and manage the streaming server. While expenses like server cost and technical staff are built into the provider's charges, they're spread over multiple customers and may be cheaper than self-hosting. If you plan to broadcast 24/7, you're essentially in the live streaming business, and it may make sense to host your own server. At the other end of the spectrum, if you plan to broadcast one or two events a month, one of the other options is probably a better choice.

Production Role

The production role involves multiple elements, including the purchase of camera, audio and lighting gear; the purchase and installation of hardware and/or software encoders; as well as that of transmission gear if the outbound bandwidth at the site isn't sufficient to deliver the compressed live streams to the remote streaming server. The production role also includes the staffing required to drive all this fancy equipment.

If you're self-hosting your live event, you're in charge of production. While you can hire outsiders, you'll have to ensure that they have the requisite experience in producing live events. The same holds true for most live streaming hosting providers and online video platforms, which typically don't have in-house production teams. In contrast, the largest LSSPs either have in-house production teams or experienced third-party contractors knowledgeable in live event production.

Player and Landing Page Creation

To watch your live event, viewers will navigate to a landing page and watch from a player. If you're hosting your own server or using a hosting provider, you're in charge of creating both the landing page and player, which requires technical resources that can program in Flash. For a simple player in a window, this shouldn't be much of a problem even if you have to bring in outside resources, although even this minimum level of player creation is complicated if you plan to support one or more mobile platforms.

More importantly, a simple player in a window is a pretty sterile environment for a live event. The trend in live event production is toward players that simulate "being there" as much as possible. For consumer-focused events, like concerts or festivals, typically this means access to live chat via Twitter, Facebook or other facility (see Figure 10-6), and perhaps the ability to upload images or even videos from the event. In a business or educational environment, this could mean the ability to issue polls or quizzes, as well as chat. All events benefit from ratings and the ability to leave comments.

In addition, for many events, you want the broadest possible distribution, which means the ability to embed the player in other websites, or Facebook or Twitter. These types of features would be expensive for occasional broadcasters to reproduce, which is why many such broadcasters migrate toward OVPs and particularly LSSPs, which typically offer more evolved social-media-related features in their players than OVPs.

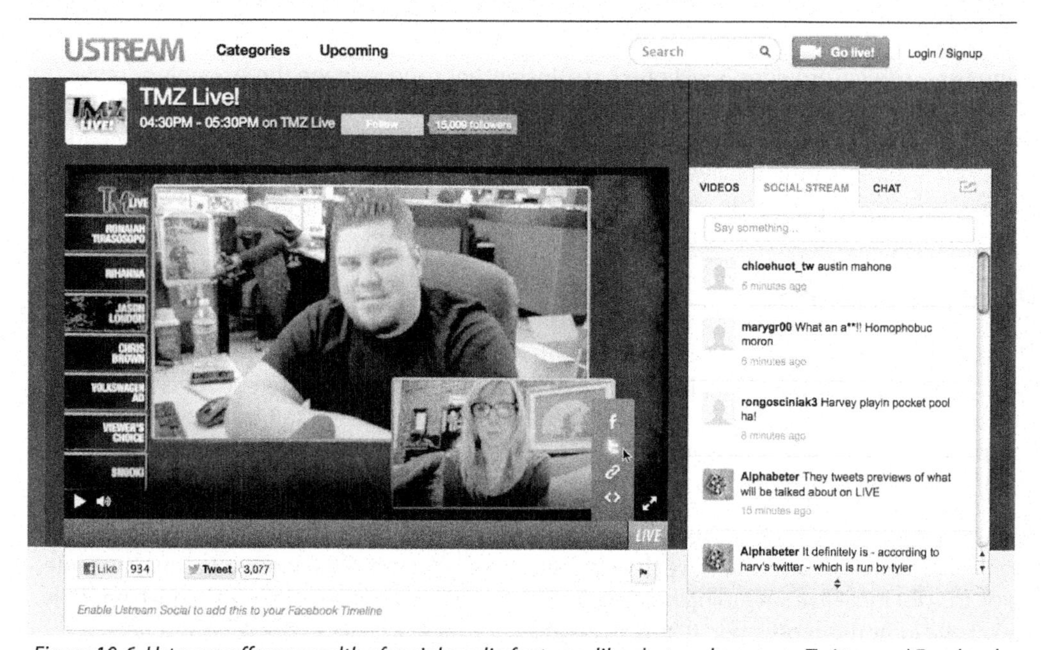

Figure 10-6. Ustream offers a wealth of social media features, like chat and access to Twitter and Facebook.

Platform Support

Most live event producers want to reach the widest possible audience, including desktop and mobile-based players. If you're hosting your own server, most servers you can buy—like the Adobe Flash Media Server, RealNetworks Helix Server or the Wowza Media Server—can repurpose Flash-encoded video to iOS and Android devices. However, this adds a level of expertise that you may not have in-house if you're hosting your own server.

In contrast, all three classes of service providers should be able to distribute your Flash-encoded video to multiple mobile platforms and provide templates and/or encoding specifications to help you produce streams that play on all target platforms.

Distribution

If you're rolling your own system, you're in charge of distribution as well, both externally and internally if you're targeting internal viewers. If you're serving even relatively few external viewers from your own server, you need massive pipes to the Internet, a co-located server or a server installed in the cloud. As simultaneous viewers increase, you may also need a content delivery network (CDN) to ensure a high quality of service to all viewers.

This is an area where all three classes of service providers can help. All will have relationships with CDNs for external delivery; once the live video stream is transmitted to their streaming servers, they're in charge of delivering the video to the end users. Some OVPs can also optimize delivery within the enterprise using technologies like peer-to-peer delivery on the corporate LAN.

Eyeballs

Finally, primarily for B2C producers, if you're looking for additional viewers to watch your event, this is one of the primary benefits of LSSPs, which have massive numbers of viewers and various programs to market to them. If you're looking for eyeballs, this is the best category.

Hopefully, the foregoing has clarified the pros and cons of the different alternatives for the server function and helped you decide which is the best for your live events. If you're looking for additional information, note that I covered distribution via OVPs in Chapter 9, and I will cover choosing a live streaming service provider and streaming server in Chapter 12.

Now let's get back to the last two elements of the live experience.

Landing Page

As mentioned above, the landing page is the web page your viewers will navigate to watch your video. The landing page provided by Ustream is shown in Figure 10-6. Note that this is the page on the Ustream site; in addition, TMZ embeds the player on its own website. Irrespective of which server option you choose, you'll need to provide a landing page.

Delivery to the Viewer

See Distribution in the section above. Again, delivering the video to your viewer is the final component; if done poorly, all the other steps go to waste.

Conclusion

OK, now you know all the components to live streaming. Kind of a shortie chapter, but I didn't want to dive into the encoding and distribution issues without providing some background to readers who may be unfamiliar with the concepts. While there may be a bit of repetition in the upcoming chapters, hopefully it won't be too onerous. That said, let's jump into Chapter 11: Distributing Your Live Video, where we take a deeper look at LSSPs and streaming servers.

Chapter 11: Distributing Your Live Video

> ***Not to nag, but:*** You can download a PDF file with all figures from this book at **bit.ly/Ozer_multi**. Since this chapter contains lots of screens, now would be a good time to do so.

In Chapter 10, I detailed the components of a live streaming workflow and identified some technology decisions that you need to make, including choosing how you access the streaming server features required for live distribution (see "Streaming Server" in Chapter 10). In this regard, your options included buying and hosting your own server; choosing a live streaming hosting provider like PowerStream or Streaming Media Hosting; signing on with a traditional online video platform provider (OVP) like Brightcove or Kaltura, which has expanded to offer live streaming; or using a live streaming service provider (LSSP) like Livestream or Ustream.

Coming into this chapter, I'm going to assume that you've decided to use an OVP. If you've decided to buy and host your own server, check out the section titled "Distributing Your Own Videos" in Chapter 9, where I discuss features to consider when buying a streaming server. All of the servers discussed there handle live and on-demand video, so the discussion is equally applicable.

In this chapter, I'll discuss issues relating to choosing an LSSP and demonstrate the workflows for streaming live with Livestream and Ustream. Since many producers need to create live experiences that incorporate PowerPoint presentations, polling, quizzes and other elements, we'll also take a look at a category of products called rich media presentation systems that deliver a much more robust experience than any LSSP.

In this chapter, you will learn:

- Factors to consider when choosing an LSSP
- The high-level workflows for streaming live with Livestream and Ustream

- How rich media presentation systems differ from LSSPs
- Factors to consider when choosing a rich media presentation system.

For the record, I'm presenting this chapter, "Distributing Your Live Video," before the next chapter, "Choosing and Using a Live Encoder," because you need to decide which distribution option you will choose and which LSSP you will use before selecting an encoder. There will be a good deal of cross-referencing between this chapter and Chapter 12; I apologize for any duplication or confusion. Thanks, and let's get started.

Choosing a Live Streaming Service Provider

Briefly, LSSPs are the simplest and cheapest way to start live broadcasting because the service provider pays all infrastructure costs and provides all necessary system components, from live encoding tool to the embedded player. As your audiences get larger and your broadcasts more frequent, you may want to cut over to your own live streaming server with a totally custom player or investigate other, less integrated services from content delivery networks (CDNs) or other third-party service providers, but LSSPs are a great place to start.

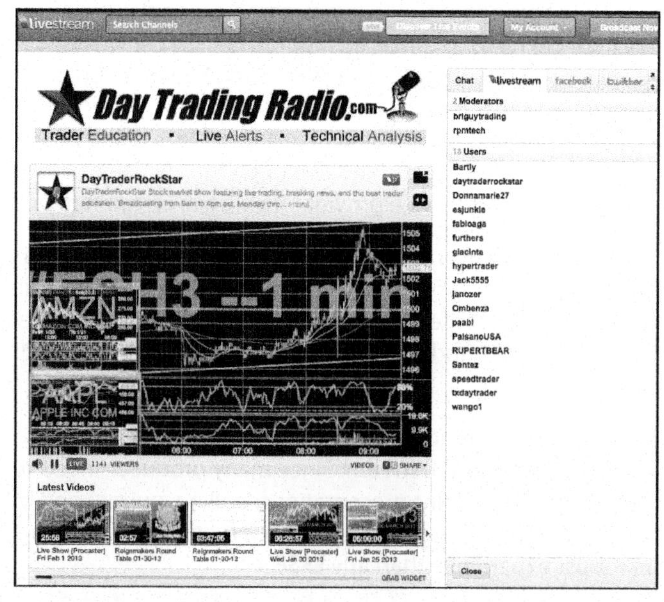

Figure 11-1. The Livestream channel page for Day Trading Radio. Note the chat and other social media-related features (and 1141 viewers!).

Like any technology or service, choosing an LSSP involves an analysis of multiple features. As an overview, let's discuss the broadcast process. Your first consideration is where the broadcast will play, which can either be the page the LSSP creates for your broadcasts on its site, called the channel page, or on a player embedded into your own and/or other websites. So, we'll start by covering the types of features to consider when comparing the channel pages and embedded players offered by the various services.

Figure 11-2. The same video embedded in the producer's website, free from any Livestream branding.

Once you've got the channel page and player set, it's time to broadcast. You accomplish this by connecting your camera or video mixer (or a webcam) to an encoder and transmitting the encoded stream over the Internet to your LSSP. From there, streaming servers hosted by the service provider distribute the video to your viewers, whether they're watching from the landing page or an embedded player.

Most service providers offer a browser-based encoder and traditional applications for encoding and transmitting your stream, or you can use third-party programs like Telestream Wirecast. I discuss these options in Chapter 12. Many service providers also allow you to broadcast from mobile devices; it's features like these that I'll cover during the second half of this chapter.

Preliminary questions: SD or HD?

Some free plans limit resolution to SD and some don't, although all of the for-fee services offer HD at this point. If you're shopping for a free service and have the equipment and outbound bandwidth to produce and transfer an HD stream, you'd better check this out early. If you're looking for a paid service and want HD, make sure you understand the size limits. Most at this point are 720p, but I'm guessing this will change over the next few months. If you want 1080p, you'd better check this early.

Single or Multiple Streams?

Most LSSPs are in the process of transitioning to adaptive streaming. For desktop computers, this typically means manual stream selection, called YouTube style adaptive streaming since that's how YouTube offers its multiple streams. For mobile viewers, typically this means HTTP Live Streaming (HLS) support, with automatic stream switching.

The bottom line is that you want true adaptive streaming; it's the simplest and most trouble-free way to support a range of viewers watching over a range of connection types on a multitude of different devices. So when you're shopping, determine:

- If the LSSP plans to support adaptive streaming
- How it plans to support it (YouTube or true) for all relevant platforms
- The encoding requirements to produce adaptive streams
- How much extra (if any) adaptive streaming will cost.

Eyeballs or Platform?

With these basics out of the way, ask yourself two key questions. The first is, Are you looking for eyeballs or just the platform? One of the great values of sites like Justin.tv, Ustream and Livestream is that they are destination sites for viewers seeking entertainment—Justin.tv for the gamer crowd, and Ustream and Livestream more for general audiences.

Like posting on-demand videos on YouTube, broadcasting on these sites can bring you plenty of viewers. If you're looking for these eyeballs, then you want to choose a site that matches the demographics of your target audience.

If eyeballs are your goal, you should assume that most viewers will watch the broadcasts from your channel page on the LSSP site and focus more on the features of the channel page than the embedded player. On the other hand, if you're looking primarily for live streaming technology to leverage within a player embedded within your own website, you care more about the features of the embedded player.

The second question is, Can you live with advertising on your channel page or even embedded player? Many services offer advertising-supported free versions of their services with some limitations discussed below. This may be acceptable to many smaller broadcasters. If you'd like to drop the advertisements, many offer for-fee "white label" versions without the third-party advertising or any branding from the LSSP.

As an example, Figure 11-1 shows the channel page for Day Trading Radio in the Livestream site; note the obvious Livestream branding. In contrast, if Figure 11-2, the same player is embedded in the Day Trading Radio website and there's no Livestream branding, so it's a white-label version. Later, I'll discuss the differences between the new Livestream and the old Livestream player. One key difference is that the old Livestream player, shown in Figures 11-1 and 11-2, offers a white-label version, while the new version doesn't.

To be clear, the new Livestream is completely advertising-free, so you don't have to worry about miscellaneous ads popping up on your video. However, there is no white-label version without the Livestream branding, although this may be coming. This probably isn't a big deal for most companies, but if you're a large broadcaster who doesn't want to let the world know you didn't build out the live streaming infrastructure yourself, it may be a problem.

Does the Channel Page Support My Monetization Strategy?

Again, the channel landing page is the page on the LSSP website where potential viewers go to watch your live and on-demand broadcasts (all services archive live broadcasts for subsequent on-demand viewing). The features of this page should be your next consideration.

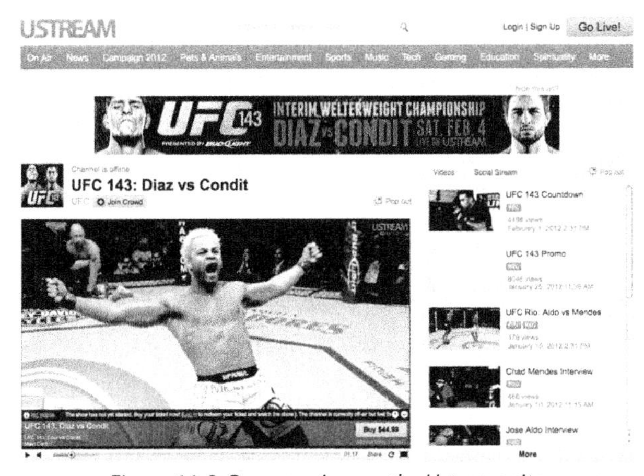

Figure 11-3. Pay-per-view on the Ustream site.

If you're seeking to monetize your content, you should ask about the monetization capabilities offered by the site. These can vary from a share of revenue for advertisements shown on your site to pay-per-view or subscription access to videos. This is illustrated in Figure 11-3, which shows the UFC broadcasting pay-per-view fights on Ustream. (I know it's an old fight, but Ustream stopped showing previews of its PPV events, so current fights show a boring empty screen.) Also on Ustream, you can pay $0.99 to learn how to bake cakes from the Cake Boss, Buddy Valastro, or $4.99 to watch Ron Volper tell you how to increase sales in the new economy. All LSSPs have different monetization offerings and different LSSP/producer splits, so investigate this issue early in your analysis.

Next consider the experience you're seeking to deliver to your viewers. Figure 11-3 is the current prototype: a video player with social media links, a related library of content on the right and a social stream—consisting of chat, tweets and other content—on a separate tab.

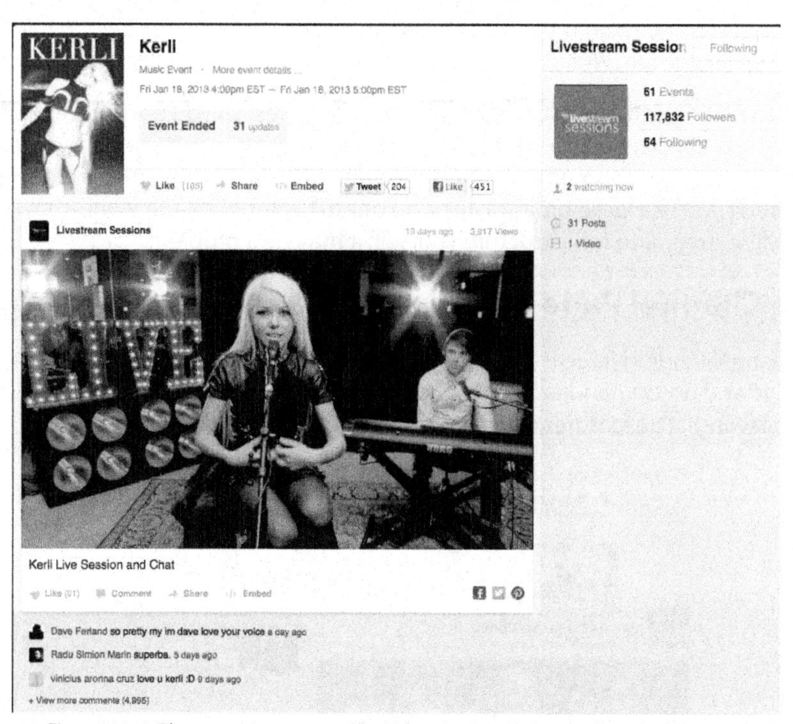

Figure 11-4. The new Livestream. Note the 4,995 comments during the event.

Although Livestream's old offering (shown in Figure 11-1) is very similar to that of Figure 11-3, with its new product, the company is attempting to change the paradigm with an event-oriented longitudinal presentation that includes pictures, video and chat from before and after the event. This is shown in Figure 11-4, with a long page full of pictures, chat, and comments with the actual concert just another stop in the stream (you can visit the stream at **bit.ly/Livestream_Kerli**). Real-time viewers can watch the band as it sets up, then watch the pre-concert chat and then the concert, and then share in the post-concert cool-down and break-down. On-demand viewers can share in any or all of the experience.

Imagine a training session hosted by your organization, or the Sunday sermon. Under Livestream's new paradigm, the live video is presented with pictures of the rapt audience, the guest speaker chatting with the CEO after the discussion, and tweets and comments from remote listeners. The experience is much more rich and nuanced for both live and on-demand viewers.

That's the Livestream vision, anyway. Whether it's the old paradigm or new, when you create your channel page, there are lots of switches to set and lots of controls to configure. Most but not all LSSPs let you customize your landing page with personalized headers, logos, overlays, colors and other options. Beyond these appearance-oriented issues, the next concern for many enterprises relates to access to content.

Can I Protect my Content and Brand?

For example, using the options shown in Figure 11-5, Justin.tv lets you password-protect your videos, prevent others from embedding your content, and remove your channel from the directory so only viewers you send to the page will find the content. (Note that these controls are actually on two different screens on the Justin.tv site.) Most LSSPs offer some versions of these; some also let you identify which URLs can embed your videos for more fine-tuned control.

Your channel is public. Adding an access code will make your channel private. This will make all your past broadcasts inaccessible to anyone without the access code. This will not take effect until you stop broadcasting. Access code must be 25 characters or less.

Add Access Code:

Control how your viewers use your video:

- Don't allow my channel to be embedded
- Don't allow anyone to view my past broadcasts
- Don't allow users to create highlights from my past broadcasts
- Automatically add highlights of my past broadcasts from viewers to my list of videos
- Remove my channel from the directory

Figure 11-5. Access protection and other ways to make your content more private on the Justin.tv site.

Other potential causes for concern are comments and chat. Some LSSPs allow you moderate all comments, so you can prevent spam or negative comments from appearing. Regarding chat, some LSSPs let you identify multiple moderators to monitor chat, while others let you block certain words or even certain chat participants (Figure 11-6). If you have concerns about how potential viewers can access and use your content, investigate each LSSP candidate's capabilities in this regard early in the process.

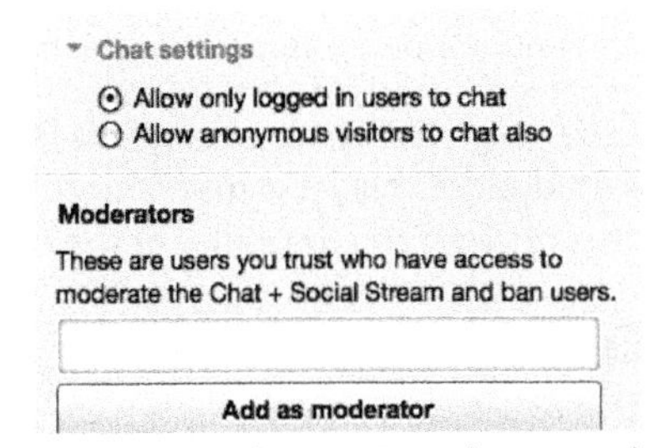

Figure 11-6. Ustream lets you turn chat on and off and identify moderators who can ban abusive users.

How Extensive are Social Media Links?

All LSSPs provide links to social media sites like Facebook and Twitter so you can easily post a link to the video to these services when you go live. In addition, Livestream does a great job integrating Facebook and Twitter chat into its live presentations in addition to its own native chat facility. That way, comments posted via Facebook and Twitter appear on its viewers' walls or pages, increasing the potential buzz around the event.

It's also useful to consider what visitors to your page see when you're not broadcasting live. In this regard, all LSSPs present libraries of previous broadcasts among which your visitors can select and play. Ustream lets you create playlists of videos that appear when a visitor opens the channel page so you can control your message to all visitors. Livestream lets you build sophisticated presentations from previous broadcasts, uploaded on-demand files and even files imported from YouTube to create rich presentations to engage your viewers.

If you're selling products on iTunes or Amazon, one great feature offered by Ustream is the ability to use "extensions" to post links to these products under your live and on-demand videos (Figure 11-7). Folks watching the video about your brand-new widget can then click on Amazon and buy it, another feature you can use to help monetize your video.

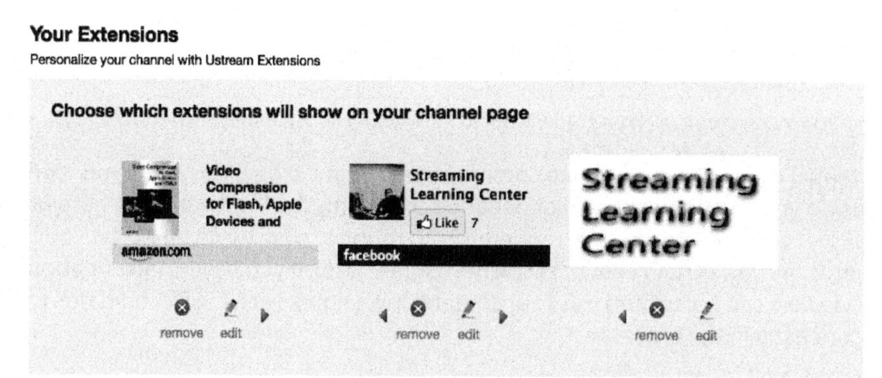

Figure 11-7. These Ustream extensions provide e-commerce links to products that I'm hoping to sell.

So far, of all the features discussed involve the channel page on the LSSP's site. For many broadcasters, the features of the embedded player is much more important. So let's spend some time looking at these features, starting with the flexibility of the embedding options.

How Flexible are My Embedding Options?

Embedding a live video stream from an LSSP is very similar to embedding a YouTube page into a website, or a video from another UGC site or online video platform. You can see this in Figure 11-8, from LSSP Bambuser, which does a nice job allowing me to embed a compact player with access to my video library and social media links. On the right, you see there are two configurations—expanded (shown) and integrated—and that you can customize the size of the player. Then you copy the embed code shown just below these options, and paste it into the HTML in your website.

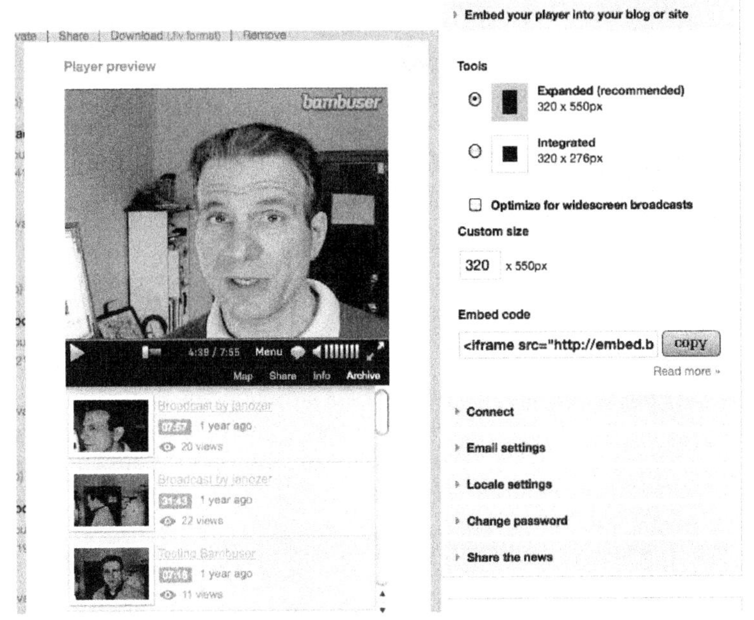

Figure 11-8. Customizing the Bambuser embedded player.

Bambuser does a nice job with its embedded player because its target customers are those who primarily want to use the embedded player, not the channel page. Sites like Ustream and particularly Justin.tv, which see their primary value in the ability to deliver eyeballs to your channel page, offer a different range of options. For example, Ustream lets you include libraries of content in your embedded player, but when users click to play these videos, they jump back to your Ustream channel page.

In addition, some sites, including Livestream and Ustream, let you embed video playback into your Facebook page or wall, which is great if you're trying to draw viewers to your Facebook pages. Many others simply let you post links to the video in Facebook, which may not be as effective.

Where does the Video Play?

Another consideration involves which platforms the videos play on and at what cost. Of course, all LSSPs offer Flash-based playback on the desktop for both the channel page and embedded players. Once you consider devices, however, compatibility is all over the map, and may be different for the channel pages and embedded pages.

For example, Bambuser's major value proposition is mobile support for both playback and broadcasting. However, while its channel page played fine on my iPad, the embedded player didn't even appear on my iPad. Another example is Justin.tv, which relies primarily on apps to play back on mobile devices, rather than browser-based playback.

The compatibility-related permutations are too complex and fluid to present in a features table. Before choosing a provider, you should sign up for free accounts at the LSSPs you're considering and test the playback compatibility of the channel page, embedded player, or both across the relevant matrix of players and platforms.

What are My Encoding Options?

A few points here on encoding options. First, a quick word on the browser-based encoders offered by all LSSPs. Basically, they use the Flash Player's older codec, VP6, which is much lower-quality than H.264 and isn't compatible with mobile devices. So either use a different application provided by the LSSP or a third-party product.

Second, you can use commercial encoders with most LSSPs, but in some instances, you may not have access to the entire feature set. For example, although you can send multiple streams to Livestream via most of the company's own software and hardware encoders, you can't from many third-party programs. So if you plan on distributing multiple streams, make sure you understand which encoders can support that.

Third, Livestream in particular has been very aggressive in offering its branded encoding products for its user base, including the on-camera Livestream Broadcaster (**bit.ly/Ozer_broadcaster**), which is much cheaper than most competitive products. For those seeking a mixer/encoder, Livestream offers the Studio HD500 (**bit.ly/Ozer_HD500**). At some point, Livestream will offer a software-only version of the Studio program, although pricing has not yet been set. Obviously, these products integrate well with the Livestream service, and often are much more affordable.

In contrast, Ustream sells custom versions of Wirecast (with one free edition) that enable multiple-camera switching. Ustream does a nice job certifying third-party products with its platform (**ustream.tv/platform**), including the very cool Logitech Broadcaster, a webcam that can also encode your video and transmit it directly to Ustream, removing your computer from the equation (**bit.ly/Ozer_logitech**).

As mobile devices gain more power and higher-quality optics, the ability to broadcast live using your mobile phone becomes much more important. Most mainstream LSSPs offer Android and iOS apps that provide this capability, but smaller providers may not. Overall, before choosing an LSSP, it's worth reviewing the encoding-related products they offer or support to ensure your encoding needs will be met.

Are Ancillary Services Available?

Larger LSSPs like Livestream and Ustream offer a full suite of production services—including camera, mixing, encoding, and player design and development—to make sure that your events go off without a hitch. If you don't have these capabilities in house and have a high-profile event in the works, these capabilities are certainly worth considering.

What's it All Going to Cost Me?

And finally, cost. Although there are some similarities, each LSSP has its own unique pricing model. For example, as shown in Figure 11-9, with the new Livestream service, there are never any advertisements, and the free service can stream unlimited HD video to unlimited viewers, one event at a time (**bit.ly/LS_pricing**). All broadcasts are converted into on-demand video files that can be embedded into your own website. However, videos expire after a month, and your viewers have to be logged in as Livestream users to view the event.

For $42 a month, your archive of past broadcasts is preserved, viewers can watch without logging in, and you get Google Analytics integration, but you can't embed the live video into your own website; your viewers have to come to Livestream to watch. With the Premium account ($333 a month), you can embed the player into your own website and Facebook, disable social media and get full phone support. You still don't have a completely white-label player, but you're getting pretty close.

Compare Plans	Free	Basic	Premium
Price (yearly subscripion prices shown)	$0	$42	$333
Unlimited Ad-Free Streaming	✓	✓	✓
Unlimited Events (1 live at a time)	✓	✓	✓
Unlimited Viewers	✓	✓	✓
Unlimited Email Support	✓	✓	✓
Unlimited On-Demand Embedding	✓	✓	✓
Unlimited Archive	-	✓	✓
Local Publishing Point *Streams based on your location for increased reliability and quality	-	✓	✓
Viewers can watch without logging into Livestream	-	✓	✓
Event/Account Vanity URL	-	✓	✓
Google Analytics Integration	-	✓	✓
Ability to disable Event Viewer Count, Post Viewer Count, and Comments & Chat	-	-	✓
Livestream on Any Website *Unlimited Embedding of the Live Event Page and Live Player	-	-	✓
Livestream on Facebook	-	-	✓
7 Days/Week Phone Support	-	-	✓

Figure 11-9. Pricing for the new Livestream.

In contrast, Ustream offers the more traditional pricing model where you pay to lose third-party advertising, gain your own branding and acquire additional viewer minutes, plus additional embedding control (Figure 11-10, **bit.ly/US_pricing**).

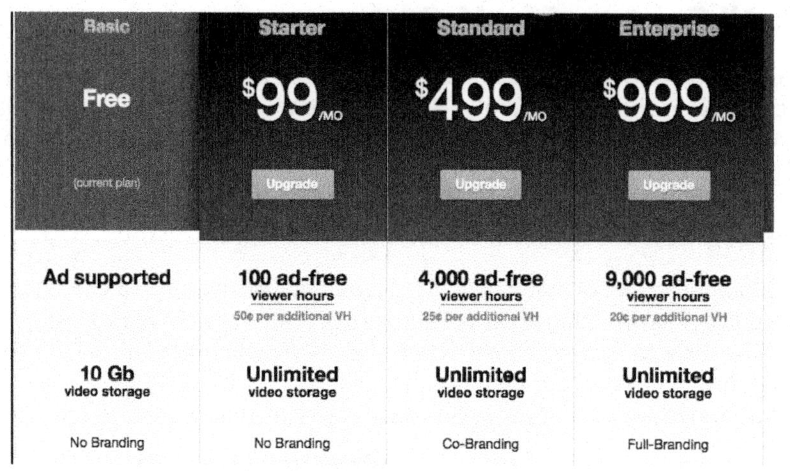

Figure 11-10. Ustream's pricing model.

Bambuser offers a similar model that you can view at **bambuser.com/premium**, where you pay for viewing hours, API access and customization options (Figure 11-11). You can also get multiple accounts, which allows you to broadcast multiple streams simultaneously. With the other services, you'd need multiple accounts (or an unlimited account) to accomplish this.

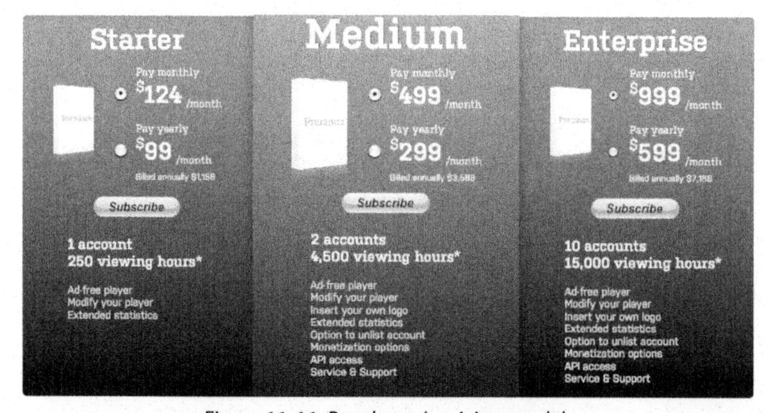

Figure 11-11. Bambuser's pricing model.

To close the loop on the LSSPs we've been describing, Justin.tv recently began offering a premium broadcasting service that offers live transcoding (Figure 11-12, **bit.ly/JT_pricing**). Note that transcoding on the desktop means manual switching, while custom HTTP streaming means automatic switching via Apple's HTTP Live Streaming (HLS) within the browser, not the Justin.tv app. You also get multiple accounts with the Storm, Blizzard and Hailstorm plans, again enabling simultaneous streaming, although there are storage limits that other LSSPs don't impose.

Remove all the ads from your channel and give all your viewers the full ad-free experience!	FLURRY $99/MO FOR 100 VIEWER HOURS SELECT	STORM $199/MO FOR 500 VIEWER HOURS SELECT	BLIZZARD $399/MO FOR 3000 VIEWER HOURS SELECT	HAILSTORM $999/MO FOR 9000 VIEWER HOURS SELECT
	$0.30 per additional viewer hour	$0.25 per additional viewer hour	$0.20 per additional viewer hour	$0.15 per additional viewer hour
STORAGE	10 GB ($0.30 per additional GB)	100 GB ($0.30 per additional GB)	500 GB ($0.30 per additional GB)	1000 GB ($0.30 per additional GB)
TRANSCODING	N/A	N/A	Live Transcoding	Live + archive Transcoding
BRANDING	Co-branding	Co-branding	White label	White label
IPAD/STREAMING	Justin.tv app	Justin.tv app	Custom HTTP streaming	Custom HTTP streaming
PRO USER ACCOUNT	1 Pro user account	2 Pro user accounts	4 Pro user accounts	6 Pro user accounts

Figure 11-12. Justin.tv's pricing model.

Before Justin.tv started offering these programs, the website wasn't relevant for most corporate broadcasters because there was no way for the publisher to eliminate the advertisements. With the new plan in place, it's definitely worth a look.

If you're running lots of events, you'd better do the math, because these plans can get quite costly, although you end up with a fully white-label player with only your own branding.

Try It—You'll Like It

Hopefully, the foregoing identified the questions to ask yourself when comparing LSSPs. From here, the best course is to narrow your focus to two or three LSSPs and then give their free services a try.

The more time you invest up front identifying your unique requirements, the smoother these trials will proceed and the more effective they will be in helping you select the best service. This includes determining if you hope to monetize your videos, how you need to protect the content, whether you want single or multiple streams, whether you care more about the channel page or embedded page, and which playback platforms are critical to your live streaming offering.

A Live Event Walkthrough

We just had a look at the various factors you should consider when choosing an LSSP. Now let's take a quick technical walkthrough of how they actually work. I'll use Livestream because it is the LSSP I've used the most for my live events.

1. ***Create the event.*** The first step is to create and schedule the event, which notifies your followers that an event is coming and gives you a page to send potential viewer to.

A short wizard walks you through the steps: first naming the event (Figure 11-13), then setting a date and time, and uploading a photo to serve as the poster for the event. You'll see the picture a few screens from now in Figure 11-19. Note that you can create an event and stream live from a computer, or from an iOS or Android device via Livestream apps.

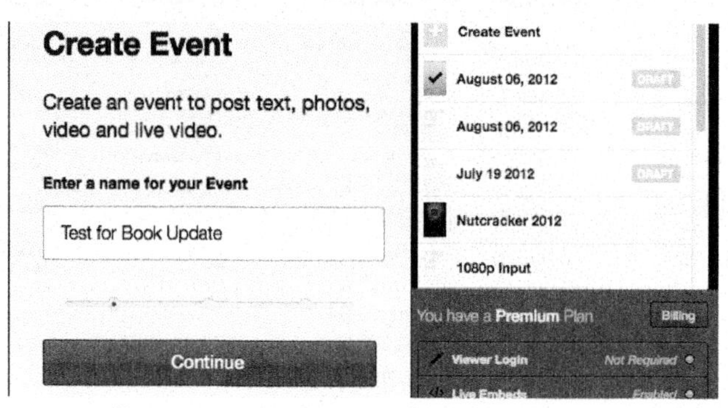

Figure 11-13. The first step of creating the event.

2. ***Configure your encoder.*** A few minutes before the event, you want to start the live stream. I published this one via Livestream for Producers, which is the free, cross-platform (Mac/Windows) program you can download from the Livestream site. In the program, you first choose the event so the stream knows where to play.

Figure 11-14. Choosing the event in Livestream for Producers.

3. ***Choose your encoding templates.*** Livestream offers a number of templates at different data rates, including single stream, multiple stream, and multiple stream with a mobile stream. Typically, you choose the highest combination that your outbound bandwidth will support. With my crummy outbound DSL, I can't support streams over the Normal quality, so that's what I use. I've produced multiple events that use the High + Medium + Mobile stream, which provides good coverage with a good-quality stream for those watching on platforms with sufficient power and bandwidth.

Note that you can customize the presets, but I never do. The various producers and encoding gurus at Livestream (and Ustream) have produced many more live events than I have, and I trust their judgment.

If you have multiple video devices on your computer, you select the active device via the drop-down list beneath the Go Live button (currently showing Display iSight). If you have multiple audio devices, click Preferences to choose the correct device.

Figure 11-15. Choosing your encoding templates.

4. ***Start the stream.*** When you're ready, click the Go Live button and you'll see the screen shown in Figure 11-16. Type the name you want to appear above the stream—I used the name of the event, but you can enter any title you would like. If you have multiple publishing points associated with your account, you'll see the IP Address field, although most users have only a single publishing point, so they won't see the field. If you do, note that the software automatically chooses the best publishing point, so you shouldn't have to adjust this in any way.

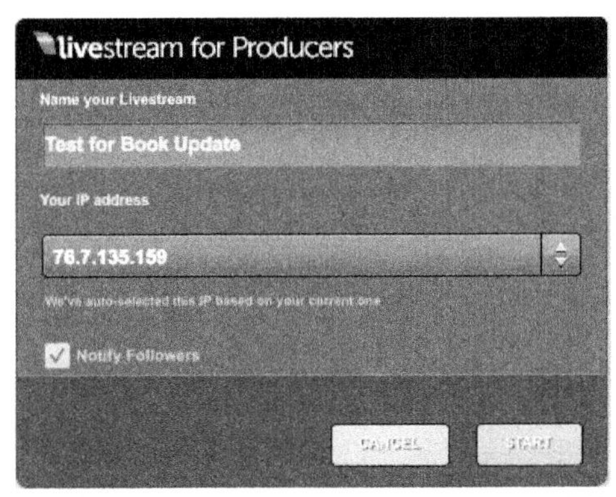

Figure 11-16. Type the event name; don't sweat about the IP address, which should be correct.

5. Click Start to go live. Careful what's on camera, since you'll be streaming live once you click Start. It's a good idea to start streaming to a different test event to get your various configurations nailed, and then cut over to the real event once you have. On the other hand, if you're showing the band setting up and checking the sound, some viewers might be interested in watching.

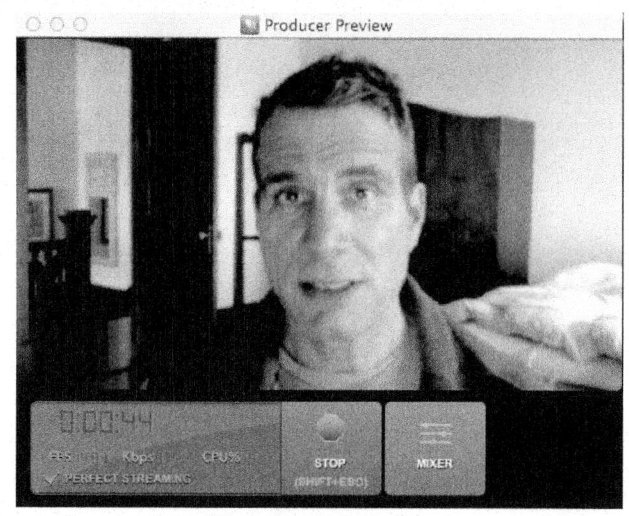

Figure 11-17. Going live.

Once you start to stream, you'll see the controls shown in Figure 11-18. I'm streaming from my bedroom Mac, since that has the best webcam, although I definitely don't recommend a background like that for your live events. Sorry about the bed in the back; I guess I shouldn't (wait for it) air my dirty laundry in public.

6. Adjust audio volume. You almost certainly will have to adjust audio volume via the Mixer, accessed by pressing Mixer on the bottom right in Figure 11-17. I cover factors to consider when controlling audio volume in Chapter 13. It's not exactly as simple as you might think. For example, should you stay clear of the red gauges between +6 and +12 dB? Shoot for 0 dB like you might when editing audio in an audio editor? I didn't know either, so I asked the Livestream producers and share the results in Chapter 13.

Tip: *Note that audio is exceptionally tough to adjust in a loud auditorium, even with headphones on. It's always good to have someone watching the live stream in a quiet place who can give you the real skinny about what's going on with the audio.*

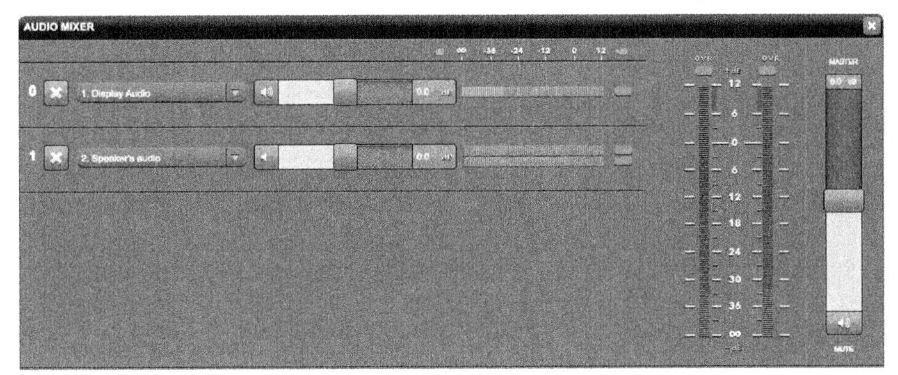

Figure 11-18. Adjusting audio volume.

7. ***Monitor the live signal.*** Go to the Livestream site (or the embedded player on your own website) and make sure things are going swimmingly. You should always have the Livestream page up on a computer while broadcasting, since this is the quickest way to sense any problems. If you're encoding on a computer (as opposed to an on-camera or hardware device), bring a notebook along to monitor. It's absolutely essential.

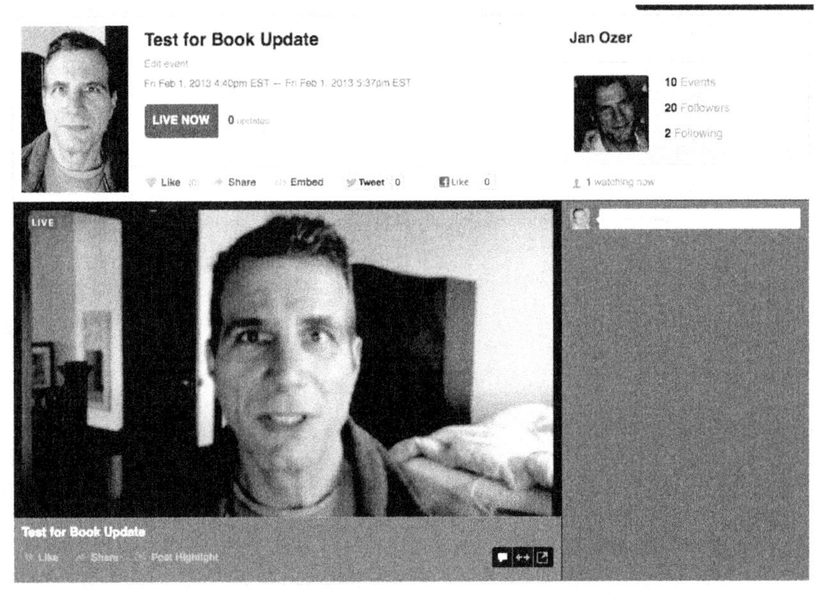

Figure 11-19. Monitoring the live stream.

So that's it. The high-level point is that producing a live event, irrespective of which LSSP you choose, is very simple. Basically, if you have a camcorder or event mobile phone/tablet computer at an event, there's really no reason not to stream it live.

Now let's touch on a separate but related market called rich media communications.

Rich Media Communications

The rich media communications category includes presentations with PowerPoint slides and other graphics, as well as chat, polling, Q&A, and other similar features. If you're running a conference or training seminar and want your live video stream to incorporate all the elements actually used by the presenter, plus viewer interactivity, you'll need a rich media communications system.

As an example, one of the most prominent rich media communications companies is Sonic Foundry. Figure 11-20 is a screen grab a presentation I gave using its flagship product, Mediasite—say hello to Sonic Foundry VP Sean Brown in the picture-in-picture. You can view this presentation, titled "Encoding Best Practices for the Enterprise," at **bit.ly/Ozer_SF**. This is an archived presentation, and you can see some of the various rich media components: video on the bottom right, clickable links to different slides on the bottom left, and slides dominating the frame. The icons on the lower right provide access to additional information, resources and email link. In a live presentation, there would be windows for chat and polling. Overall, it's a pretty impressive presentation for training, corporate communications and the like.

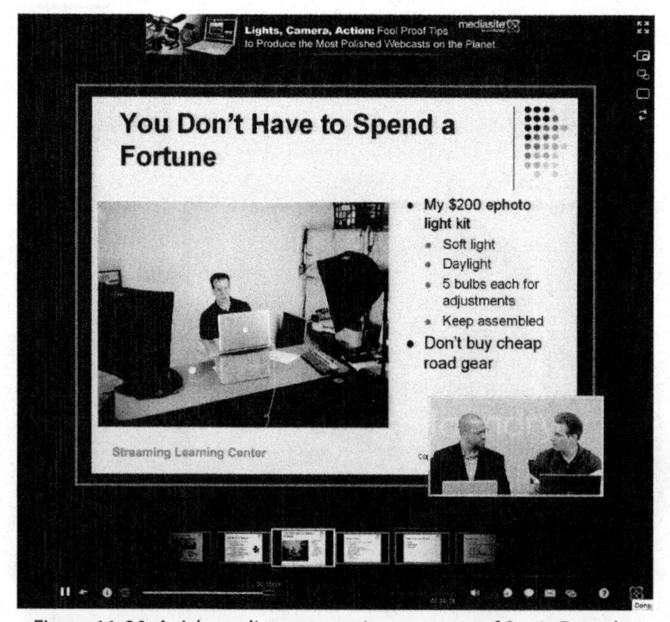

Figure 11-20. A rich media presentation courtesy of Sonic Foundry.

There are three prominent players in this space: Polycom (formerly Accordent), Sonic Foundry and MediaPlatform. Although the technologies seem to do the same things—mix PowerPoint with video—they have unique focuses. Here's a quick overview of each with a link to my review of the product or service.

- ***MediaPlatform WebCaster.*** A software-only solution—either SaaS or behind your own firewall—for producing live events incorporating PowerPoint, video, polling and quizzes. The system manages the entire event process, from invitation to analytics, and can support multiple speakers at multiple locations with no hardware investment. It's suitable for both casual and frequent use (see bit.ly/MP_review, also excerpted at the end of the chapter).

- ***Polycom.*** Polycom RealPresence Media Manager is a very deep, enterprise-oriented system that integrates into your existing network and security infrastructures to transform the way you use video in the enterprise. It is not a casual product; it wants to change your life, and will require substantial resources and capex to implement (bit.ly/PC_review). For a screencam overview of the system, check out bit.ly/PC_tutorial. Note that when I reviewed the product, it was owned by Accordent and named the Accordent Media Management System.

- ***Sonic Foundry Mediasite.*** Mediasite is closer to Polycom, but more user-friendly and easier to implement. Unlike Polycom, Sonic Foundry has a division that will produce your events for you, which is great for those producing one or two major events a year. So if you want the rich media functionality during the event and after, Sonic will send gear and personnel to produce the event, and will host the content after the event is over.

With this as an overview, let's jump into factors to consider when choosing a rich media presentation system. Although I haven't reviewed the system, note that Qumu (qumu.com) also has multiple highly regarded products in this space.

Choosing a Rich Media Presentation System

Again, there will be a great deal of similarity between all the systems at a very high level. That is, there will be an environment for announcing events; customizing the look and feel of the player; integrating PowerPoint, video, polls, quizzes and other content; and reporting on who watched what and for how long. Beyond this, there will be a long list of relevant differences that these questions will help you ferret out.

At the end of this chapter, I'll share the story of a presentation that I gave with MediaPlatform WebCaster, which not only details how to use the technology, but also gives tips about producing the best possible webcast with the largest possible attendance. Unfortunately, you'll be learning from my mistakes, since I did neither, but there you go.

Where Does the System Live?

Some products are available for purchase and installation within the enterprise firewall, while some are available only as a software as a service (SaaS). There are also hybrid products that incorporate purchased software and a service aspect. If you have a strong preference either way, be sure your candidate products meet that preference.

Internal or External Focus?

Some products, like Polycom RealPresence, are designed for heavy integration into the enterprise with an obvious focus on enterprise viewers. Others, like MediaPlatform WebCaster, are more targeted toward external viewers, with a range of features discussed in a moment that help you gain attendees from external sources. Obviously, you want to acquire a system that matches your own needs.

How Integrated is the System?

There are multiple aspects here, including network, authentication, and distribution integration, which is primarily of interest to organizations seeking an enterprise delivery system as opposed to an outward-facing system.

On the network side, if your focus is enterprise distribution, you need to understand how the system will integrate with your existing infrastructure. For example, Polycom RealPresence's viewer routing feature lets you route groups of viewers by location (called Location Profiles) to servers that are closer to their location, like an edge server that's at their facility. For example, in Figure 11-21, the Laptop location profile is sent first to the 2k3-Edge1-Acc Media server, with failover to the AMMS – Acc Media server. This improves the quality of service for the viewer because the data is closer, while decreasing WAN traffic and reducing bandwidth costs (if any) associated with video playback.

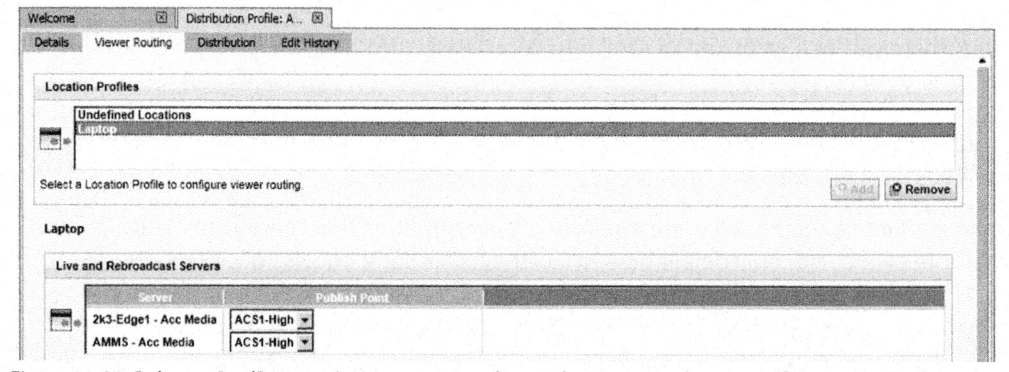

Figure 11-21. Polycom RealPresence's viewer routing directs the viewer to the most efficient source of content.

Another feature of the Polycom system is the ability to distribute content to the servers to which you're routing your viewers—no sense sending the viewers to the server if the content isn't there. In this regard, Distribution Profiles let you create rules that control which content gets delivered to (and later removed from) the various on-demand and live streaming servers in the enterprise's network delivery infrastructure, including edge servers from Blue Coat, Cisco, Riverbed and Microsoft. This feature ties together disparate network resources and automates a manual distribution task that would otherwise be exceptionally time-consuming.

On the authentication side, you'll want to make sure your candidate systems work with existing network authentication databases like LDAP or SAML; otherwise, you'll have to create and maintain two databases. On the distribution side, if you have an existing portal, you'll want to make sure your candidate products and services can work within these portals, both to present content to your community and to help deliver it efficiently.

What's the Cost?

Big differences here, particularly between systems (like Polycom RealPresence and Mediasite) that use their own dedicated hardware capture devices and others (like MediaPlatform WebCaster) that use the software-only approach. You should definitely understand your intended usage pattern before you begin shopping, and make sure you understand how pricing works for your intended usage. For example, if you want to use the system to create a portal for enterprise viewing, but have multiple divisions to serve, ask how many portals the base system can create.

Also ask about how you can scale should usage grow beyond your projections. If the system is built upon proprietary technology, you may need to purchase additional units to scale. If it's built upon standard database and OS technologies, like Microsoft SQL, you may have the databases in-house and the expertise to scale yourself.

What are the Video Specs?

From a video perspective, HD isn't such a big deal, since most of the time the content will be watched in a small window. However, adaptive streaming clearly is, since you'll likely be attempting to reach viewers connecting via a range of connection types, from 3G to LAN.

What Content Does the System Work With?

As mentioned, all the systems should be able to incorporate video and PowerPoint into a presentation, as well as quizzes, polls and the like. For enterprise systems, determine if the system can actually administer and grade exams, which can be a huge benefit in heavily regulated industries where competence testing is a requirement.

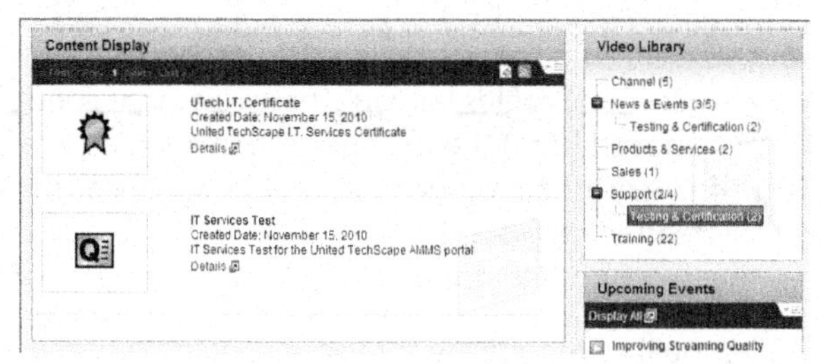

Figure 11-22. Polycom's system can administer tests and award certificates and certifications.

If you're a government, educational or related facility that needs closed captions, get that on the table early, since not all systems manage these to all target platforms. In Figure 11-23, which is the Sonic Foundry Mediasite presentation configuration screen, you can see that Mediasite can handle closed captions, polling and links.

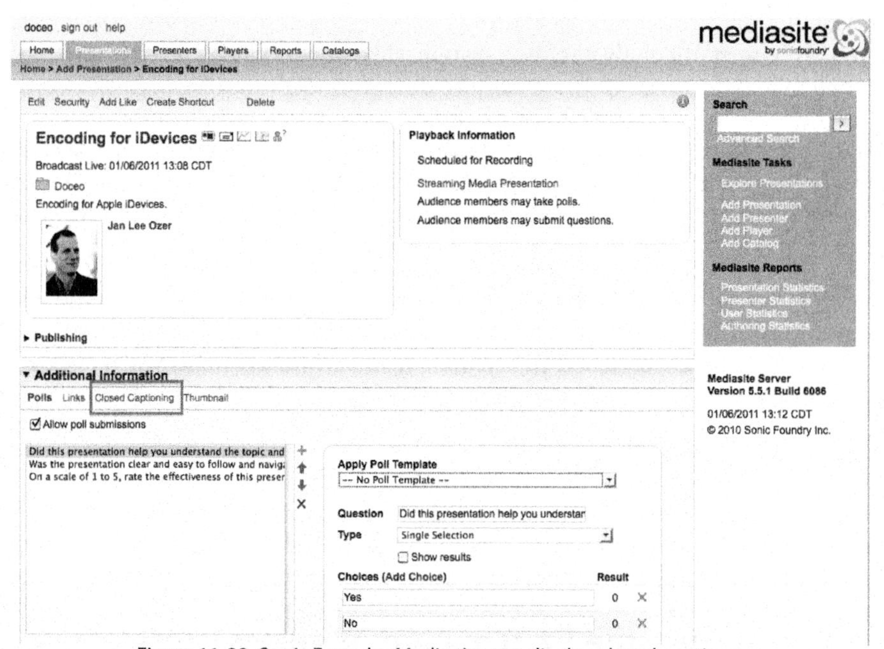

Figure 11-23. Sonic Foundry Mediasite can display closed captions.

While we're on the subject of geeky network stuff, it's a good time to ask which protocols the system can use to distribute content within the enterprise. Using available network resources is nice, but technologies like multicasting and peer-to-peer can be even more efficient in some distribution environments.

Which Player Technology and Where Does It Play?

The two predominant desktop players are Flash and Silverlight. Both are fine for the desktop environment where you can dictate which players are installed on the system, but I prefer Flash for primarily outward-facing systems. If you're looking at a new system, be sure that features like chat and links to social media sites are available, as well as the ability to post content like slides for viewers to download.

Obviously, mobile playback is important and will only become more so. You should understand which mobile platforms each candidate system will support and how that support compares feature-wise with the desktop player. For example, does it duplicate the experience completely, with polling, Q&A, adaptive streams and the like, or is it solely playback within a window?

Can the System Help Me Get Attendees?

This will differ for enterprise and outward-facing systems. On the enterprise side, with the Polycom RealPresence system, once an event is scheduled, the event appears in the Upcoming Events section of the Polycom RealPresence portal (Figure 11-24). Once the event goes live, viewers can click a link in the portal to view the presentation. Viewers can also search through content trees for other content.

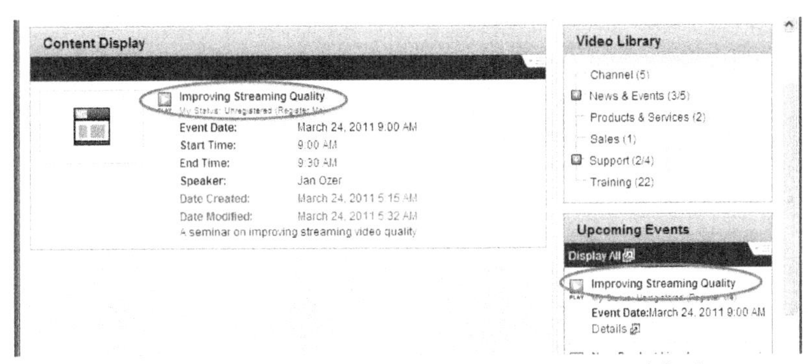

Figure 11-24. A scheduled event appearing in a user's portal display.

As you'll see in the closing story below, the MediaPlatform system provides a complete outward-facing lobby system to help users sign up for your webinars, with an email system for sending reminder and thank you emails, and the ability to send Outlook and iCal reminders.

Does the Reporting Meet Your Needs?

Again, this will differ depending upon whether your system is primarily inwardly or outwardly focused and will vary greatly by vendor. The Polycom RealPresence system collects almost scary amounts of detail, including reports that identify which videos a particular user watched and for how long, and which tests and surveys they took and which certificates they earned. Again, in a heavily regulated environment, this information can be crucial. The Polycom system also gets down and dirty with network statistics that only an IT person could love, including playback metrics like bit rate served and packets lost by portal.

Figure 11-25. A presentation report showing which sections of the presentation were viewed most.

Sonic Foundry breaks its reporting into five categories, as follows:

- ***Presentation statistics.*** Provides data showing how many of your presentations are being watched and by whom during any given time period.

- ***Presenter statistics.*** Provides data showing how many of a presenter's (or a group of presenters) presentations are being watched and by whom.

- ***User statistics.*** Provides data showing a specific user's (or a group of users) viewing activities over any given time period.

- ***Authoring statistics.*** Shows the presentations authored as a whole and by each presenter or groups of presenters.

- ***Server statistics.*** Provides data on usage and utilization of the server.

Reports can all be customized by presentation, presentation folder, presenter, user or IP address, and all combinations can be saved as custom reports. One Mediasite report that particularly caught my eye was titled "Presentation Viewing Trends" (Figure 11-25), which shows how long individual viewers watched your presentation, with a color-coded bar showing which sections the viewers visited most often. Color coding can be particularly useful with on-demand presentations, since viewers can choose which content to watch and often don't watch linearly. Like most vendors, Mediasite allows its reports to be exported to Excel for further massaging.

While each rich media platform offers literally hundreds of features, these are the highlights that should be considered when comparing the various systems.

As (I hope) you can tell from the very frequent bit.ly references to articles I've written (over 100 so far), I've resisted the urge to use these articles en masse, and most of this book is new or significantly updated information from the first book.

But I've decided to add the MediaPlatform article largely intact, since it captures the technical workflow of producing a live webcast, as well as many of the ancillary considerations, like acquiring attendees. You'll learn how to produce a webcast with the MediaPlatform system, which will be similar to most of the other outward-facing systems. I hope you'll get a sense that the technical side is fairly simple, so that the marketing is really the critical element.

The review is about 18 months old, so I apologize in advance if the screens are outdated, although from demonstrations I saw at Streaming Media West in December 2012, the general workflow is still similar. You should also verify the pricing data shown at the end of the review, which is shared here for historical purposes only.

So, without further ado, I bring you:

MediaPlatform WebCaster—Review and Webinar

Webcasts are a great tool for acquiring and retaining customers, as well as communicating to and training employees and partners. If you're considering presenting webcasts for any of these reasons, MediaPlatform WebCaster should be on your short list of platforms or platform providers.

Why? A number of reasons. First, the basic webcasting functionality is comprehensive, and can include live streaming audio and video, on-demand audio and video, PowerPoint slides, Flash animations, and screen demos. The system also enables audience-participation features like polling, surveys, and Q&A.

The system itself is available as a software as a service (SaaS) for companies that don't want the hassle of installing and maintaining a webcasting server, but it can also be licensed and installed onsite for behind-the-firewall usage. With the SaaS model, your webcasts are delivered by Akamai or other CDNs, ensuring highly scalable delivery, plus the licensed system can efficiently deliver within corporate networks via Adobe's multicast fusion. You also get rare features that only a cloud-based solution can inexpensively provide, like the ability to integrate audio and video feeds from multiple speakers at different physical locations.

Plus, the system is template-based and totally easy to use. How do I know? Because to test WebCaster, I produced my own webcast—"Encoding Best Practices for the Enterprise"—from soup to nuts (**bit.ly/WC_best**). What, you missed it? Well, although I won't share the pitiful attendee numbers precisely (only 20 attended), let's just say that missing it did put you in the overwhelming majority of the world's connected population.

Although I've been the speaker in plenty of webcasts, this was my first time as producer, and some self-induced rough edges during the webinar made me glad that more people didn't attend. Still, my experience does highlight that having the technical ability to produce a webcast—which MediaPlatform provides admirably—doesn't guarantee bountiful attendees or a smooth webcast; you're in charge of both the latter activities.

So in this review, I'll devote 98% of my attention to how WebCaster works, and 2% on the marketing and performance aspects of the webcast. You'll learn what WebCaster can do and how it works, plus pick up a tip or two about webcast production. Hey, no reason you shouldn't avoid the mistakes I made as a first-time producer.

Getting Started

Getting started in WebCaster is simple. You log in, then choose your project type. If you're working on an existing project, you click Open Existing. If it's a new project, you can clone an existing project, or create a new live or on-demand audio, or audio/video project. My project type was live audio/video, which you can convert to on-demand after the show.

Then you chose a template from the Theme Explorer. As you can see from the tabs on the window on the right in Figure 11-26, a template controls multiple aspects of the presentation, from the "Lobby," which contains information about the presentation, to the Registration page to the presentation itself (shown in Figure 11-26) to how the archived presentation looks.

Figure 11-26. Choosing a template from the Theme Explorer.

After choosing a template, you input the basic presentation data like name, time, date and duration, and upload any marketing-related graphics like speaker image and company logo. This gives you the basic information you need to save and publish the presentation, which gets it on the calendar and provides a URL for the Lobby and Registration page so you can start inviting attendees.

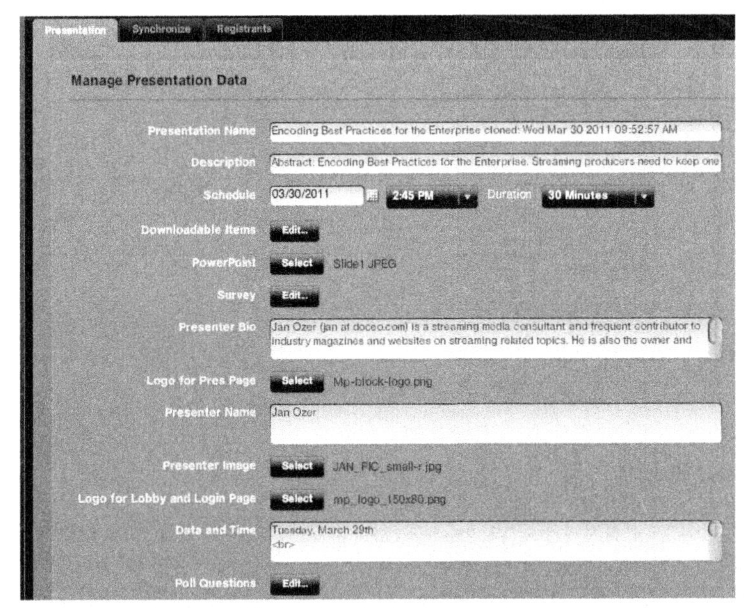

Figure 11-27. Entering the basic presentation data.

You can also upload PowerPoint slides at this time, and create polls and surveys (more on these later), but you can do all this at any point before going live with the presentation. Basically, you need enough information to create the Lobby page so you can start inviting folks, and that's it—the rest you can add later.

Note the Downloadable Items button that's four slots down from the top, which is where you upload and manage files to be downloaded by the viewers. My first rookie error was to forget about this detail until about 10 minutes before go time, when it was too late to upload the handouts—a definite bummer because they contained so many technical details. You definitely already knew this, but you should start thinking about what you want the viewers to download from the start.

Once you click Save & Publish, the system processes the uploaded information and creates the Lobby page shown in Figure 11-28. MediaPlatform could make it easier to format the description, perhaps with a WYSIWIG editor for adding bullet points and the like. You can do this, but you have to manually insert the necessary HTML in one long line of text entered into the Description text box in Figure 11-27. I'm sure I would have gotten more than 20 attendees if I was able to put the three major topics in bullet points so that invitees could easily discern the value in the presentation. Just kidding, but since this screen is the face of the presentation to the folks you invite, MediaPlatform should make it simpler to add basic formatting (and may already have done so).

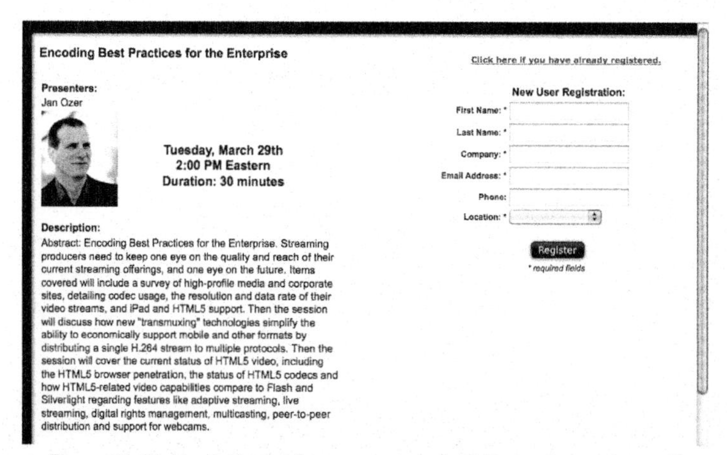

Figure 11-28. Here's the Lobby page, ready for folks to sign up (or not).

System Overview

Now that you've got your presentation set up, let's take a quick look at the WebCaster interface, shown in Figure 11-29. The tabs on the top left walk you through the various stages of the presentation, from Overview (shown in Figure 11-29), which provides links to the Lobby and Presentation pages; to Manage, where you manage presentation data and registration information; to Email, where you create emails to send to your registrants and production team; to Present, where you actually present your webcast; to Analytics, where you analyze the results.

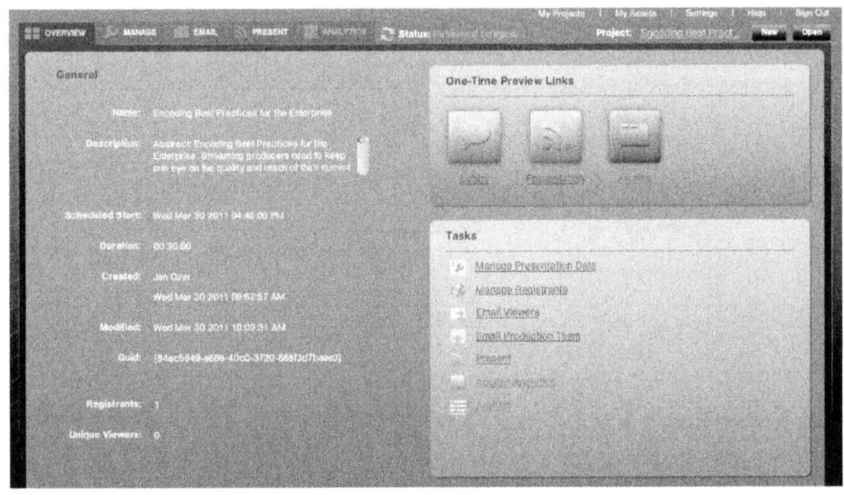

Figure 11-29. The control center for the WebCaster system.

Getting Registrants

Let's start with getting registrants. You can add users manually for small-scale internal events, with an API available for large-scale insertion, although the system isn't designed to be your first point of contact with your targets. Rather, the preferred workflow is to use your existing email system to invite your target viewers with a link to the Lobby page where they can sign up for the webinar. Once they enter their data, they're in the system, and you can easily communicate with them via email.

In this regard, WebCaster contains four editable templates for Registration Confirmation (shown in Figure 11-30). These include Outlook- and iCal-compatible reminders; a reminder note (Scheduled) for sending to registrants before the event; Missed Event, which sends the URL of the archived event to those who registered but didn't actually attend; and a thank you note with the archived event URL for those that did attend. The Registration Confirmation is sent immediately upon registration, and you can schedule the others within the system.

Rookie mistakes No. 2 and 3 were failing to include the Outlook/iCal reminder in the Registration Confirmation email, and failing to send an automatic notification sometime before the webcast. In my experience with other webcasts, about half the registrants actually attend, while my attendance was slightly below 33%. No doubt I could have improved this result by using the tools MediaPlatform provided.

If you're marketing your event through multiple sources, you can create a custom URL for each and track their effectiveness. The directional URL is the same, but the reference can be different and will be reported as such by the program. For example, the URL I used in the link from my website (StreamingLearningCenter, or SLC) ended with the phrase "?ref=SLC," as shown below:

http://a4.g.akamai.net/ ... /Lobby/default.htm?ref=SLC

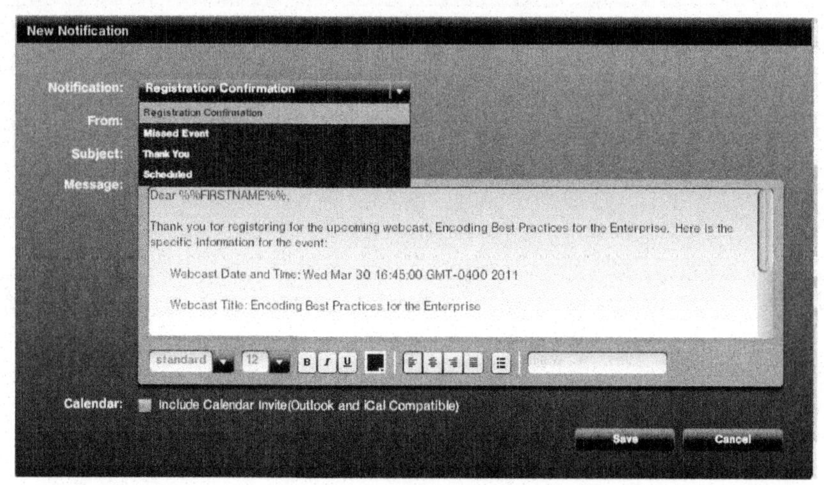

Figure 11-30. WebCaster comes with these four emails, which you can modify, or create your own.

I was fortunate to have two other organizations help announce the webinar—MediaPlatform and Wowza—and each had its own custom reference in their Lobby URL. You can see how this appears in the registration form in Figure 11-31; obviously, this information is critical in determining which sources produced the best return.

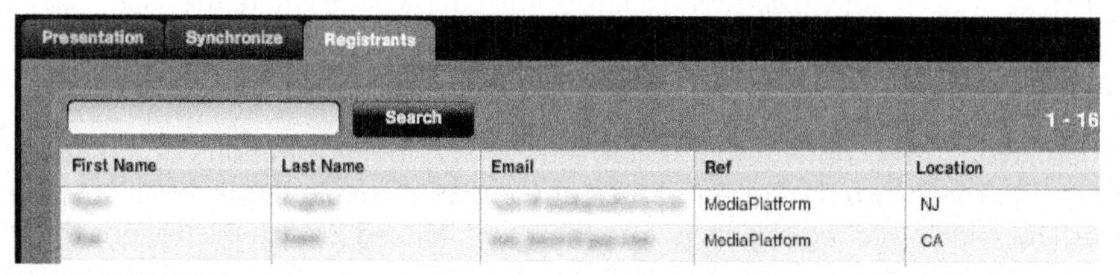

Figure 11-31. The Registrants screen let me know which registrants came from which sources.

Team Building

Although this wasn't relevant to my presentation, WebCaster also includes the ability to create production groups and easily communicate with them via email. You create the group as shown in Figure 11-32, then name the group, choose their functions and permissions, and add emails on the bottom. Once the group is saved, you can email all members simultaneously by selecting that group and creating your message.

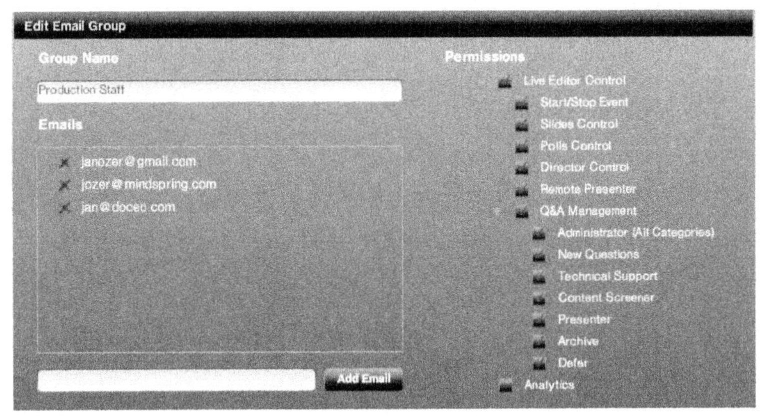

Figure 11-32. Creating a production group for easy communication.

During the event, the assigned permissions will determine which functions the group can perform. For example, if you want to restrict access to the analytics for some groups, make sure the Analytics checkbox is empty for those groups.

Staging the Event

Let's spend a moment on the physical setup for the event. I had decided to use my MacBook Pro to produce the webcast because its webcam and internal microphone work well for web conferences, and because its 3.06 GHz Core 2 Duo CPU was more than sufficient for the 320x240 stream I was going to broadcast. I also wanted to keep things simple, so a separate camcorder and lavalier system, which I definitely could have used, were out. Using a webcam, however, meant that I would have to be very close to the camera and microphone and would need lots of light.

As you can see in Figure 11-33, I have one light blue wall in my office that I keep blank to use as a background for video shoots. I keep an inexpensive light kit (the ePhoto INC SFT2KIT, under $200 at Amazon) with two 1,000-watt soft boxes assembled and ready to go for these occasions.

The kit uses 5500k compact fluorescent bulbs that you can control individually if you're trying for a shadowed, three-point lighting look. I was trying for flat lighting, so I powered all the lights on in both soft boxes and placed them equidistant at about 2:00/10:00 location just slightly above my head and pointed down. Technically, I should also have a back light shining down on my head and shoulders, but the contrast between my solid dark blue shirt and solid light blue background was sufficient without it.

Those solids were by design, of course, since pinstripes and other fine patterns quickly turn to mush under low-bandwidth, low-resolution compression. I also didn't wear my black turtleneck because the contrast ratio, or the difference between the brightest brights and darkest darks in the frame, would have been too high for the webcam and compression technology to preserve.

As it was, as you'll see in Figure 11-35, the blue shirt was a bit too dark, and turned into a dark blob (but at least an artifact-free dark blob) during the webcast.

Figure 11-33. Here's my physical setup, including the fancy notebook stand.

Returning to the scintillating narrative, I also muted all my phones and turned the heating way down so it wouldn't start up loudly in the middle of the webcast. Couldn't do anything about the dogs in the neighbor's backyard, so I crossed my fingers and hoped for the best.

The Software Side of Things

By default, the WebCaster application uses the webcam interface from the Flash Player to encode the audio and video. In my conversations with MediaPlatform, the company said that I could get higher-quality results by using the free Adobe Flash Media Live Encoder (FMLE), so I did. To tie WebCaster and FMLE together, I clicked the Settings button on the left side of the WebCaster interface (see Figure 11-34), which opened a window containing the Flash Media Server URL to plug into FMLE, which I entered in the FMS URL shown on the right of Figure 11-34. Note the stream name "Webcast," which I'll refer to again in a moment.

Figure 11-34 also shows the encoding parameters I used for the webcast, as recommended by MediaPlatform. I would have preferred to use the H.264 codec, and to produce at 30 fps, but first time out, I was went with MediaPlatform's recommendations. Note that audio was a frugal 48 kbps mono, which is really all that you need for a talking head webcast.

Figure 11-35 shows the MediaPlatform WebCaster command center, which has separate widgets for each type of content or interaction, like the PowerPoint slides or Viewer Questions. Starting to the left of the Video 1 window, note the Settings button that I referred to a moment ago, which contains the Flash Media Server credentials, and the Test Event button that lets you take a practice run that's published to a private URL that no one else can view. When you're ready to go live, click Start Event, which then toggles into the Stop Event button.

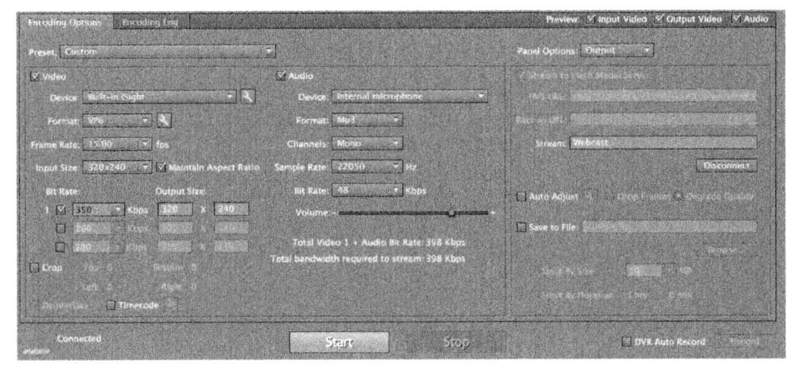

Figure 11-34. I encoded the live stream in the Adobe Flash Media Live Encoder.

We started our event about 10 minutes before the scheduled start time. To fill the dead air before the event actually started, we played a video available in the On-Demand tab beneath the video window. You upload these into the system before the event, and then open the tab and click the video to start playback as needed.

When you're ready to go live with the speaker, you click back to the Live tab, and you would see all speakers who have signed into the webcast, identified by their stream name. In this case, it was just me, identified by the name "Webcast," which is what I plugged into FMLE on the right in Figure 11-34. Click the little head icon, and you're live.

Operation of the other widgets is straightforward: For example, to change slides, you choose Next in the PowerPoint window. To show a poll, you click the poll in the Audience widget and then click Show.

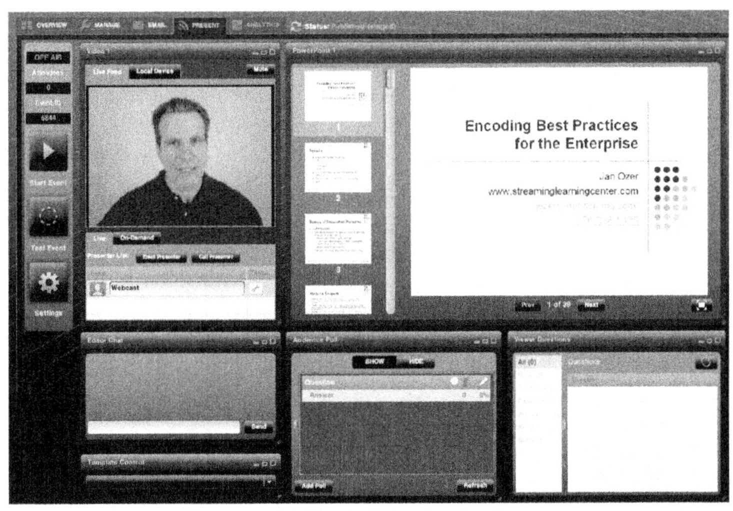

Figure 11-35. Here's the command center.

Attendees watch the presentation in the Window shown in Figure 11-36, with tabs to access the different kinds of content included. Most of the tabs are straightforward, although a couple could use some explanation. For example, the difference between a survey and a poll is that poll results are shared with the audience in real time (beneath the video window), while survey results are not, with survey results appearing only in the Analytics window.

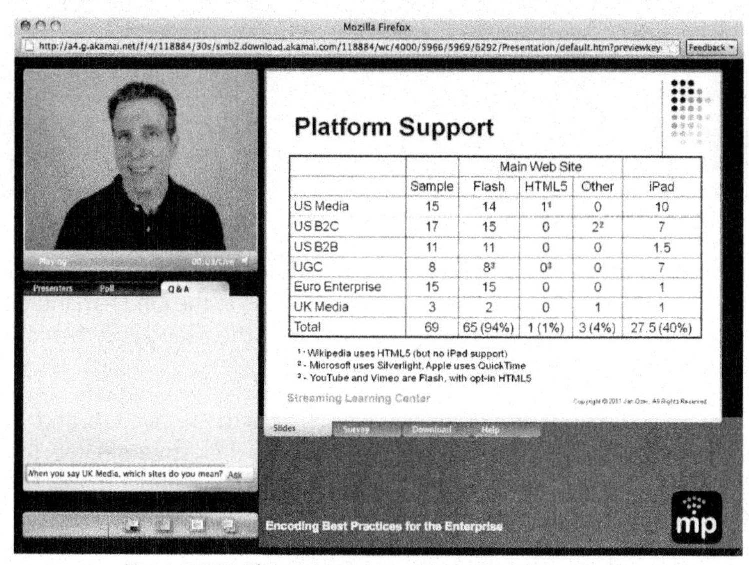

Figure 11-36. This is the presentation seen by the viewer.

Otherwise, moving from left to right, attendees click the Presenters tab to see the presenters' bios, the Q&A tab (shown in Figure 11-36) to ask questions, the Download tab to download materials (when available), and the Help tab for minimum system requirements. On the bottom left are four icons that let the viewer change the presentation display, enabling large-screen video, large-screen PowerPoint, swapping of the video and PowerPoint windows, and the default view shown in Figure 11-36. Presenters have a similar control over what the viewer sees via the Template Control widget, on the bottom left of the presenters' interface, shown in Figure 11-37.

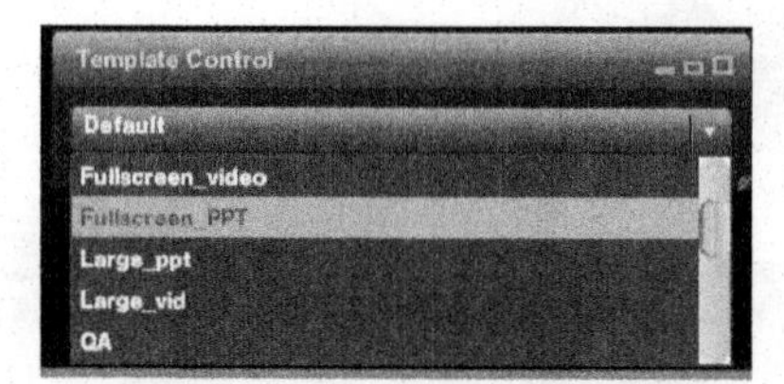

Figure 11-37. Controlling the screen in the viewer's presentation window.

What mistakes did I make during the presentation? Well, I attend most of the StreamingMedia webinars, and one of the first things that Streaming Media moderators Dan Rayburn and Eric Schumacher-Rasmussen always say is, "Ask questions during the presentation, and we'll answer them at the end." I forgot to say that, and had no questions when the time came, which was embarrassing to say the least. With only 60 or so registrants, I should have anticipated this issue, and created a few questions to prime the pump ("Hey Jan, when will your new book be available?"), but I didn't do this either.

Basically, before producing your own webcast, attend a few StreamingMedia webcasts and pay attention to what pros like Dan and Eric say at the start, and make sure you have that material covered at the start of your presentations. Probably an even better general rule is to divvy up the presentation and production functions to separate individuals or groups. As a presenter, you're always concerned with getting your materials together and rehearsed, so many of the presentation details can fall through the cracks.

The WebCaster software is simple enough for one person to operate during the presentation, but having a moderator to set the stage and perform triage on any questions takes a huge burden off the presenter, and makes for an overall better webcast. I'm sure most readers knew all this, but hopefully the reminder is helpful.

Finally, at least for marketing-oriented webinars, don't underestimate how hard it is to acquire attendees. Ultimately, I learned that Brightcove held a webinar at the same time as mine, and that it covered many of the same topics that I did, which no doubt poached at least some of my potential viewers. Although I loved the WebCaster software, you should recognize that there's a big difference between being able to successfully produce the technical aspects of a webcast and producing a webcast that achieves the desired marketing results.

Synchronizing and Archiving

Once the event is complete, it's time to polish up the presentation and archive it for on-demand viewing. This process is called synchronizing (Figure 11-38), and if you're using the same PowerPoint presentation that you originally uploaded, you'll probably only use the function to trim video from the front and back ends of the presentation. For example, to delete the dead air at the start of the presentation, I dragged the faint green box on the upper left of the timeline to the actual start of the video, which is why there are no slides (the little numbered callouts) beneath this area.

Otherwise, the synchronization process is simple: You play the video and then drag the slides on the timeline to the desired location in the video. You can see the content of the slide by hovering the pointer over the callout. Behind the slide announcing my upcoming book is a list of slides by PowerPoint title, which you can add and delete as needed. When you're done, click Save and Publish, and WebCaster archives the presentation where it can be viewed at the same URL as the live presentation.

As mentioned, you can easily send emails to your registrants to let them know that the archived version is up. In my case, this was important, because it was the only way for the attendees to download the slides. If you make the on-demand presentation generally available, potential viewers will have to register to watch it.

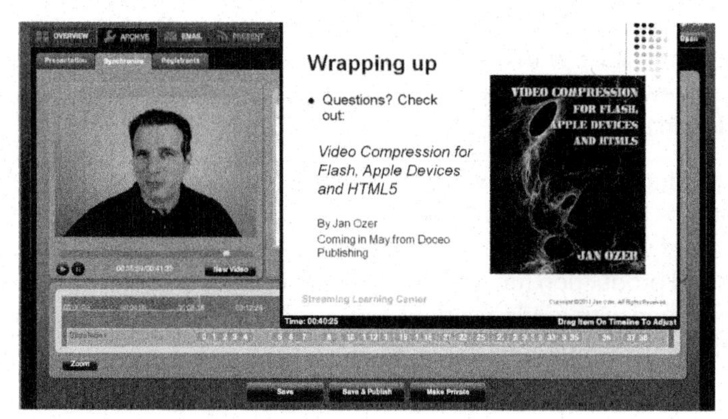

Figure 11-38. Synchronizing the video and slides.

And they'll probably be glad that they did. The archived presentation is impressive as you can see in Figure 11-39. Beneath the video window in the Slide Index are the titles of all the PowerPoint slides, so viewers can click to the desired topics and quickly access the desired content.

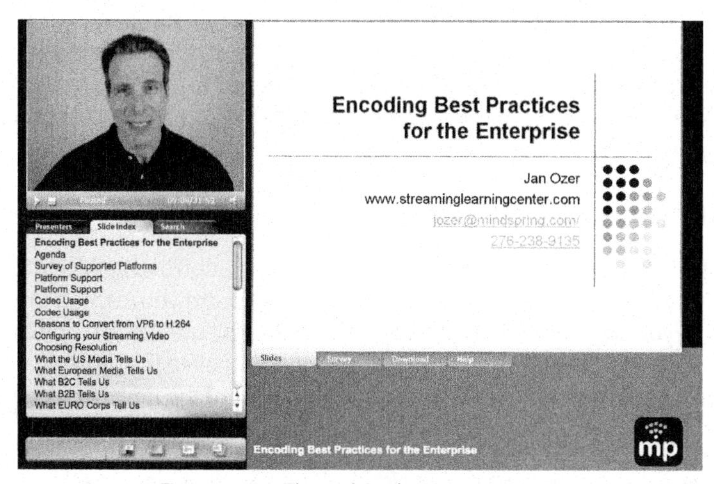

Figure 11-39. The archived presentation.

Or, you can click the Search tab and search for any text item in the original PowerPoint slides. For example, to find where I mentioned Wowza, type "wowza" in the text box, click Search, and all locations will appear in the results box. Click the red Wowza to jump to that slide.

Scanning the rest of the archived presentation, you can see the Download tab, where folks can (finally) download the PowerPoint slides. The Survey tab will show the survey questions asked during the presentation, and if there were any polls, they would appear in a separate tab beneath the player window. On the bottom left of Figure 11-39, you see the same icons that let viewers configure the presentation window to their liking.

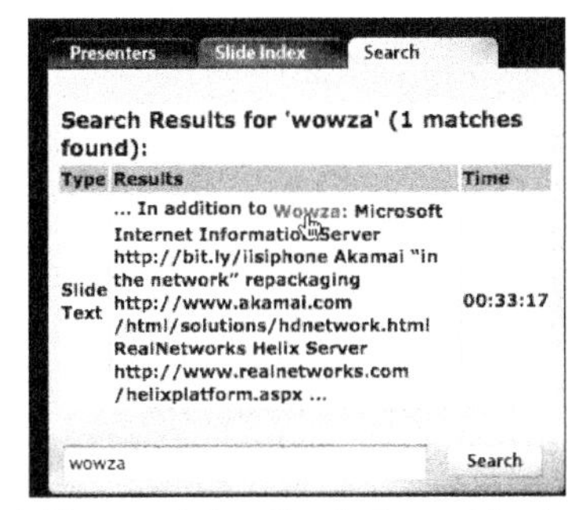

Figure 11-40. The search function in the archived presentation.

Overall, the archived presentation is an efficient way for viewers to find and watch exactly the content they're interested in. In my view, the index and search function adds a lot of value, because if potential viewers know they won't have to slog through 60 minutes of video to watch the five minutes they really care about, they'll be more likely to have a look.

Analytics

WebCaster presents analytics in two basic categories as shown on the left in Figure 11-41. At the top is Viewers, which shows where they watched from geographically (Map Overlay), their identity (Unique Viewers), where they came from (Campaign Tracking, shown in the figure) and how long they stayed (Length of Visit). Note the Export button on the upper right, which lets you export these reports in either PDF format (for graphic-oriented reports like Campaign Tracking) and Excel or PDF for more data-oriented reports, like to create a list of registrants or attendees.

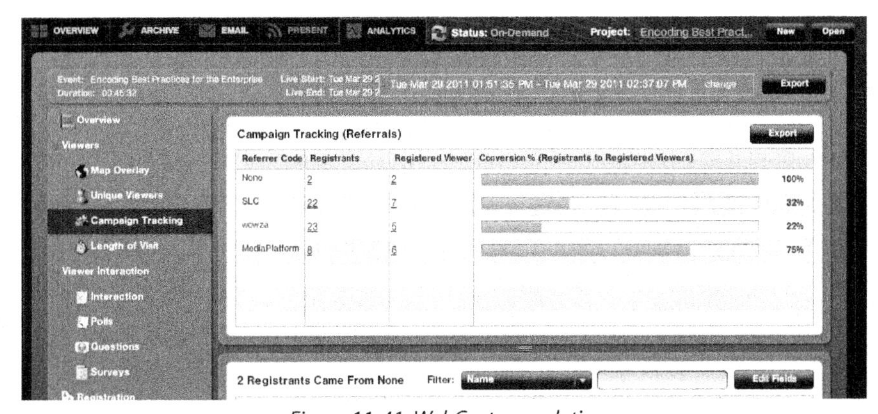

Figure 11-41. WebCaster analytics.

The second category details all the interactions each user had with the system, including when they signed on, how they responded to polls and surveys, and whether they asked any questions. In most categories, you can see this in list view for all participants, or zoom into the view shown in Figure 11-42 for any single registrant.

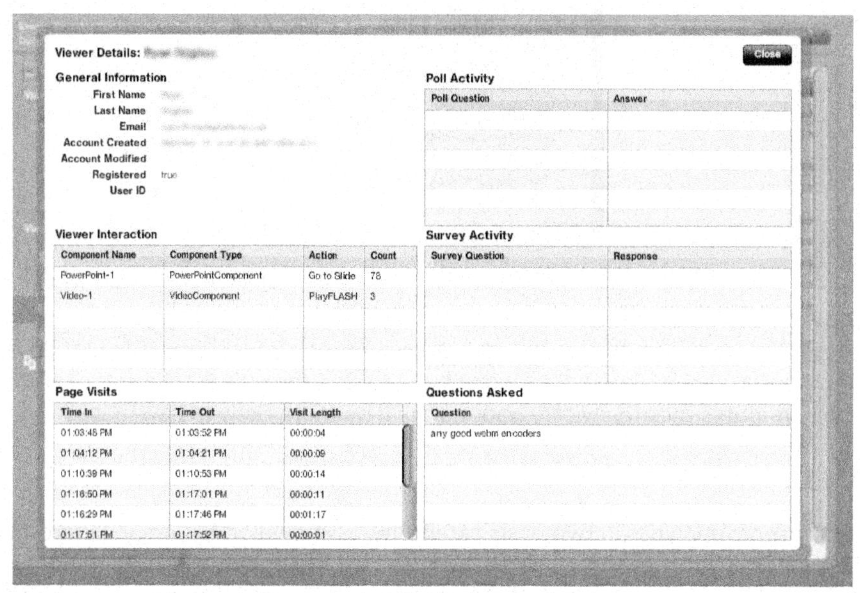

Figure 11-42. Individual viewer details.

What's it Gonna Cost You?

MediaPlatform pursues several business models, predominantly entering into long-term contracts with their various customers. When I asked about pricing, this is what I got back:

> Pricing for WebCaster can vary significantly depending upon the size of the company and the breadth of usage within the company, but a typical range is $25,000 to $75,000 per year for a mid-size company, and $75,000 to $175,000 per year for a large company, for unlimited webcasting. The price range is based on usage, allowing the client to start small and grow as needs dictate.

In addition, MediaPlatform offers simple Pay Per Event pricing for service providers and for enterprise customers that do not require unlimited webcasting. The Pay Per Event pricing varies based on a number of factors (e.g. audio vs. video, size of event, length of event and bit rate), but a typical 60-minute video event for a few hundred viewers might cost about $1,200.

What's my take? I love the web-based operation, and found the software exceptionally easy to use with some great functionality throughout. To present, all your speakers need to do is click Next in their PowerPoint deck, and you can even do that for them. Unless you want a behind-the-firewall version, there's no software to install or maintain, and back-end operation is far less technical than those of other products that I've seen. Heck, a tech-savvy marketing person should be able to handle 99% of it, freeing IT folks to conquer more serious issues.

I should also say that the MediaPlatform WebCaster manual comes with lots of checklists you can use to avoid the problems that hindered my efforts, and to help streamline your webcast. Overall, if you're considering presenting your own webcasts, you've got to give MediaPlatform WebCaster a serious look.

Conclusion

Now you know all about choosing an LSSP and rich media webcasting system. In the next chapter, you'll learn your capture and encoding options for feeding video into these systems.

Chapter 12: Choosing and Using a Live Encoder

Lots of screens in this chapter, too! You can download a PDF file with all figures from this book at **bit.ly/Ozer_multi**. Since this chapter contains lots of images, now would be a good time to do so.

As you've learned in previous chapters, in order to stream live, you must encode your video and send it to a streaming server. In this chapter, you'll learn to choose between the various categories of live encoders—hardware, software, and cloud—and the questions to ask before choosing a tool within each category. You'll also learn how to set up your encoder to connect to your streaming server and encode to the required streams. Specifically, you will learn:

- An overview of the relevant categories of streaming encoder, including software, portable, and cloud
- How to choose the right category for your productions
- How to choose a product within each category
- How to configure your encoder to connect to your streaming server, and where to find the configuration options
- The typical encoding parameters available in live encoding tools and how to configure them.

You will leave this chapter will clear direction as to the type of encoder you should be considering and how to configure it. Big promises, eh? Let's get started.

The Key Functions of the Live Encoder

Let's start with an overview of the key functions of the live encoder. There are three: connecting to your camera or mixing gear, connecting to your streaming server, and encoding the stream(s) in the required format(s).

Let me reiterate my invitation from Chapter 10. If you're unfamiliar with the process of live streaming, you may want to start by watching a tutorial I produced in 2010 titled "Producing Live Webcasts with ViewCast's Niagara 2100" (**bit.ly/introtolive**). While few producers are using the Windows Media codec any longer, otherwise the setup and configuration steps are identical to units streaming H.264 today, and you'll get a useful overview of the process in about 10 minutes.

Connecting to Your Camera or Mixing Gear

Connecting to your camera or mixing gear is job No. 1. As you can see in Figure 12-1, which are input specifications for various models of Digital Rapids' TouchStream streaming appliance, different models offer different connections. Using the legend shown on the upper right, a solid dot means the feature is provided, an open dot means it's optional, and if the space is empty, it's not available.

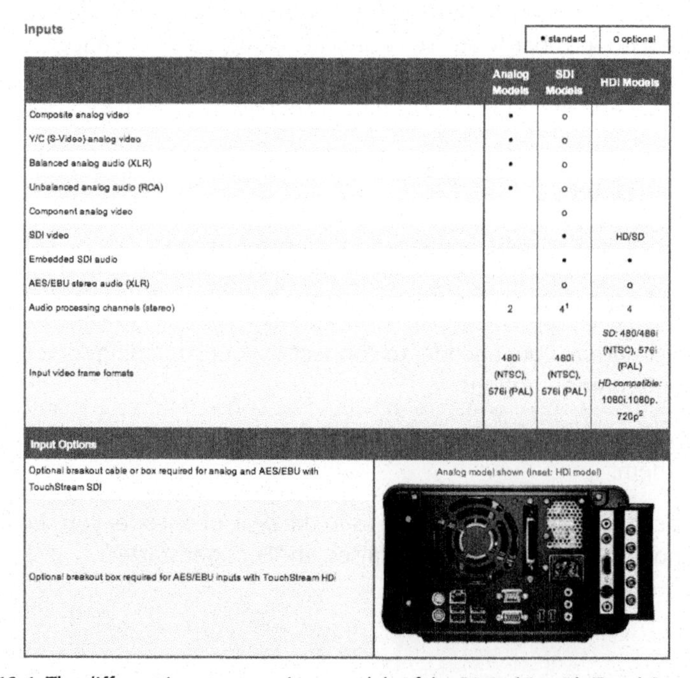

Inputs

• standard | o optional

	Analog Models	SDI Models	HDi Models
Composite analog video	•	o	
Y/C (S-Video) analog video	•	o	
Balanced analog audio (XLR)	•	o	
Unbalanced analog audio (RCA)	•	o	
Component analog video		o	
SDI video		•	HD/SD
Embedded SDI audio		•	•
AES/EBU stereo audio (XLR)		o	o
Audio processing channels (stereo)	2	4[1]	4
Input video frame formats	480i (NTSC), 576i (PAL)	480i (NTSC), 576i (PAL)	*SD:* 480/486i (NTSC), 576i (PAL) *HD-compatible:* 1080i, 1080p, 720p[2]

Input Options	
Optional breakout cable or box required for analog and AES/EBU with TouchStream SDI Optional breakout box required for AES/EBU inputs with TouchStream HDi	Analog model shown (Inset: HDi model)

Figure 12-1. The different inputs on various models of the Digital Rapids TouchStream unit.

If you're currently working with SD analog gear, you have to buy an analog or SDI model and pay the cost for the optional analog configuration. If you're working with HD-SDI, you need an HD model that will cost more. Since units like these cost $3,995 and up, you need to consider both your existing gear and any foreseeable upgrades.

You also need to consider how you'll connect to your camera or mixer when using a software encoder. Three or four years ago, most camcorders were HDV or DV and featured a FireWire connector that connected neatly to the FireWire input on most notebook computers and desktops. Today, the primary interface for prosumer camcorders like AVCHD is HDMI, while higher-end camcorders use HD-SDI. No computer in the land ships with these inputs, so you'll definitely need an input device to feed the video signal into the encoder.

Note that Blackmagic Design sells a line of mini-converters that you can use to bridge the gap between camera and mixing gear and encoding gear (see **bit.ly/BM_converters**). For example, if you have an AVCHD camera with HDMI output but plan to buy a higher-end camcorder with HD-SDI output in the next year or so, the HDMI to SDI/HD-SDI converter ($295) will convert your HDMI signals to HD-SDI in real time and feed it into your capture device. Using the Blackmagic converter, you can buy an HD-SDI capable encoder now and use it with your existing gear, and not have to purchase another encoder when you upgrade your camera.

Figure 12-2. Blackmagic Design's format converters can come in handy when planning your capture gear purchases.

Blackmagic also offers SDI to Analog/Analog to SDI converters, an SDI to HDMI converter, SDI to Audio/Audio to SDI converters, and many others. In addition to future-proofing current encoding purchases, they're great for connecting existing encoders with your new camera gear.

Connecting to Your Streaming Server

When I was producing my first webcast, one primary concern was, "How do I point my encoder to my streaming server?" It's a big question, obviously—so big that all streaming service providers and encoding tool developers have conspired to make the answer quite simple. Basically, you need to know two items to point your encoder to your streaming server: the stream URL (typically an RTMP address for Flash streaming) and a stream key. If you're streaming using a Live Streaming Service Provider (LSSP) like Livestream or Ustream or a Live Streaming Hosting Provider (LSHP) like PowerStream, they will make these available to you. If you're working with your own server, ask your server administrator for these credentials.

For example, Figure 12-3 shows the URL and stream key provided by Ustream with a few digits strategically blacked out at the end. If the location of these settings isn't immediately obvious, search the service provider's support site for server address or stream ID; I'm sure you'll find the information quickly.

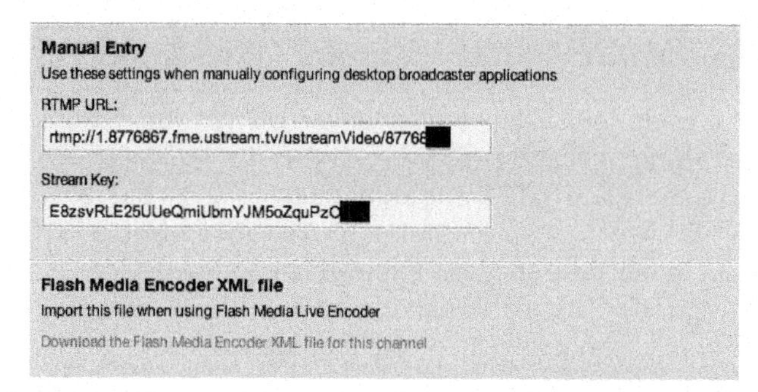

Figure 12-3. The server address and stream ID as provided by Ustream.

Note also in Figure 12-3 that you can download an XML file for the Flash Media Live Encoder (FMLE) that will set your credentials for you. Some services offer these files for popular encoders like FMLE or Telestream Wirecast. When provided, just download the XML file and import it into the encoder, and these credentials will automatically be set.

If the service doesn't offer an XML file, just copy and paste the information from the credentials page, and you'll be streaming in moments. Figure 12-4 shows where that information is plugged into the Flash Media Live Encoder (top) or Telestream Wirecast.

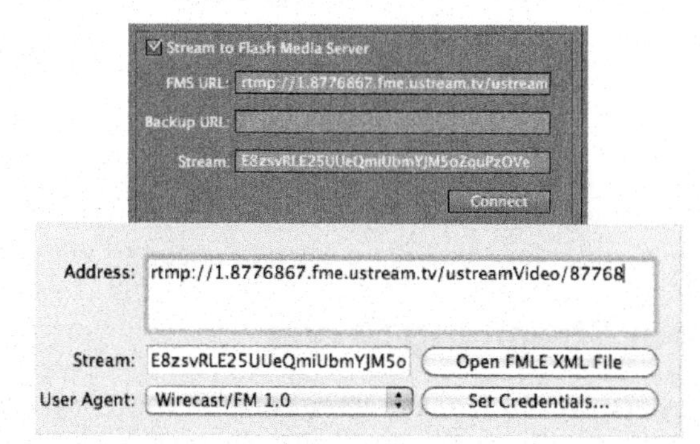

Figure 12-4. Plugging these credentials into the Adobe Flash Media Live Encoder (top) and Telestream Wirecast.

Third-party programs like Wirecast also simplify logging into popular services with a custom interface that's much less technical. You can see this in Figure 12-5 for Livestream, and also other interfaces for services like Limelight, Brightcove, Justin.tv and Sermon.net.

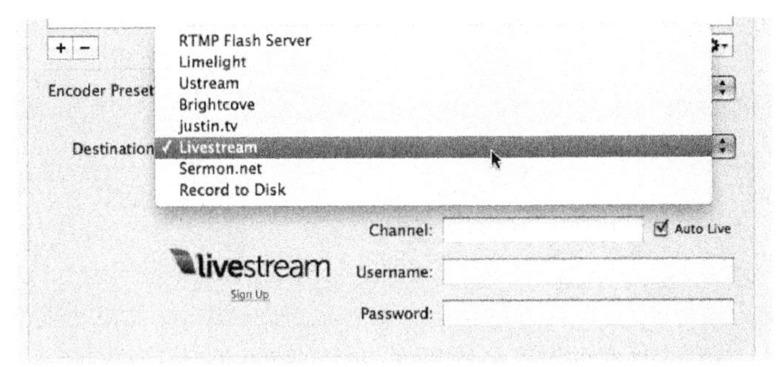

Figure 12-5. Wirecast also simplifies logging into other services.

While you should definitely get your encoder connected to your streaming service well before the live event, if you're concerned that this will require a high-level of geeky network knowledge, like I was, don't be. It's all pretty simple.

Configuring Your Encoding Parameters

The final key function of your encoding tool is to produce the required number of streams in the required configuration(s), which will obviously depend upon your unique encoding needs. Note that all products have their limits—some hard, like the Adobe Flash Live Media encoder's limit of three simultaneous streams, and some soft, which depends upon the number of streams and their respective configurations. For example, when testing ViewCast's Niagara 2120 portable streaming appliance (**bit.ly/Niagara2120**), I tried three configurations:

- Two streams at ***320x240@300 kbps and 480x360@400 kbps***. This pushed CPU utilization to 48%, which is under the 80% recommended by ViewCast.
- Two streams at ***320x240@300 kbps and 480x360@400 kbps***. This pushed CPU utilization to 63-79%.
- Three streams at ***640x480@800 kbps, 480x360@400 kbps, and 320x240@300***. This pushed CPU utilization to 100%, resulting in dropped frames.

Obviously, when working with software-based encoders, performance will depend upon the computer's CPU(s), the efficiency of the encoder, and the number and configuration of streams.

Before the event, you should know how many streams you need to produce and their approximate configuration. Once you have this information, all encoders will provide an encoding interface with most of the options shown in Figure 12-6. If you click the wrench icon next to the format drop-down box, you'll see familiar options like the profile, level and key frame interval.

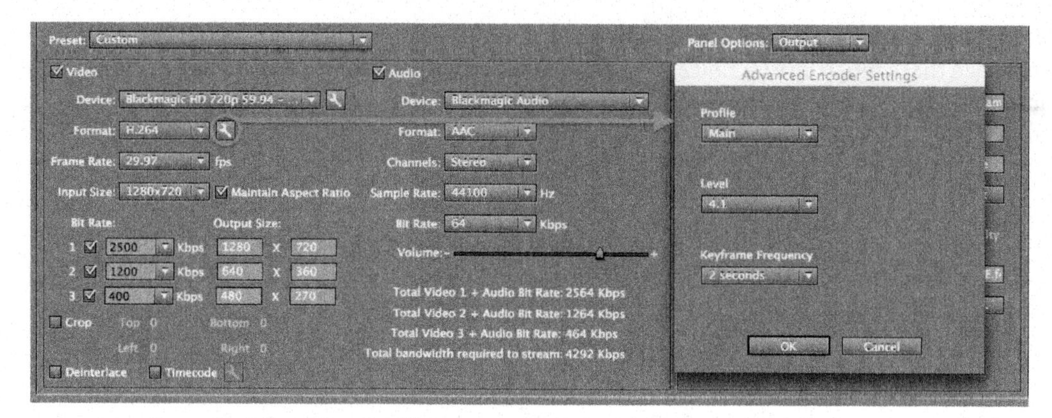

Figure 12-6. Choosing encoding parameters in the Adobe Flash Live Media Encoder.

OK, those are the key functions performed by a live streaming encoder; let's cover one more topic and then we'll go shopping.

Before You Go Shopping

Choosing a live streaming encoding tool used to be simple: You typically would encode a single stream for delivery to your desktop viewers, and budget was the most important buying criterion. When buying today, of course, you've almost certainly expanded your target viewers to include both mobile and desktop clients, with adaptive streaming preferred over single-file delivery.

You have a host of new workflow options to consider, from live cloud or server-based transrating to server or CDN-based transmuxing. You also may have several new requirements, from digital rights management to closed captions to advertising insertion. So before you go shopping for a live encoder, you really need to nail down exactly what you're looking for. Answering the following questions will give you a good start.

Question 1: Who is the service provider and what encoder does it recommend?

If you're using a live streaming service provider like Livestream or Ustream, you should first check its list of compatible software programs and hardware devices. Significantly, both Livestream and Ustream offer free software encoders that are fully integrated with their distribution platform, ensuring both ease of use and comprehensive platform and feature support. If your requirements dictate the purchase of a hardware encoder, you'll want to buy an encoder that similarly integrates into the distribution platform of your selected service provider.

If you're streaming to your own server, ask the server administrator about the required formats for input. Most streaming servers provide highly detailed information about the required formats; you can check out Adobe's tutorial on producing live video via HTTP Dynamic Streaming at **bit.ly/HDS_tutorial**.

Question 2: What platforms will you be serving and will you distribute single or adaptive streams?

Assuming that you're not using a live streaming service provider, you have to start by defining the delivery requirements for the event, beginning with the platforms you're targeting and how many streams you plan to deliver. Are you targeting desktop, mobile and over-the-top (OTT)? Of course you are; that's why you bought the book. So you have to configure the steams accordingly. Then you can work backward to identify the optimal workflow—and encoder—to produce the required streams for the required targets.

Question 3: What features do you need surrounding your video?

Beyond the target platforms and the number of streams, you also have to consider the features surrounding your video. For example, will closed caption support be a requirement, or digital rights management (DRM) or advertising insertion? If so, few stand-alone software encoders, or even low-end hardware encoders, can handle these requirements. If you need to add captions or DRM or advertising, get these concerns on the table early, because you're looking at a very much abridged collection of potential candidates.

Question 4: What functions do you need your encoder to perform?

Most hardware and software encoders are single-function encoding engines. However, several software programs can perform valuable production functions, like multiple-camera switching, title insertion or playback of disk-based files. On the hardware side, there are some multiple camera production systems that can also output a live encoded stream. If you're looking to integrate production and encoding functions, get this on the table first, since it will severely limit the hardware and software products you can consider.

Question 5: What's your overall encoding workflow?

This is where the analysis gets complicated. A few years ago, a producer seeking to distribute adaptive streams to multiple platforms—say, four streams each to Flash and iOS—had to produce all required streams for both platforms in real time. Typically, this involved multiple sets of expensive rack-mounted hardware encoders.

Around 2010, the real-time ability to convert one set of adaptive streams to meet the requirements of another delivery platform became commonplace. This process, which is called transmuxing, involves rewrapping (but not re-encoding) H.264 video streams into the container format necessary for another target platform and creating any required metadata files. Transmuxing has become a standard feature on most media servers and is also offered by some content delivery networks (CDNs).

For example, you might input four streams bound for Flash delivery via RTMP-based Dynamic Streaming into a streaming server, which would transmux this content into a set of MPEG-2 transport streams for delivery via HTTP Live Streaming (HLS) and create the necessary .M3U8 manifest files. If you build transmuxing into your workflow, you won't need separate encoders for each format; the media server can transmux the incoming RTMP streams for HLS.

In 2011, the first live transcoders hit the market, available either as stand-alone servers or in the cloud (live transcoding is discussed in detail at the end of this chapter). Live transcoders can input a single high-quality stream and then create the necessary file iterations for the adaptive streaming group. In our example, the transcoder would input one file and produce four files at different quality levels, or three files if the input file was going to serve as the highest-quality file in the adaptive group. These files could be configured as necessary for delivery to multiple target platforms—whether HLS, RTMP-based Dynamic Streaming, or any other adaptive format.

If you build transcoding and transmuxing into your live streaming workflow, you no longer need multiple-file encoding capabilities on premise. Rather, you need an encoder capable of supplying a single high-quality stream to the transcoding facility, which virtually all hardware or software encoders can do. You also significantly reduce your outbound bandwidth requirements because you're sending a single stream out of the facility, not multiple streams.

Transmux? Transcode?

Choosing whether to integrate transmuxing and transcoding into your workflow involves multiple factors. Of the two, transmuxing is more compelling; since there are really no negatives, it's hard to imagine when you wouldn't want to use it.

Live transcoding is newer and less proven, and it's unclear whether quality will suffer because the lower-bit-rate streams are produced from a previously compressed file, rather than the original source. Outbound bandwidth is obviously a key decision-making factor here; if it's limited and you need to distribute multiple streams, sending a single stream out of the facility and then transcoding that stream into your adaptive group may be your only option.

If outbound bandwidth isn't an issue, budget becomes the next consideration. If you're a quality-at-all-costs operation, buying a hardware encoder capable of producing all the required streams may be your best option. If funds are short, producing a single stream for downstream transcoding may be the most affordable option, particularly if you produce relatively few live events and can't amortize the cost of a big iron encoder over multiple events.

Overall, before even choosing a class of encoding tool, you should plot out the workflows you'll use to produce your videos to their final form, including the container formats and metadata requirements of your actual distribution files, as well as any ancillary features. You should also identify the input requirements for the selected transcode/transmux tools, since these will become the required outputs of your selected encoder.

Once you have this information, you'll have to choose an encoder from one of the following categories:

- ***Software encoder.*** In this category, there are free tools like the Adobe Flash Media Live Encoder, as well as the free offerings from LSSPs like Livestream and Ustream. Beyond this are programs like Telestream Wirecast and Combitech VidBlaster. To use a software encoder, you'll need a sufficiently powerful computer, plus some type of input device into your computer. For many older DV and HDV cameras, you can use an IEEE 1394 connection (or FireWire), but most new cameras don't support these connections.

- ***Hardware encoder.*** Hardware encoders come in several classes: stand-alone encoders installed in a server room, whether rack-mounted or smaller, plus portable devices that can be taken and operated from the road.

- ***On-camera encoders.*** These encoders sit on the camera itself, enabling total freedom of camera motion. Typically, they connect to the Internet via either Wi-Fi or 4G.

- ***Cloud or server-based transcoder.*** These are programs that input a single stream and output multiple streams, typically in multiple adaptive-bit-rate formats.

After answering the questions above, you should be pretty far along on your requirements list, understanding the number of streams that your encoder needs to produce, their respective configurations and ancillary features that it must support. This should help you focus on a category of products—whether it's a software encoder, mid-level standalone hardware encoder or industrial-strength, rack-mounted encoder. Within this category, you should identify the most affordable product that can produce the required streams, deliver all the ancillary features and connect to your relevant inputs, whether IP or video.

If you're caught between categories—say, choosing between a software encoder or inexpensive hardware encoder, or wavering between multiple inexpensive hardware encoders or a single industrial-strength product—here are some high-level considerations.

First, as for the choice between hardware and software encoders, there's a general perception that hardware encoders are more reliable. To a great degree, this relates to the platform the program is installed on, not the software program itself. That is, if you install a software encoder on your director of marketing's notebook, it may be affected by any program later installed on that computer. Since most hardware encoders are single-function devices, the platform itself is much more stable.

This single-focus functionality also makes hardware devices easier to use. So long as your non-technical users can turn the device on and connect the necessary gear, they should be fine. While the difference isn't that significant, software programs are undeniably more difficult to use. You have to know which program to load, which menus to select, and often which options to select. If you're sending non-technical users into the field to stream a live event, the edge in simplicity goes to hardware.

Finally, as between a single industrial-strength encoder or multiple less powerful and cheaper encoders, if you're producing adaptive streams, it's best practice to encode all streams in an adaptive group on a single encoder. There are certain parameters like key frames, chunk sizes and audio parameters that should remain consistent among all files, and encoding on a single encoder is the easiest way to make this happen.

Let's go shopping.

Choosing a Software Encoder

As a caveat, note that few, if any, live software encoders can integrate closed captions or any form of digital rights management. So if these are a requirement, you're probably looking at the wrong category.

Again, while we talk about software encoders, some form of hardware capture devices is always involved. That is, all software encoders need some kind of video capture device to work—whether it's a simple DV port on your computer or notebook, a higher-end Blackmagic DeckLink card or a low-end ViewCast analog capture card.

Fortunately, in most instances, commercial software programs like Telestream Wirecast support well-defined standards, like DirectShow on Windows and QuickTime on the Mac. As you would expect, most capture/encoder cards are built to support those same standards, so the software and hardware should play together well without much effort on your part. Still, if you end up choosing your software encoder first, check the software developer's website for a list of compatible hardware devices before buying your hardware. This is especially crucial if you plan on pushing the envelope, like installing multiple cards to switch four streams of video in Telestream Wirecast.

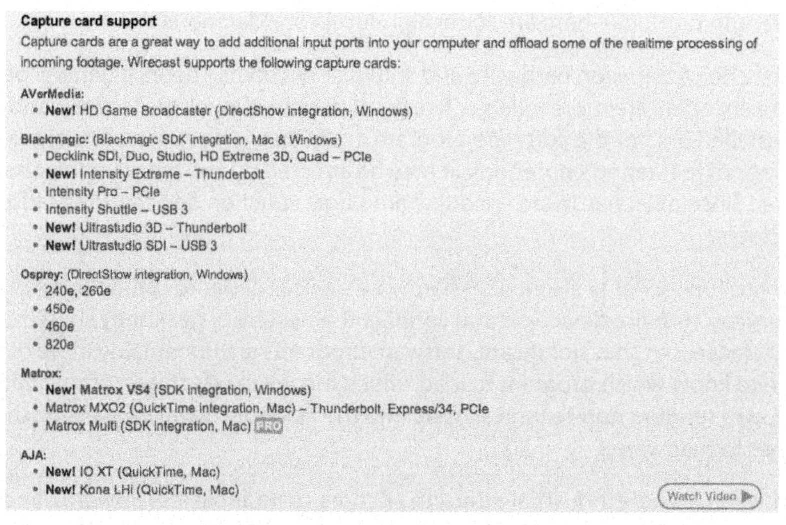
Capture card support

Capture cards are a great way to add additional input ports into your computer and offload some of the realtime processing of incoming footage. Wirecast supports the following capture cards:

AVerMedia:
- **New!** HD Game Broadcaster (DirectShow integration, Windows)

Blackmagic: (Blackmagic SDK integration, Mac & Windows)
- Decklink SDI, Duo, Studio, HD Extreme 3D, Quad – PCIe
- **New!** Intensity Extreme – Thunderbolt
- Intensity Pro – PCIe
- Intensity Shuttle – USB 3
- **New!** Ultrastudio 3D – Thunderbolt
- **New!** Ultrastudio SDI – USB 3

Osprey: (DirectShow integration, Windows)
- 240e, 260e
- 450e
- 460e
- 820e

Matrox:
- **New! Matrox VS4** (SDK integration, Windows)
- Matrox MXO2 (QuickTime integration, Mac) – Thunderbolt, Express/34, PCIe
- Matrox Multi (SDK integration, Mac) PRO

AJA:
- **New!** IO XT (QuickTime, Mac)
- **New!** Kona LHi (QuickTime, Mac)

Watch Video

Figure 12-7. The list of compatible capture devices for Telestream Wirecast.

For example, Figure 12-7 is a screen grab from the Telestream site that contains a list of hardware products that are compatible with Wirecast (**bit.ly/WC_hardware**). Obviously, if you plan on using Wirecast, you should buy a card off this list. Later in this chapter, I'll point you to tutorials on installing the Blackmagic Design Intensity Pro and the ViewCast Osprey 820e, the latter a dual-input card ideal for mixing multiple cameras in Wirecast. I'll also talk about the external options shown, including the Thunderbolt and USB 3.0 options.

This leads to the obvious question: Which should you choose first, hardware or software? For the most part, I would go with the software first. Once you've identified the software you plan to use, it should be fairly simple to identify hardware devices that perform the capture tasks necessary to produce the desired outputs. So that's the order we'll approach this analysis: first the software, then the hardware.

Why not the LSSP Software?

If you're using an LSSP, the first question you have to ask is, Why not use software supplied by the LSSP? Let me qualify that a bit; you should never use the browser-based apps supplied by the LSSPs, since these typically use the VP6 codec enabled in the Flash Player. If you're streaming to an LSSP that does supply applications, like Livestream and Ustream, consider them first. If you're using an LSSP that doesn't, check their website for recommendations.

If you're using Ustream.tv, you have three choices: Ustream Producer (free), Producer Pro ($199) and Producer Studio ($549). All are versions of Telestream Wirecast; Producer is SD-only with no titles or multiple cameras. Pro adds HD, multiple cameras and titles, while Studio adds virtual sets and support for IP cameras. You can check out the full range of features at **ustream.tv/producer**. Note that Ustream also offers iOS and Android apps for both broadcasting and viewing.

If you're using Livestream, the company offers Livestream for Producers, which encodes multiple user selectable streams, for free. While the program can also broadcast your screen, I've tried it and found it clumsy and difficult to use. Livestream for Producers is perfectly adequate for video-only productions, but I'd find another solution for screen-based output.

In late 2012, Livestream released the Studio HD500 ($8,500), a lunch pail computer with Blackmagic Design capture components and Livesteam's own Studio program. Designed as a competitor to NewTek's fabulously successful TriCaster line, Studio includes multi-camera mixing, graphics and encoding. In addition to being a stand-alone product that Livestream will continue to sell, the HD500 is also a testing ground and proof of concept for the Studio software, which Livestream will offer as a stand-alone product in 2013.

I reviewed the product for *Streaming Media Magazine* (see **bit.ly/LS_HD500**), producing a three-camera shoot of *The Nutcracker* ballet, as performed by my wife's ballet company. In Figure 12-8, the three cameras are in the middle row on the left, with the big screen on the upper left the preview window, and the screen on the upper right the live window. Overall, I was impressed with the Studio 500's stability and performance and concluded:

Figure 12-8. A boring screenshot of the Livestream Studio HD500, but I didn't want to risk a screenshot during the broadcast.

> As both a product and proof of concept, Studio HD500 is impressive. I'm sold on the lunch pail form factor as more convenient than a stand-alone CPU or mixer that requires a separate monitor, and really appreciate the use of proven Blackmagic Design and NVIDIA components as opposed to cheaper alternatives. Going forward, I hope that Livestream offers different Studio configuration options for a range of camera inputs. As a producer, I would prefer to buy a proven, self-contained unit like the HD500 as compared to building my own from existing or newly purchased components.
>
> Relating to the software rollout, it will be interesting to see how Livestream ensures the stability and responsiveness of the Studio software on a diverse range of homegrown systems. I'm guessing Livestream will release a known list of compatible hardware products with stringent CPU, RAM and graphics requirements. For best results, you should meet or exceed these requirements and install the Studio software on a clean system that's used solely for live event production.

Livestream released the software-only version of Studio in February 2013 for $1,999, and it works only with Blackmagic Design components (see **bit.ly/Studio_sw**), which should maximize stability. All Livestream users who are producing multiple-camera shoots and need a video mixer should definitely consider it. Note that Livestream configures and sells a range of

Mac- and PC-based portable and desktop systems on its website, which you should check out for serious production, particularly if you decide to use the Studio software.

Solo Encoder or Production Station?

There are two product families that combine encoding with production features: Telestream Wirecast (**bit.ly/WC_family**), which is cross-platform and starts at $495, and VidBlaster (**bit.ly/VB_family**), which is Windows-only and starts at $195. Both families offer versions with a dizzying array of features that can dramatically enhance a live broadcast, from multiple-camera switching and titles to adding disk-based files and screen-based or PowerPoint presentations to the live feed. More advanced features include live greenscreen and virtual sets, and scoreboards that you can integrate into your sports productions. If you tried to buy all these features in a dedicated hardware device like the NewTek TriCaster, you'd easily pay $15,000 or more.

Figure 12-9. Telestream Wirecast, which provides TV production-like functions at a fraction of the price of hardware.

The product families are similar in feature sets and both have their rabid advocates. I've used Wirecast for multiple live events, and reviewed it back in 2010 for *Streaming Media Magazine* (**bit.ly/WC_review**), and again in 2011 for *Sound & Video Contractor* (**bit.ly/Wirecast_SVC**). I produced a short video with and about the product that you can see at **vimeo.com/9361582**. While there have been some minor changes since then, the program still uses a similar workflow to perform similar functions. Here's my conclusion from the first review:

> Overall, Wirecast offers an exceptional blend of well-targeted functionality that's unique in the price range, and after you get settled into the interface, it's relatively simple to use.

> If you're serious about chucking those writing chores to make it on Internet TV, or if you have a real job and just want to create some highly polished live or live-to-disk Internet broadcasts, Wirecast should be on top of your purchasing list.

I like this class of product even for single-camera shoots because the ability to add titles, on-demand videos, and static slates for when you're taking a break adds a nice touch of class to any event. Note that the various versions of Ustream's producer are all Ustream-specific versions of Wirecast, so if you're a Ustream user and are considering a product in this category, check Ustream Producer first.

What's Wrong with FMLE (and Free?)

Probably the most widely used live encoder today is the tried-and-true Adobe Flash Media Live Encoder—and why not? It's functional, cross-platform and free. Probably the most significant limitation for those hoping to stream adaptively is the three-stream maximum. If you're interested in learning to use the product, I produced a tutorial titled "Streaming Live with the Adobe Flash Media Live Encoder" that's available at **bit.ly/FMLE_tutorial**.

KulaByte—High-Volume Producers

One software live encoder that's carved out a niche for high-quality, multi-stream productions is the KulaByte XStream Live encoder, now owned by Haivision. I reviewed the product for *Streaming Media Magazine* in 2009 (**bit.ly/XStreme**), and concluded, "Overall, though ease of use could (and will) be improved, from a performance and quality standpoint, the KulaByte encoder performed exceedingly well."

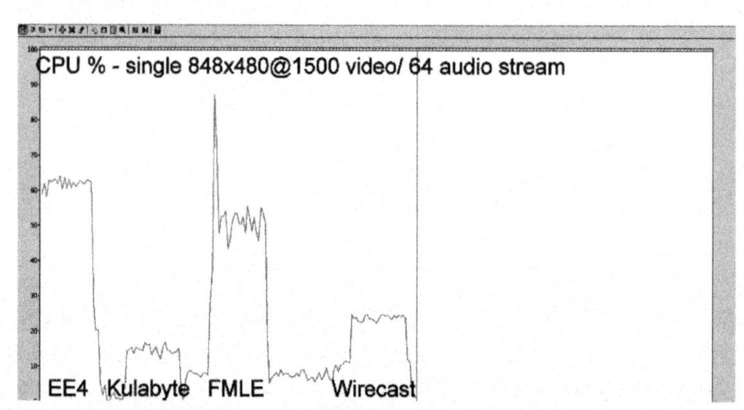

Figure 12-10. The CPU required to create an 848x480 stream using the various programs.

A couple of years later, I took another look at the product in a lengthy article, also in *Streaming Media Magazine*, titled "Live Streaming from a Notebook" (**bit.ly/liveencoder**). There, I compared streaming efficiency, quality and data rate consistency for KulaByte, FMLE, Wirecast and Microsoft Expression Encoder 4. Figure 12-10 shows CPU utilization while encoding an 848x480 stream on an HP EliteBook 8740w with a 2.0 GHz four-core (eight with HTT) i7-based CPU, running 64-bit Windows 7 Professional with 8GB of RAM, NVIDIA Quadro 5000m, 7200 rpm disk drive. Lower is better, and KulaByte was the lowest of the bunch, which means that it can create more streams on the same hardware than any other encoder.

In other tests, KulaByte ranked just behind FMLE in quality and performed competently otherwise. In the industry, KulaByte has gotten a reputation for a rock-solid software encoder that has produced some very prominent events. The software is so good that KulaByte was purchased by Haivision in 2011 and is now available from its website (bit.ly/KB_Haivision). The only downside is price; it's so high that Haivision doesn't typically disclose it. Back in 2009 when I reviewed the software, it cost $6,495, but I believe that it's in the $10,000 range now. Still, if you've got a powerful computer, that's lots cheaper than any dedicated hardware encoder performing the same function.

Expression Encoder for Silverlight Producers

The final software program I'll mention is Microsoft's Expression Encoder ($199.95), which is the only software encoder I'm aware of that can produce streams for Silverlight (bit.ly/EE_buy). Combined with Microsoft's Internet Information Server 7 (IIS7), Expression Encoder can send create a stream that IIS7 can transmux for delivery to iOS devices—a very cost-effective and efficient combination for serving Silverlight and iOS clients (bit.ly/IIS7_iOS). I've reviewed the product several times, most recently in January 2011 in a review you can read at bit.ly/Ozer_EE. The product worked well, and if you're a Silverlight producer, you need to check it out. Otherwise, since the product can't communicate with an RTMP server, EE has nothing to offer Flash producers.

Capture Hardware for Your Software

So you've selected your capture software, and now it's time to choose the hardware device that connects your camera to your computer. As mentioned above, the first place to check is the list of compatible hardware devices posted by most live streaming software producers, as shown in Figure 12-7. Beyond this, your choice will obviously diverge depending upon whether you're installing the card into an internal slot or using external connection like USB, FireWire or Thunderbolt. We'll look at both cases, starting with internal installation.

Cards for Internal Installation

Most internal capture devices install in a PCI Express slot. In desktop computers, you'll need a free slot. With notebook computers with an ExpressCard slot, you can add a device like the Magma ExpressBox 1 (magma.com/expressbox-1), which connects to the ExpressCard slot and provides a half-size or full-size PCI Express slot in the external enclosure. This is the setup I used to test HD capture in the *Streaming Media Magazine* article "Live Streaming from a Notebook" (bit.ly/liveencoder).

Although the product isn't cheap (about $800 at B&H Photo-Video), it enables access to the high-bandwidth PCI Express connection which most producers prefer over externally connected capture devices. On the HP 8740w notebook I used for those tests, for example, my alternative would have been either a USB or FireWire connector, which is much slower. Note that Magma also offers a similar expansion chassis for Thunderbolt connections, although this isn't quite so useful given that there are several capable Thunderbolt-connected capture devices available that cost much less than the expansion chassis.

Wherever you install the card, once you find a list of boards compatible with your selected capture software, make sure the vendors offer cards that connect to your camera or mixing gear. This shouldn't be a problem, since most producers offer cards ranging from composite analog to HD-SDI, with steps in between for component analog and HDMI. Then you need to find cards that enable the density that you need. For example, if you're mixing two cameras in Wirecast, you need to either install two cards or find a card that captures two streams.

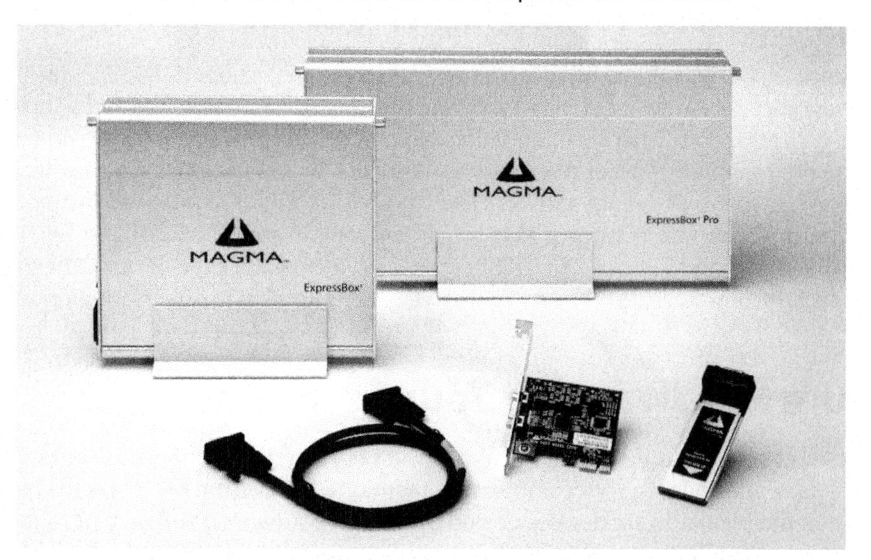

Figure 12-11. Magma ExpressBox 1 adds a PCI Express slot to a notebook.

Finally, when reviewing your candidates, check their lists of compatible computers and see if the computer you'll be using is listed. I almost exclusively use HP computers, which are very popular among the content creation community and typically supported by most capture vendors. If a capture card vendor doesn't list your computer, it's a concern; make sure you buy from a source where you can return the capture board if it's not compatible.

Once you've narrowed the candidates, make sure you purchase from a reputable vendor, which includes Blackmagic Design, ViewCast, Matrox and AJA. From my perspective, Blackmagic has the most mindshare with the user community, which means its software developers work really hard to ensure trouble-free installation and operation. I would rate the other three vendors as a close second, but second they are.

Here are factors to consider when buying a capture card/encoder.

- ***What incoming audio/video adjustments are enabled?*** This will typically depend upon input source, but it's nice to have brightness, contrast, hue and saturation adjustments for video, and volume adjustments for audio. Most devices that input digital video directly—like DV, HDV, SDI and HDMI—won't offer these features, while most that input composite, component or S-Video should.

- ***What hardware pre-processing features are supported?*** Make sure the card can scale, crop, deinterlace and perform inverse telecine in hardware, which is more efficient and usually higher-quality than processing in software.

- ***What software is available to drive the hardware?*** If you haven't chosen your software encoder, identify the programs the hardware vendor offers (if any) and ask if there is a published application programming interface (API) if you need to perform custom development for your live streaming application.

Once you narrow down your selections, I would sniff around the web to see how the board is faring in production environments. If the vendor has a support forum, check that, or Google "name of product" "name of company" "driver issues" or "problems." That should identify any major issues you may be facing.

Leave Plenty of Time for Installation

In the great panoply of things you can do with a computer, I would rate installing a hardware capture device among the most complicated. Not only do you have to get the cards and drivers installed, you have to make sure your live capture software is communicating effectively with your capture hardware. This sounds simple, but I would budget two hours to get a new card installed and working correctly, and pat yourself on the back if you can shave time off of that. Heck, I've spent two hours getting a capture card in a computer working again after not using it for a few months. It's just a pain in the butt and very persnickety.

Anyway, if you're installing any Blackmagic product, you might save yourself some time and hassle by checking out my tutorial "Configuring Blackmagic Intensity Pro for a Live-Switched Wirecast Webcast," where I walk you through installing two Intensity Pro cards into an HP computer (**bit.ly/BM_install**). If you're installing any ViewCast card, check out "Configuring ViewCast Osprey 820e for a Live-Switched Wirecast Webcast" at **bit.ly/VC_install**. The 820e is a very powerful dual-density card that worked very well in my tests.

Now let's shift our attention to external capture devices.

External Capture Devices

There are four commonly used connections on notebook computers as shown in Figure 12-12, and your capture options will vary by connection. Let's take each in turn from left to right.

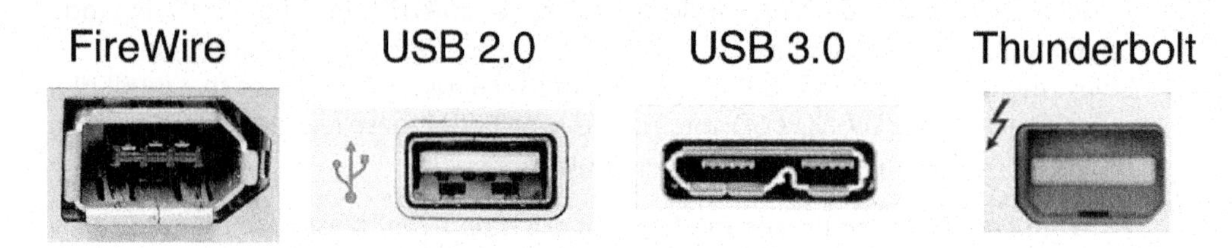

Figure 12-12. The various external connections found on notebook computers.

FireWire (IEEE 1394) Capture Devices

FireWire is the commercial name for the IEEE 1394 interface Apple acquired in 1986. There are multiple connection speeds—most notably FireWire 400, which can transfer up to 400 megabits per second (Mbps), and FireWire 800, with a bandwidth of up to 786 Mbps. Virtually all FireWire capture devices use the earlier specification, which became exceptionally popular in the late '90s and early '00s as the capture medium for DV and HDV camcorders. However, FireWire 800 connectors are backward-compatible with FireWire 400 devices, and you can purchase an adapter that allows you to connect FireWire 400 cables into your FireWire 800 port. See the article "An Essential Gizmo for Mac Video Producers" at **bit.ly/Mac_gismo** for more details.

As shown in Figure 12-13, there are two connection types for FireWire 400: On the left is the 4-pin connector, on the right the 6-pin connection. The 6-pin connector can power an external device, like a hard drive, while the 4-pin connector can't.

Figure 12-13. Two form factors for a FireWire 400 connector: 4-pin on the left, 6-pin on the right.

When you buy a cable to connect your camera or mixer with your computer, you obviously have to purchase a cable with the right connections. Most camcorders use 4-pin connectors, as do most notebooks, so you'll need a cable with 4-pin connectors on both sides. These are widely available, but not common, so if you forget your 4-pin-to-4-pin cable on your way to a shoot, you probably won't find one at the local RadioShack or Walmart. Cables.com sells them for about $9, so you should buy two to have an extra around. The most commonly available cable is a 4-pin-to-6-pin, which connects camcorders to most full-sized computer connectors.

If you're working with a DV or HDV camera and have a notebook with a FireWire port, you won't need any additional hardware, just the required cable. On the other hand, if you're working with a camera that doesn't have a FireWire port and you want to capture via the FireWire port on your computer, you'll need an external device to convert the camera input to FireWire.

Note that a key benefit of external FireWire devices is that they convert the incoming signal to DV video, so the devices look like a camcorder to the computer. This means you typically don't have to install a driver to use the converter, which simplifies installation.

If you're capturing SD analog video, I recommend a Canopus product like the ADVC110 Converter (**bit.ly/can_ADVC110**), which costs around $200 at Amazon. I've used one for many years for multiple purposes, including live webcasts, and the quality, reliability and stability has been great. You can find cheaper analog-to-FireWire converters, but I don't recommend going the cheap route for live webcasts.

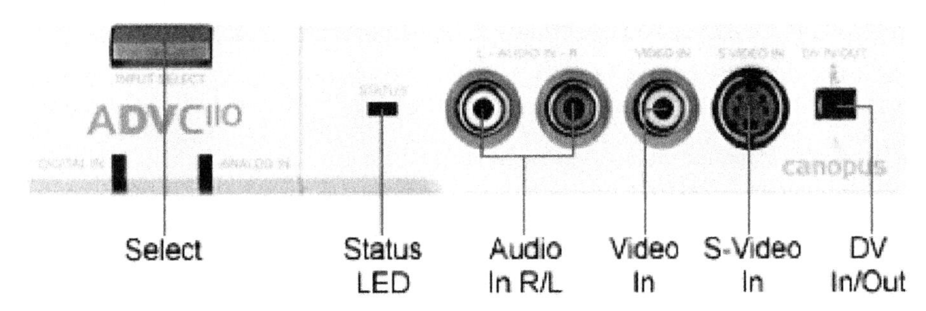

Figure 12-14. The connections on the Canopus ADVC110 FireWire capture device.

Figure 12-14 shows the analog connectors on the ADVC100. What about if you're shooting in HD and want to connect via component analog—HDMI or HD-SDI ? Operationally, all FireWire capture devices convert the incoming signal into SD DV video for input into your computer via the FireWire connection. For this reason, there's no reason to input an HD signal into a FireWire device; you get the same result from composite or S-Video input.

Although Blackmagic offers an SDI to Analog converter that can connect HD-SDI output to a FireWire converter (**bit.ly/BM_converters**), you're still inputting SD video. Basically, if you want to capture HD video, FireWire isn't an option. So let's move to the next category of capture devices.

USB 2.0 Capture Devices

USB 2.0 was released in 2001 and has a top-end transfer rate of 480 Mbps, slightly higher than that of FireWire. However, FireWire involved both a connection standard (400 Mbps) and a video format standard with a bandwidth of 25 Mbps. Since FireWire has a transfer rate of about 16 times faster than the bandwidth of DV video, connection speed was never an issue. And, since DV is a standard video format, there was never any compatibility issue with editing or live streaming programs.

In contrast, although USB 2.0's bandwidth is slightly higher, there is no standard capture format, so it's kind of like the Wild, Wild West out there. Different products use different codecs at increasingly higher bit rates to produce better quality at HD resolutions, at the risk of connection speed issues and potential compatibility issues. The bottom line is that I've never liked USB 2.0 capture and have never really used it for a live production. However, numerous producers have, so the key takeaway is that while DV capture should be exceptionally stable and hassle-free from the start, you'll have to budget more time when working with USB 2.0 devices to iron out any wrinkles.

If you Google "USB 2 capture devices," you'll see a multitude of options ranging in price from $24.99 to $499 and higher. As mentioned before, you need a unit that connects to your camera or mixing gear and is also listed as compatible with your selected capture software. Interestingly, none of the devices listed on the Telestream compatibility list are USB 2.0 devices.

Livestream does a nice job listing devices compatible with its capture software, and the only USB 2.0 device it includes is Blackmagic Design's H.264 Pro Recorder (**bit.ly/LS_capture**). I reviewed that product as a real-time H.264 converter back in March 2012 (**bit.ly/H264Pro**). My conclusion was:

> Overall, I'm an unabashed supporter of Blackmagic capture cards, and have recommended DeckLink cards many times and use them frequently in my work. Unfortunately, I can't say the same about the H.264 Pro Recorder. Unless your current application fits perfectly into the existing sweet spot—encoding perfect 29.97 HD footage to 1080p or 720p output—I would stay away from this product until Blackmagic adds output and H.264 configuration options and works through the technical issues described above.

Interestingly, encoding live video footage to 1080p or 720p input is a sweet spot for the device. A few days after the review came out, I received an email from a contact at Livestream, saying:

> It does a nice job since you can do HDMI/HD-SDI input on any laptop via USB 2.0. We take the 50 Mbps full-resolution feed from the device over USB 2.0 then Procaster decodes and re-encodes using our own codec/profile. It's a little more CPU-intensive than getting a feed from a PCI , Thunderbolt or USB 3.0 and a slightly lower-quality result (but you have to be an expert to see any difference) but it saves the user from buying new laptops or carrying workstations at event. So we do like it. We are not using their software though; it shows in Procaster as a certified device.

So the Pro Recorder works well with the Livestream Procaster software (since renamed to Livestream for Producers). Unfortunately, this doesn't mean that it works equally well with other software. Here's a thread (**bit.ly/H264Pro_WC**) that talks about issues with Wirecast, and here's a thread (**bit.ly/H264Pro_FMLE**) that talks about unstable results with FMLE. So, if you must connect via USB 2.0, the Pro Recorder may work for you. I would Google "H.264 Pro Recorder" and "live capture program" to see if any users have any good or bad news to report.

USB 3.0 and Thunderbolt

USB 3.0 has transfer speed of up to 5 gigabits per second (Gbps), which is more than 10 times faster than USB 2.0. Thunderbolt doubles that connection speed with a maximum transfer rate of 10 Gbps. These connection speeds are ideal for high-quality video transfer, so it's not surprising that there are multiple options using both connections. Unfortunately, I have no direct experience with any of them. So I'll repeat the general guidance that I've related before: find a list of compatible devices provided by the producer of your selected live encoding software or your LSSP, and find a device that connects to your camera gear or mixer.

Get a Lunch Pail Computer

One other option is a "lunch pail" computer like the Next Computer Radius, which I reviewed along with the Roland VR-3 video mixer in an article that you can read at **bit.ly/next_VR3**. The Radius has three PCI Express slots behind the LCD, into which I installed my Blackmagic DeckLink Studio capture card, into which I fed the output of the VR-3. Although the lunch pail form factor won't win any style points, it's wonderfully convenient for live event production—so much so that Livestream used the same form factor for the Studio HD500 I discussed last chapter.

Figure 12-15. The Next Computer Radius has three slots in back, making it perfect for live event production.

The article discusses a concert that I produced with the two products, and you can see a short video review of the VR-3 and computer on the second page of the article.

Hardware Encoders

Hardware encoders come in many shapes and sizes for multiple uses. For the purposes of this discussion, I'll divide them into two categories: portable encoders and big iron encoders, which are my shorthand for rack-mounted encoders typically installed in server rooms. Let's start with the portable ones.

Portable Streaming Appliances

Again, note that few, if any, live software encoders can integrate closed captions or any form of digital rights management. So if you need these, you're likely looking at the wrong category.

If these aren't required, why use an appliance rather than a notebook with streaming software? Several reasons. First, since they can be pre-configured back at the home office, all the onsite producers have to do is connect the camera to the device and press the streaming button.

For example, Figure 12-16 is the Digital Rapids TouchStream appliance, which I reviewed at **bit.ly/touchstream**. If you configure the TouchStream back at the office, field operation is exceptionally simple. When you boot the appliance, the TouchStream software loads automatically, and you can "arm" the system so that all the user has to do is plug in the Ethernet cable and audio/video connectors, and press Start when ready to begin streaming. You'll want to check your work by logging into the landing page for the live stream, but if there's video in the embedded preview monitor and no error messages, you should be streaming.

Figure 12-16. The touchscreen controls on the Digital Rapids TouchStream portable streaming appliance.

In addition, unlike software encoders that vary in performance depending upon the platform they're installed upon, appliances should always perform up to their rated specifications. Even though they're usually Windows-based computers in a box—the TouchStream runs an embedded version of Windows XP—users don't load games, shareware and other funky applications on appliances like they do their own notebooks, so the appliances should function more reliably over the long haul.

Driving a Portable Streaming Appliance

Let's take a little time to discuss the experience of operating the TouchStream. The unit itself is slightly larger than a breadbox (about 16"x9"x6" tall), and weighs about 8 pounds, so is fairly transportable, and the 800x480 touchscreen tilts up and out for easy viewing. It comes in three basic hardware configurations—analog, SDI and HD-SDI—with connectors well documented on the Digital Rapids' website, and above in Figure 12-1.

I tested the analog model—with VC-1, H.264, and iDevice formatting and segmenting—which would have retailed for $6,140. The CPU was a 2.4 GHz Intel Core2Quad with 1 GB of RAM and an embedded Intel Graphics controller. Analog capture and pre-processing is provided by an embedded Digital Rapids DRC-500 capture card that you can buy separately for $895 and higher, depending upon formats supported. Basically, Digital Rapids built the TouchStream around its own proven board designs, adding the portable case, convenient touchscreen and the additional software that pulls it all together.

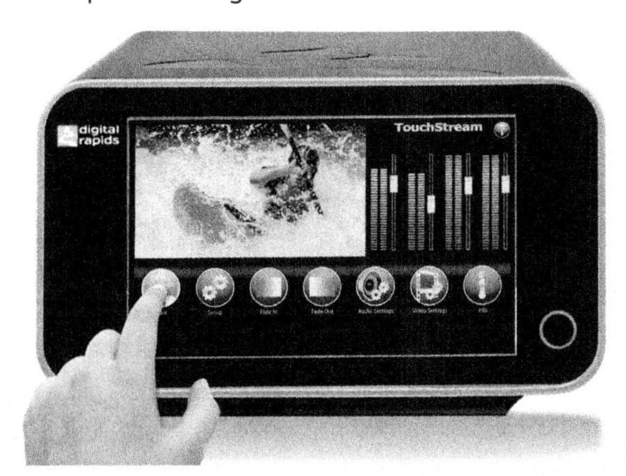

Figure 12-17. Here's the full glamour shot of the TouchStream.

Once you're up and running, you can click the Video Settings button to adjust brightness, contrast, saturation and tint, as well as deploy both spatial and temporal noise reduction—all while you're encoding (Figure 12-18). You can also display a graphic overlay that you can fade in and out. You can't crop while encoding, however; you have to do that before you start. Of course, you can also adjust audio levels while encoding.

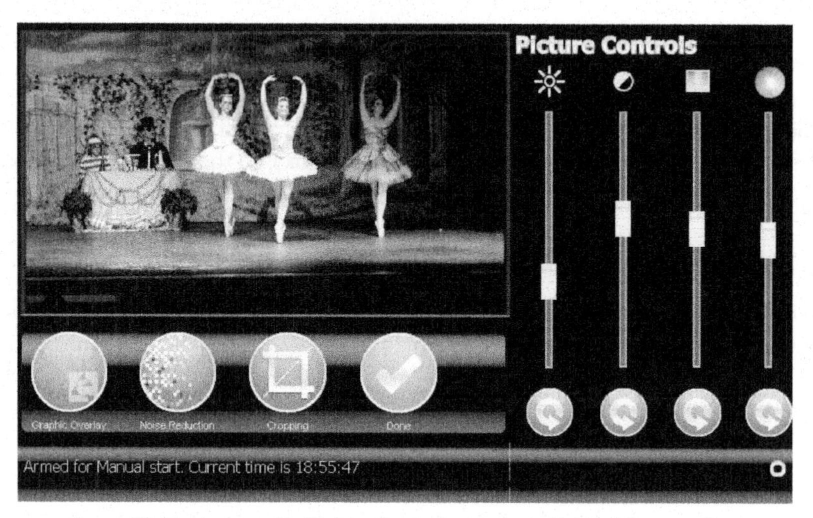

Figure 12-18. Here are the Picture Controls you can adjust while encoding. Don't you love high school stage lighting?

Digital Rapids ships the TouchStream with dozens of project presets, including single and adaptive streaming, which vary by stream count, resolution and data rate. If you can use one of the canned presets, life is simple: You simply choose a preset and move to the next step.

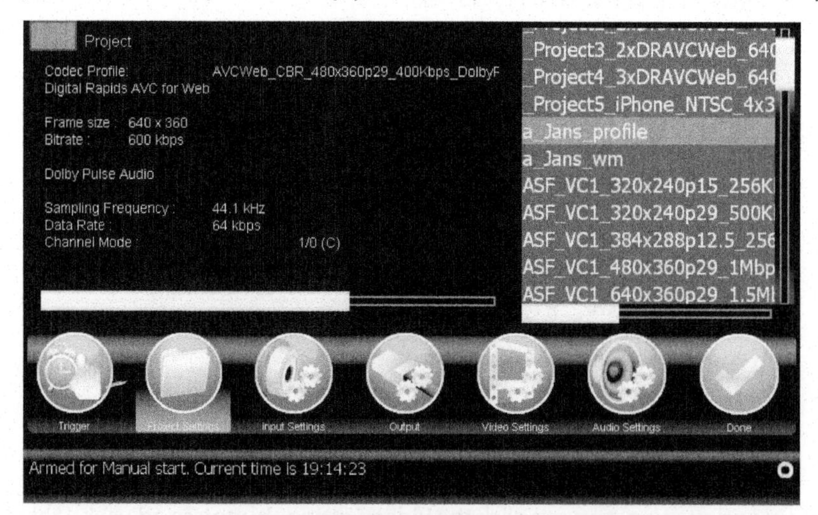

Figure 12-19. Choosing an encoding preset. Pretty simple stuff, eh?

If you can't, you have to exit the TouchStream software, run the TouchStream configuration program, and either customize an existing preset or build one from scratch. If you've read Chapters 2 through 4, the encoding controls shown in Figure 12-20 should look instantly familiar. Once you customize a preset, it shows up as a preset that you can easily reuse.

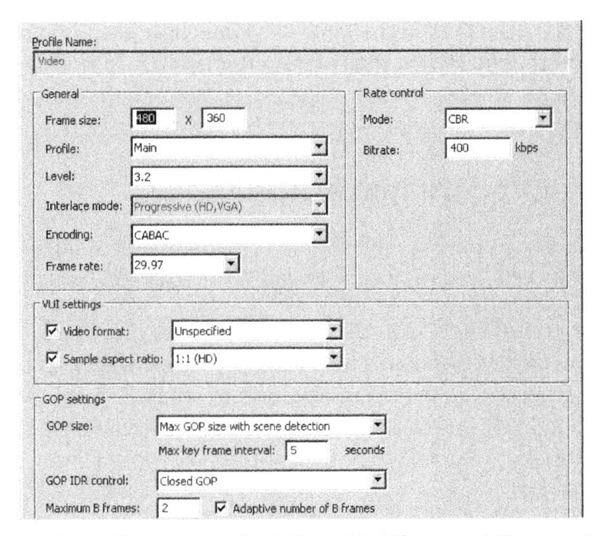

Figure 12-20. The controls should be old hat if you read Chapters 2-4.

After choosing the preset, you select your audio and video inputs, and then enter your server credentials, including the server URL and stream name (Figure 12-21). You learned where to get this information and where to put it at the start of the chapter.

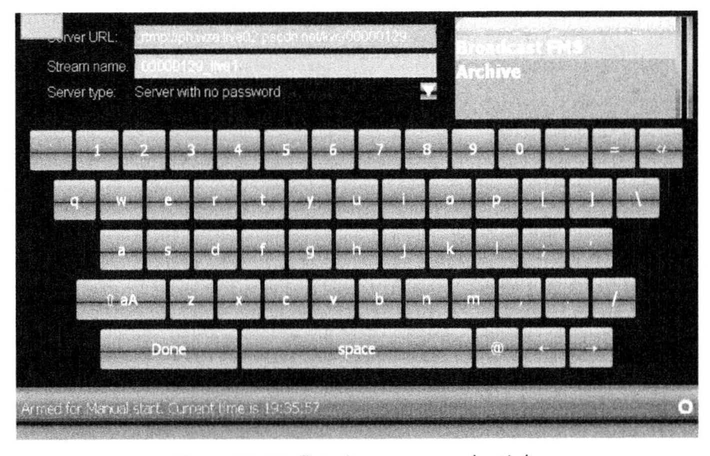

Figure 12-21. Entering server credentials.

As is the case with many portable appliances, you can start and stop encoding manually, or schedule triggers that allow you to start the unit according to a fixed schedule. Either click the Start button shown in Figure 12-17 to get started, or make sure something interesting is in front of the camera at the scheduled start time.

Basically, every hardware encoder works the same way: You choose your inputs and encoding parameters, enter server credentials, and then press Live when you're ready. Again, you can simplify this by choosing all these parameters in your office so that inexperienced users can simply plug the unit in, connect the audio/video/Ethernet cables, and press the Start button to start the show.

Choosing a Portable Streaming Appliance

When choosing a streaming appliance, you should consider many of the same factors identified previously for capture/encoder cards, along with a few unique ones, which I'll cover first.

- ***Touchscreen or not?*** Although the units are more expensive, Digital Rapids' line of TouchStream appliances give you a full preview of streaming operation with touchscreen controls. It's nice to see that video preview, and to adjust controls directly on the unit if things go wrong. With devices that lack a touchscreen, you have to bring either a notebook to connect to the unit over the LAN, or a keyboard, mouse and monitor to drive the unit directly.

- ***How tough is it to create my settings?*** I've reviewed both ViewCast and Digital Rapids appliances (see the ViewCast review at bit.ly/Ozer_2120) and I favor the ViewCast software for creating your encoding settings because you create and implement the settings in the same program. With Digital Rapids, you use separate programs to create settings and drive the hardware, and because of some funky license restriction, you can't have both programs running at the same time, which is frustrating. The interface for creating encoding settings is very obtuse, as well.

- ***How can I access it remotely?*** If you'll be sending the unit out with non-technical users, find out how easily you can control the software remotely. For example, the ViewCast software is totally browser-based, so you drive it locally or remotely via the same interface, which is a nice paradigm.

- ***How noisy is it?*** Both the Digital Rapids and ViewCast units we've discussed are full computers with high-powered CPUs and noisy fans required to cool them. You typically can't use them in closed spaces; they're just too loud. In contrast, the Vitec Optibase MGW Nano is about the size of a large deli sandwich and has no moving parts, so it runs silently. If you're streaming a deposition from a closed conference room or from a similarly quiet environment, this is a unit you want to check out. You can read my review at bit.ly/Ozer_nano.

Beyond this concern, you have many of the same considerations discussed above, including:

- ***Does it produce the required streams?*** You've planned your broadcast and know how many streams you need and in what formats and configurations. Make sure the appliance you select can output those streams. Not all devices can produce outbound HD streams, so if that's a requirement, make sure it's supported.

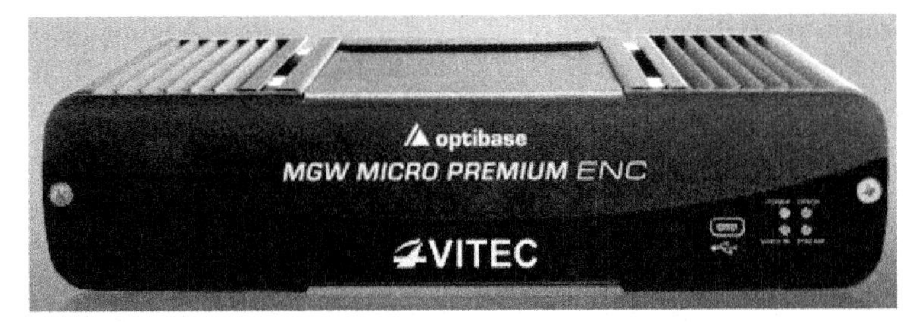

Figure 12-22. The exceptionally quiet MGW Micro.

- ***Can it connect to my input feeds?*** This is a big one as well. Inexpensive appliances typically support either FireWire or composite/S-Video input, with more expensive devices supporting component and SD-SDI, and the most expensive devices supporting single or multiple HD-SDI inputs. Make sure the appliance supports the feeds you'll be working with.

- ***What incoming audio/video adjustments are enabled?*** This will typically depend upon input source, but it's nice to have brightness, contrast, hue and saturation adjustments for video, and volume adjustments for audio. Some appliances that input digital video formats—like DV, HDV, SDI and HDMI—won't offer these features, while most that input composite, component and S-Video should.

- ***What hardware pre-processing features are supported?*** Make sure the appliance can scale, crop, deinterlace and perform inverse telecine in hardware, which is more efficient and usually higher-quality than processing in software.

Since appliances tend to be very expensive, I'd definitely check for user reviews before buying. If the vendor has a support forum, check that, or Google "name of product" "name of company" "driver issues" or "problems." That should identify any of the major issues that you may be facing.

Choosing a Rack-Mounted Live Encoder

Rack-mounted systems are for high-volume applications and typically cost $20,000 and higher. For this, you get real-time encoding of multiple streams with full support for both closed captioning and DRM. For full service streaming, this is the category you need to buy from.

Unfortunately, while I've reviewed my share of on-demand enterprise encoders, my experience with live big iron systems is minimal. However, for an article I wrote for *Streaming Media Magazine* in 2010 (**bit.ly/webcastmasses**), I asked Major League Baseball Senior VP Joe Inzerillo—who's charged with broadcasting 2,430 regular season games, plus pre-season and postseason—for some tips.

For Inzerillo's work with MLB webcasting, reliability is critical. Before choosing his encoder, he held a "bake off" between multiple vendors, running them for a 5-hour pre-game/game/post-game period and analyzing factors like CPU use and dropped frames.

While an occasional dropped frame may not be a huge deal on an iPod, an increasing number of his customers are watching the 720p streaming feed on widescreen television sets, and dropped frames are very noticeable. He recommends that if you can't get a system in for simulated testing, ask the vendor to put you in touch with clients whose applications and run times are similar to yours.

Inzerillo also noted a significant disparity in the breadth and depth of H.264-related controls made available by each encoding tool. This is significant because he uses different encoding parameters depending upon factors such as whether it's a day or night game. In addition, his organization provides services to other networks during baseball's off season, and encoding parameters that work well for the center field camera in baseball may be totally inappropriate for the side-to-side action in soccer or basketball. So work through the preset creation process to make sure you're comfortable that it provides access to all relevant encoding controls.

Beyond these considerations, Inzerillo advised to make sure that any encoder you choose can be controlled remotely, preferably with multiple encoding stations that are simultaneously viewable from a remote monitoring station. Also look for features such as scheduling, automated failover and alert notifications.

Beyond these concerns, you have many of the same issues discussed above, including:

- ***Does it produce the required streams?***
- ***Can it connect to your video feed?*** Make sure the system can connect to your current video feeds—and, since the systems are so expensive, the feeds you plan to be using over the next two to three years.

On-Camera Encoders

The next category to consider is on-camera encoders, which provide ultimate portability when you need to move around when you shoot. They are all small and battery-powered, and they usually can transmit in 4G as well as Ethernet and Wi-Fi. I covered the 4G angle in a section titled "Gee? No, 4G" in Chapter 10. Here I'll cover the H.264 encoding end. As with our software and portable encoders, none of the on-camera encoders I've looked at can incorporate closed captions or DRM into their streams. If these are a requirement for you, these devices may not be suitable unless you can implement captioning and DRM downstream using the encoded streams.

How They Work

Again, you learned at the top that there are three key functions of an encoding: connecting to your camera or mixer, connecting to your server, and encoding the streams. I've looked at three products: Livestream Broadcaster (**bit.ly/Ozer_broadcaster**), Teradek Cube (**bit.ly/Bond_cube**) and LiveGear AirStream from Vislink (**bit.ly/AirStream_ozer**). The first two units are true on-camera units, while the AirStream is a lunch pail configuration with worthy of note because of functionality (including touchscreen control) and battery life. Irrespective of the form factor, all three products get all the key functions done, albeit in slightly different ways.

First, recognize that you're working with a unit with limited screen space and configuration controls, as you can see in Figure 12-23, which shows the four-line OLED display and joystick controls. Although you can drive most systems via these controls, most encoders also let you configure them via iPhone and/or Android apps, and also via web controls. In Figure 12-23, you see how Livestream connects your device to your Livestream account: You power up the unit and connect to the Internet, then enter the code shown into your web page shown. That links the Broadcaster to your Livestream account—no muss, no fuss.

Figure 12-23. Connecting to your Livestream account with Livestream Broadcaster.

Teradek lets you drive the unit via hardware controls, browser controls or an iPhone app, with Figure 12-24 showing how you select your server in a browser (note that the Bond unit only works with the old Livestream, not new Livestream accounts). To connect to the unit via a browser, you connect the unit to your LAN, and then any computer on the LAN can run TeraCentral, which logs into the unit via the shared LAN connection. In the field, you can configure the Cube's Wireless connection into Master mode, which sets up the MIMO Wi-Fi connection that your iPad or iPhone can connect to, enabling control via apps on these devices. I used my iPad in my remote 4G tests, and it's a great convenience.

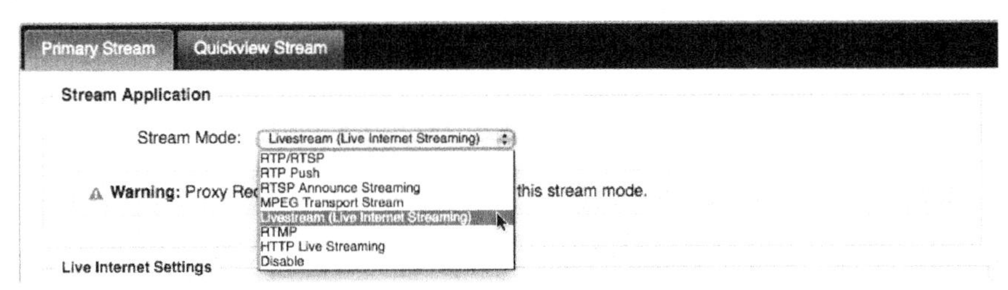

Figure 12-24. Choosing the Livestream preset.

With the AirStream unit, you typically control settings via the touch screen, shown in Figure 12-25, or using browser-based controls.

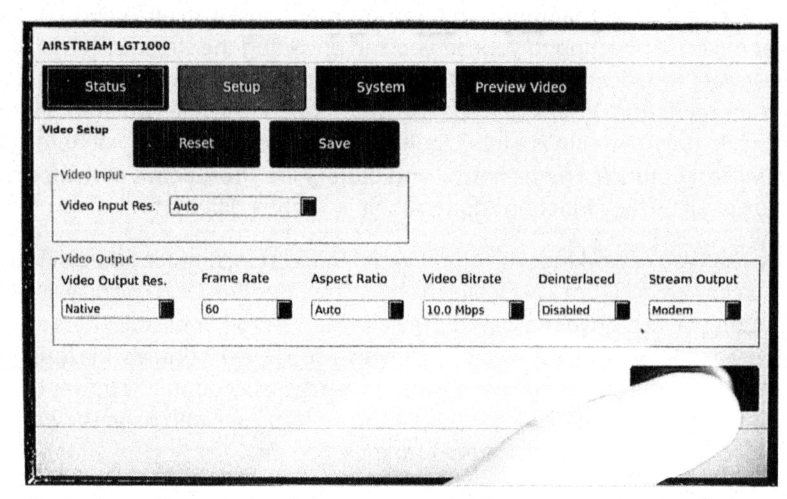

Figure 12-25. Choosing encoding settings via the AirStream touchscreen controls.

Figure 12-26 shows how you change compression presets for the Broadcaster via iPhone controls. Basically, you run these units just like any other live encoder, using direct, browser-based or app-based controls.

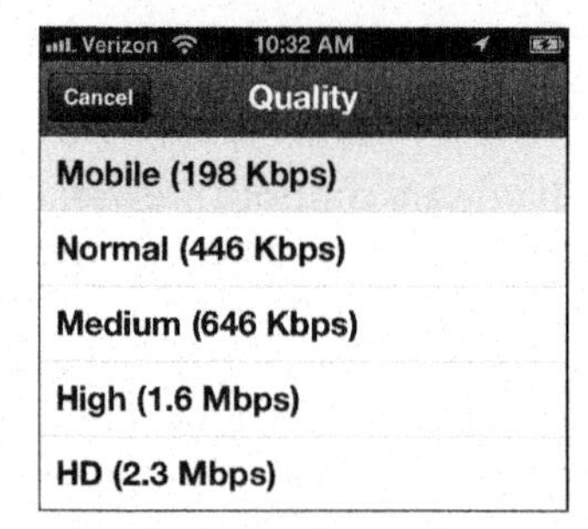

Figure 12-26. Choosing an encoding preset for Broadcaster in the iPhone app.

Choosing an On-Camera Encoder

Now that you know how these units work, let's discuss how to choose the right one.

- ***Hardware connection type.*** Obviously, the unit has to connect to your camera/mixer. Here, Livestream is the most limited, offering only an HDMI version. Though you could use a Blackmagic Design converter to convert your signal to HDMI, these AC-powered units would limit your portability.

- ***Connection to your server.*** Due to Livestream-imposed limitations, the only on-camera encoder that works with the new Livestream service is its own Broadcaster unit.

- ***Available as encoder/aggregator.*** As you undoubtedly recall from Chapter 9, I prefer encoder/aggregator pairs if you're serious about 4G. This combination enables adaptive control over the outgoing bit stream, which maximizes signal quality over all connection speeds. Aggregation also means more than one 4G modem, which can improve overall connection speed and, by installing modems from difference services, ensure a high-speed connection when one service isn't available or is working slowly. The AirStream is an encoder/aggregator, and the Cube can be paired with sister product Bond, which is a 4G aggregator. In contrast, the Broadcaster offers a serial port for a single 4G modem, which offers less speed and less stream flexibility, and limits you to one service at a time. On the downside, note that aggregators add complexity and cost to the broadcast, since you'll need a server at the back end to combine the disparate streams (as explained in Chapter 9).

- ***Form factor.*** On-camera devices are nice from a portability perspective, although battery life is often an issue.

- ***Battery life.*** If you plan on broadcasting very long events, find a unit that can be externally powered or has hot-swappable batteries (Figure 12-27), which means theoretically unlimited broadcasting with no interruption to switch batteries. For shorter events, I prefer units with internal rechargeable batteries since it eliminates the need to buy new batteries.

- ***Number of modems and configuration.*** Most systems max out at five modems. Some are configured internally, like those on the AirStream Video TX, which is more convenient because you don't have to carry them around separately.

- ***The availability of forward error correction.*** This helps ensure signal integrity and video quality.

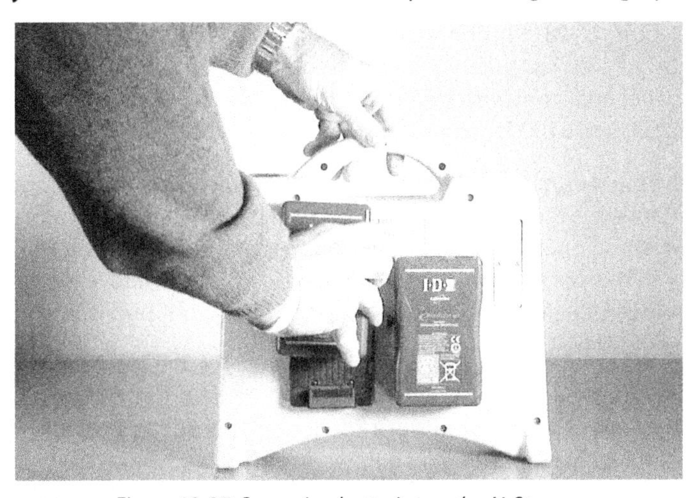

Figure 12-27. Swapping batteries on the AirStream.

Mixers with Encoders

Another hardware category worthy of note are mixer/encoders. The poster child for this category is the Livestream Studio HD500 shown in Figure 12-8, which can mix your camera input and send and encoded stream to your new Livestream account. Note also that most recent versions of the popular NewTek TriCaster can also encode to H.264 format and connect to RTMP or Windows Media streaming servers.

Let's turn our attention to cloud encoding.

Live Transcoding in the Cloud

We've discussed how to encode onsite and transfer the streams to the LSSP or other streaming server. However, what are your options if outbound bandwidth at a live event is limited, or if you don't have a spare $20,000 or so for a live encoder capable of producing the required streams in multiple formats in real time? Why, transcode in the cloud, of course.

Here's how it works. Rather than producing multiple streams in multiple formats on-site, you encode a single high-quality stream and transmit that to the cloud transcoding facility. There, the single stream is re-encoded into the required adaptive streams, which are properly formatted for each adaptive streaming technology. This obviously reduces both the cost of on-site encoding and the bandwidth requirements for getting the stream out of the building.

How much does this service cost? Figure 12-28 shows a cost estimate from Haivision HyperStream Live for an hour-long presentation with five streams (outputs) and two targets (Flash and HLS). The estimate is for 658 credits; at $0.15 per credit, that's $98.70.

As you can see in Figure 12-28, HyperStream is a web-based service. Another service option is from Zencoder, a Brightcove subsidiary that is currently beta testing its live transcoding services. While pricing is not yet final, the Zencoder rep I spoke with estimated a cost of around $10 per stream per hour, or around $100 for our 10-stream, hour-long presentation.

If you'd like to own your own software—either installed in the cloud or within your own service center—your best option is the Wowza Transcoder, which costs $495 as an add on to the Wowza Media Server. Wowza also offers several leasing options, which cost as little as $2 per channel per day. Obviously all this pricing seems very attractive if your alternative is a $20,000 hardware encoder, and perhaps a satellite truck to ship your streams to your publishing point.

This category is set to explode in 2013; at Streaming Media West in November 2012, it seemed like every company I spoke with—including Elemental Technologies, iStream Planet and many others—were planning to release a product or service in early 2013. From my perspective, it's one of the most useful technologies to come down the pike over the last few years.

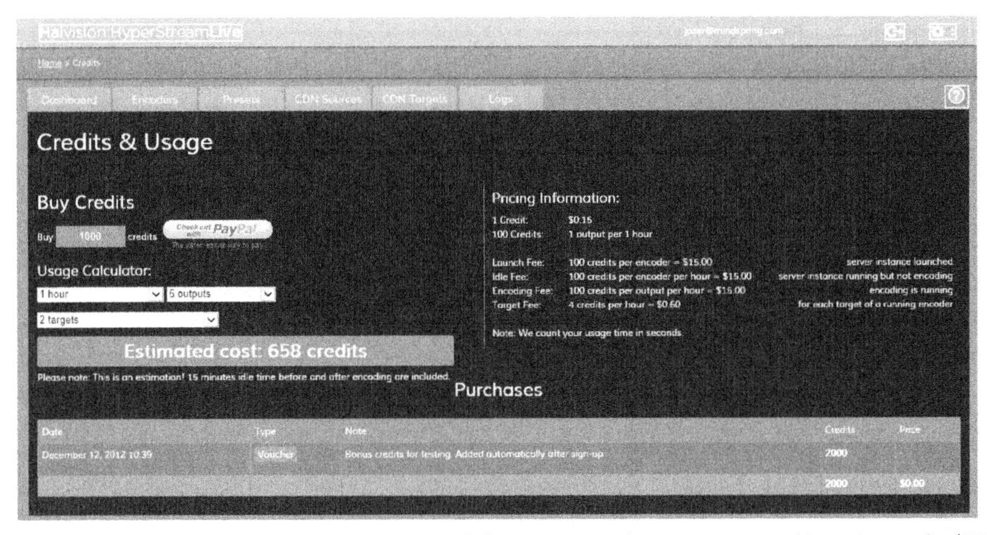

Figure 12-28. A one-hour-long live presentation with five streams and two outputs would cost just under $100.

Conclusion

That concludes our look at choosing and using a live streaming encoder. The next chapter is kind of a hodge-podge, stuff-you-should-know-before-attempting-to-stream-live kind of chapter that should prove much more fun.

Chapter 13: Producing Live Events

> ***Note:*** You can download a PDF file with all figures from this book at bit.ly/Ozer_multi. Since this chapter contains lots of images, now would be a good time to do so.

By now you've learned enough to technically produce a live event, at least from the encoding and distribution perspective. However, in the context of a live event, most streaming producers end up performing double or triple duty in multiple roles—whether driving a camera, helping with the sound and lighting, or simply assisting with overall planning. In addition, it's frustrating to do a competent job with the encoding, but have quality degraded because of poor decisions in other areas, or to produce the event and find that only a handful of viewers watched because inadequate attention was paid to the marketing aspects.

So in this chapter, I'll walk you through the planning stages of a webcast and touch on the other disciplines involved. While an entire book can be written on these topics—a book I plan to write, by the way—even a little bit of knowledge goes a long way. Then I'll take your through two discrete topics: driving a camera in a seminar scenario, which will affect those working as a camera-person in a live event, and managing audio levels, a seemingly simple topic that is actually quite complicated.

Without question, this is mostly entry-level stuff, so if you've produced a few events in the past, feel free to skip on. However, if you've never produced a live event before, you should find it both illuminating and useful. Specifically, in this chapter, you will learn:

- How to plan for a live event, touching on issues like bandwidth, lighting, camera gear, sound, backgrounds and clothing, and staffing
- How to properly frame a speaker or speakers at a seminar or similar production
- How to properly set and adjust the audio levels for a live event

He Who Fails to Plan Is Planning to Fail

This general statement is attributed to Winston Churchill, and it's very true when it comes to live event production. While there are aspects you can correct on event day, there are many that you can't, so if you don't plan ahead, the show may not go on.

A lot of the information in this planning section comes from materials prepared for a webinar I gave with Sonic Foundry in mid-2012 titled "Lights, Camera, Action: Fool Proof Tips to Produce the Most Polished Webcasts on the Planet." The webcast is available on demand at **bit.ly/Ozer_webast**, although you will have to register to watch it. Many of the resources that I refer to in the webcast and in this chapter, including the original slides, are available on my website at **bit.ly/webcastresources**.

Resources for Sonic Foundry Webcast

Written By: Jan Ozer 5-24-2011 Categorized in: Live Production, Streaming production

Here are some resources discussed in my Sonic Foundry Webcast.

Presentation in PDF format

Planning mindmap

Background checklist

Clothing checklist

Flat lighting tutorial

Three-point lighting tutorial

Shooting skills tutorials (including controlling exposure)

Figure 13-1. Resources available at **bit.ly/webcastresources**.

Note that many of the pictures I use in this chapter are in the presentation. So if you're reading the book in black and white and want to see how things look in color, try downloading the presentation in PDF format. I can't guarantee that the picture will be in there, but if not, others demonstrating the same or a similar principle will be.

I'm not quite sure if there's a difference between a live event and a webcast, although the latter certainly seems to connote a smaller event. I've worked events from three-camera concerts with thousands in the stands, to one-camera seminars with hundreds in the seats, to church productions with dozens in the seats, to webcam-driven webinars of just yours truly. If you're looking for tips on how to produce the next presidential inauguration, you probably should look elsewhere. Otherwise, if you'll be producing events similar to mine, the following synthesizes the lessons learned in the many events I've produced. Still with me? OK, let's get started.

Long-Term Planning

There are four items you should look into once it's been decided to produce a live webcast: stream definition, Internet connection, streaming service provider, and recruiting viewers. Let's start with bandwidth.

Defining Your Streams

The critical decision that drives many others is the number of streams, their configuration, and the computers and devices that you want to play on. Often this decision is made in conjunction with other known realities; for example, if outbound bandwidth is fixed and limited, perhaps only a single stream can be broadcast. On the other hand, if you can get a single high-quality stream out, you may be able to transcode that to multiple streams using a service like the Haivision live transcoding service mentioned at the end of Chapter 12.

In addition, if device support is deemed essential (as it almost always is), it may dictate your choice of live streaming service provider. Similarly, if digital rights management, monetization options, registration, closed captioning or other features are involved, they need to be identified as early as possible because they impact nearly all downstream decisions.

You also need to understand the content of the streams. For example, will you need to incorporate PowerPoint into your streams? Do you need polling, Q&A or quizzes, or chat? If your answer is yes to any of these, you may need a rich media presentation system like that discussed in Chapter 11 rather than a traditional streaming server. Will all the speakers be in one place, or will you have remote contributors? This may direct which rich media presentation system you will need to select.

How Much Outbound Bandwidth?

Outbound bandwidth is the signal you'll use to send the encoded streams to your streaming server. If you're new to live event streaming, I can't make this point strongly enough: If you don't have sufficient and preferably dedicated connectivity, you don't have live streaming. Pay attention here, because this is make-or-break kind of stuff. If a hotel or similar facility tells you there is Wi-Fi that you can access, or even direct Ethernet, ask about speeds and whether it's dedicated. If other users can access it, you have no assurance that you can get your signal to the server.

Fortunately, you are probably not the first producer with this concern, so most facility directors can direct you to someone at the local phone company or other service provider that can supply you with dedicated connectivity. This can take anywhere from two days to two months, however, so you should jump on this as soon as possible.

How much bandwidth do you need? Obviously, that depends upon the number of streams you intend to produce and their respective bandwidth. Typically, you should allow for a 20-to-30%

overhead in your calculation. For example, in my office, I have 866 kbps outbound, so typically I stream at no more than 550 kbps. If you have four streams totaling 2.5 Mbps, you'll need 3 Mbps+ to comfortably deliver the streams to the server.

How do you know how much bandwidth you have? The live event webcaster's best friend is speedtest.net, a website that lets you benchmark ping time, download speed and upload speed. For example, Figure 13-2 shows the results from the site from a telephone-company-sponsored concert that featured a 1 Gbps connection. Despite the fabulous 25.13 Mbps outbound connection speed, because the connection was shared, there were short periods during the event where the signal had trouble getting through. You truly can't be too skinny, be too rich, or have too much outbound bandwidth when producing a live webcast.

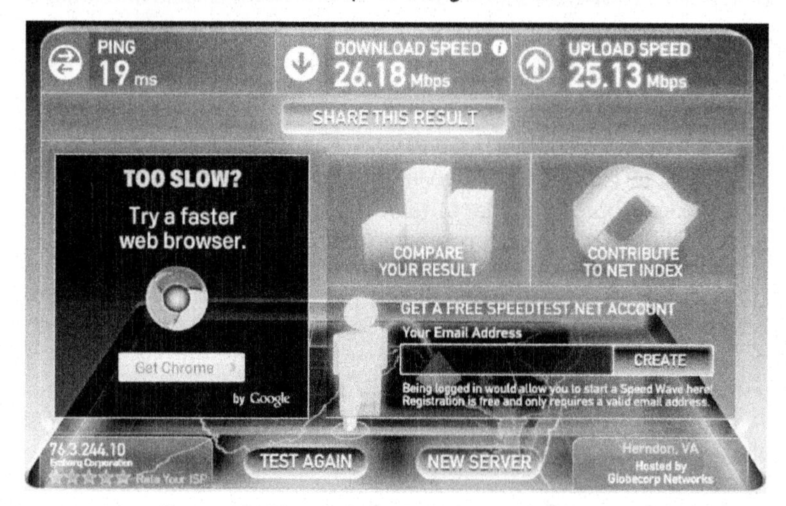

Figure 13-2. The results from a scan at speedtest.net.

If you're webcasting from your own site, surf on over and check your connectivity. If it's inadequate, either upgrade, reduce the size and number of your outbound streams, or cancel the event. In addition, if you're working in a facility shared with others, figure out if there's a way to reserve the bandwidth for the sole use of the webcast; otherwise, if an officemate decides to upload their vacation video to YouTube during the event, you'll have serious issues.

If working with a potential client to produce a webcast at their location, ask them to test the bandwidth during your initial conversation. If it's inadequate or can't be reserved, you have some hard decisions to make. If the event is at a separate facility, explain the issue to the client and make sure that they are on board to pay for decided outbound bandwidth.

If you're working with 4G, are you using a single 4G modem or a 4G aggregator? Have you tested the connection speed at that location? If the answer to the last question is no, you could be in for an unfortunate rude surprise come event day.

Choosing Your Streaming Service Provider

You'll need a streaming server to stream a live event, and in Chapters 10 and 11 I describe the various alternatives and how to chose between them. Often, whether you're working with an internal group in your own enterprise or a consulting client, you'll have to explain the pros and cons of each alternative—whether hosting your own streaming server, or using a live streaming hosting provider (LSHP), traditional online video platform (OVP) or a live streaming service provider (LSSP). I find Table 10-2 helpful in this regard. For some events, you may also have to consider a rich media presentation system.

Whatever option you choose, you need to choose it as early as possible. If you're using your own server or a hosting provider, you have a landing page and player to create. If you're using a live streaming service provider or OVP, you may want to create a page and start getting invitations out. If you choose an LSSP, you also have to decide whether to send viewers to the channel page on the LSSP site or to an embedded player on your own site.

While not as time-sensitive as the outbound bandwidth decision, it's critical to choose your approach as soon as possible, since you'll need a fixed URL to send viewers to and may be using features provided by your service provider to help market the event.

Marketing: Getting Viewers

In my experience, sometimes webcasts are almost afterthoughts, as in, "Hey, we're having a conference, why don't we stream it live?" In this context, almost no marketing is dedicated on the streaming aspect and you end up with suboptimal web viewing numbers.

I'm not qualified to tell you how to market an event, but I will say that most event marketers need at least eight weeks to market the event properly, so if you missed that timeframe, you're already behind the eight ball. I will also say that the technical aspects of webcasting, while seemingly magic to non-technical users, are actually pretty mechanical after you've done a few. The hard part is getting viewers to watch the event.

If you're the person writing the check for the event, ask, "How are we going to market this event?" Any plan should be formal and written, and should involve both multiple pitches to both internal and external lists. It should start early and end late, continuing right up until the event.

Production Planning

Once the high-level plans are in place and being implemented, it's time to turn your attention to the production aspects. If you've never produced a broadcast before and don't have access to production gear, you're probably better off hiring a production team to do the job for you. I'm going to assume that you have access to production gear and want to produce some or all of the event yourself.

There are multiple aspects of the event that you have to consider; let's take them each in turn.

Background and Clothing

Scout the background early, as some backgrounds are simply unworkable. For example, if you're shooting against a background with lots of fine detail, like a paisley or herringbone wallpaper, or a rattan curtain, the detail will turn to shifting piles of artifacts that will be distracting and look unprofessional. Ditto for trees or other shrubbery—particularly on a windy day, which adds detail and motion.

You should avoid very dark colors, or very bright colors. These can be workable, as we know from all the Apple commercials shot against the infinite white background, but they both add a level of complexity you don't need with a live event. Avoid very bright lights behind the subject, whether natural light or artificial, since these can cause backlighting.

What about things to seek rather than avoid? Seek simplicity: solid, non-reflective colors that will contrast easily with appropriate clothing worn by the speakers. Black curtains are very common, but many speakers wear black suits, or have black hair, or both, which makes the contrast tough. You can see this in Figure 13-3 from a webcast I helped produce in Washington, DC, in 2012.

When I have control over the background, I prefer light grays or blues—you'll see the background I use in my office in a few pages. I also tell my speakers what the background color will be and ask them to wear colors that will contrast well with the background.

Figure 13-3. Black backgrounds are very common, but tough when your subjects wear black.

Other clothing and appearance-related advice includes:

- Avoid fine detail like pinstripes, plaids or other patterns. These can cause artifacts, just like backgrounds with fine details. My favorite example of this is an old shot from the *Wall Street Journal* website. If you're viewing this in grayscale, you're missing the moiré pattern on the blue-and-white pinstripe shirt on the left, but the artifacts from the fine patterns is clear on both shirts.

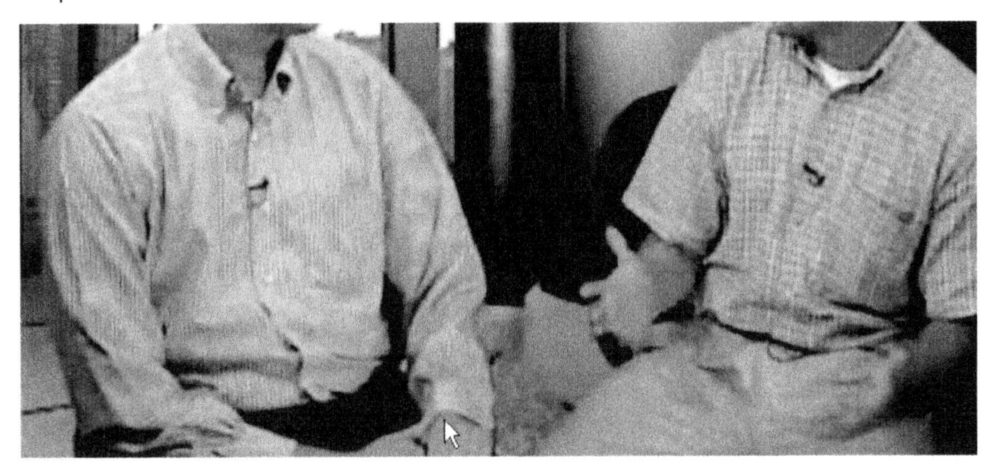

Figure 13-4. Too much detail causes very noticeable artifacts.

- Avoid clothing with extremes in brightness and darkness, like a black suit and white shirt. Extremes in brightness can cause contrast ratio issues that result in detail loss at either or both extremes. A brown coat and light blue shirt works much better than a black or very dark blue suit with a bright, white shirt.

- Avoid highly saturated colors like bright red, green or blue. Highly saturated colors tend to bloom and bleed, both of which look awful.

- Be conservative with hair grooming. Frizzy, curly or otherwise wild hair seldom compresses well, and often creates noticeable compression artifacts.

- Avoid glittery jewelry. This can cause artifacts during compression, or glint and shine unnaturally.

- Wear contacts if possible. Glasses add complexity to the lighting since lights reflecting in the glasses are distracting. If the subject can wear contacts, that's simpler. Otherwise, light from the sides so the lights aren't reflected in the camera.

These tips are all available on the clothing checklist you can download from **bit.ly/webcastresources**.

Lighting

Low lighting is the absolute enemy of compression. As a camera person, you have two options: keep the video dark, which doesn't work, or boost the camera gain, which adds noise to the video and degrades quality. How can you tell when lighting is sufficient? It's a bit complicated, but my definition is when exposure on the face is between 70 and 80 IRE with no gain adjustment.

Let's take this one step at a time. Figure 13-5 shows my buddy Gary on the right and a waveform monitor on the left, courtesy of the scopes in Adobe Premiere Pro. The waveform monitor measures the brightness of pixels along the IRE scale (which stands for Institute of Radio Engineers). As you can see from the values to the left of the waveform, the scale starts at 0, which is true black, and ends at 100, which is true white. Any pixels darker than 0 or brighter than 100 are captured as solid black or white, with no discernable detail. If you've ever heard someone warn against crushing the blacks or blowing out the whites, that's what they're describing.

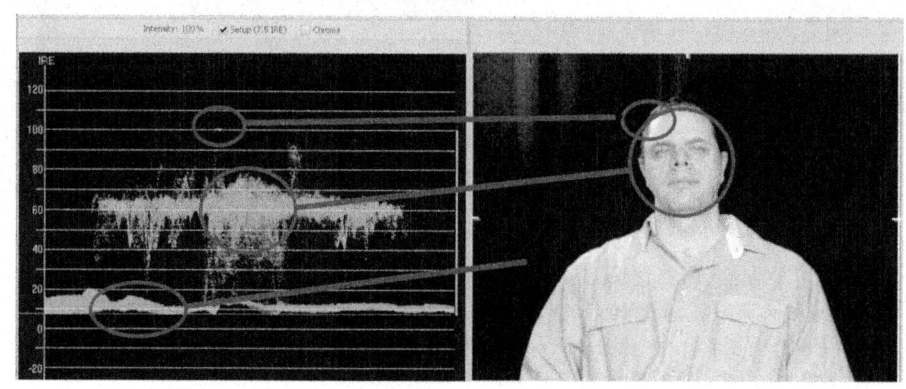

Figure 13-5. A waveform measures brightness on the IRE scale.

When setting exposure with your camera, a procedure that I'll describe below, the goal is to make sure the brightest regions in the face are between 70-80 IRE. If you have a waveform monitor, it's pretty simple. But most producers don't, so you have to use an alternative technique involving zebra stripes.

One more piece of background, then I'll describe the technique. In the waveform, note that the brightness of the pixels are shown at their horizontal location in the frame. For example, the band of pixels at or near the 0 IRE mark is the black background. The blob of pixels in the middle is Gary's face, which peaks just above 75 IRE. The blip at the top is the bright spot on Gary's forehead. While I would attempt to eliminate the bright spot, otherwise the values look good.

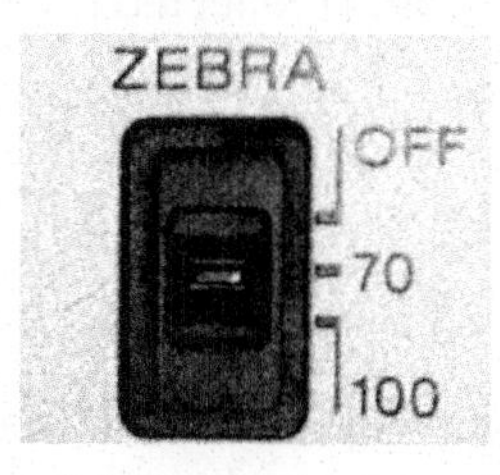

Figure 13-6. Zebra settings on my retired Sony DCR-VX2000.

How do you gauge brightness on a shoot? Using a tool on most prosumer and professional cameras called zebra stripes. Figure 13-6 shows the external control on my retired Sony DCR-VX2000, one of the world's first three-CCD DV camera and quite the camera back in 1997. As you can see, you have three settings: off, 70 IRE and 100 IRE. On most newer camcorders, the controls are in the menu, and you can set any value that you'd like.

When the face is the most important element in the picture, as it is in Figure 13-7, set the shutter speed to 1/60th of a second (on the left in the frame), and set the gain to 0dB (left center frame), and then adjust the aperture until the zebra stripes are apparent in the brightest portions of the face. In Figure 13-7, it's mission accomplished: There are zebra stripes on the lead singer's forehead, but not the entire face, which would be too bright.

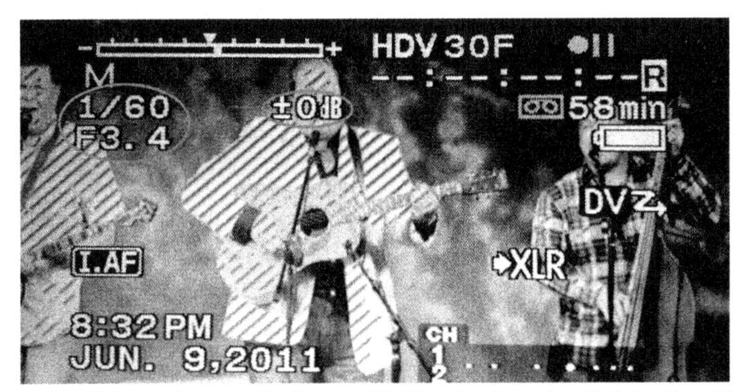

Figure 13-7. Zebra stripes on the face and 0 dB gain show that lighting is sufficient.

Let's return to my initial statement when I said that lighting is sufficient when exposure on the face is between 70 and 80 IRE with no gain adjustment. In Figure 13-7, gain is set to 0dB and I've got my zebra stripes, so lighting is adequate. If I were in charge of lighting and I couldn't get the stripes to appear at 0 dB, I would add lights. If you're not in charge of lighting, you have to boost the gain until the zebra stripes appear.

When would you set zebras to 100 IRE? When the face isn't the most critical detail to be preserved in the scene. A great example is a ballet performance, where the whites in the dresses shouldn't exceed 100 IRE or the ballerinas will look like blobs of white dancing across the screen. For these shoots, I set the zebras at 100 and adjust exposure so that just a hint of zebras appear, as you see in Figure 13-8. Obviously, the face is much too dark, but if I exposed the face at 70 IRE, all the detail in the dress would be lost. To retain my shooter's credibility, I should point out that I truncated the frame shown in Figure 13-8 to save some space on the page; when originally shot, the framing included both the hands and feet, which is the cardinal rule of shooting ballet.

When shooting interviews, meetings, conferences and speeches, set the zebras to 70 or so and make sure the faces are adequately lit. For other performances, identify the most critical element that must be preserved and set the zebra accordingly.

Figure 13-8. Set zebras at 100 IRE here so you don't blow out the tutus.

Whew, what a detour. The key point in this section is to make sure that lighting is adequate and how to determine when it is. What can you do if lighting isn't adequate? Depends upon the situation. If you're in charge of lighting, you can just add more light, or move it closer to the subject; more on this below.

If you're solely the shooter/encoder, advise the person in charge of the event that they need to add more light, or focus the light more on the speaker. This is usually pretty simple to resolve if you're working in a studio or on a stage, but can take hours to accomplish if you're working at a hotel or similar location where they may have to bring in additional light. Again, if you can't increase the lighting, you'll have to boost the gain on the camcorder until zebra stripes indicate that lighting is adequate.

Let's cover a few other lighting-related points.

> **Tip:** *The shooting skills tutorials document that you can download at* **bit.ly/webcastresources** *goes into a bit more detail about the settings to use while shooting and how to monitor your zebra stripes.*

Don't use Overhead Lights for Small Webcasts

One common issue with small webcasts is using the overhead fluorescents as the sole source of light, as shown in Figure 13-9, which causes the bottom of the face to be noticeably darker than the top. Sometimes you can offset this by using a bounce card on the subject's lap that reflects light to the lower regions of the face, but I prefer to shut the ceiling lights and light exclusively with compact fluorescent soft boxes that you'll see in a moment. See **bit.ly/bounce_card** for more on bounce cards.

Figure 13-9. Using solely overhead lighting makes the bottom of the face too dark.

If you're working in a hotel ballroom or similar environment with predominant top-down lighting, request spotlights pointing at the speakers from around eye level to light the bottom of the faces. Make sure that the color temperature of this additional lighting matches the temperature of the overhead lighting (see the next section).

Don't Mix Lights with Different Temperatures

If you haven't downloaded the presentation at **bit.ly/webcastresources** yet and you're reading in black and white, here's one that you have to see in color. Both pictures are lit the same way with incandescent lighting with a yellowish tint from the front, and daylight with a bluish tint streaming in through the window. Gary's shirt is actually white.

Figure 13-10. If you mix lights on the set, you can't get the color temperature correct.

On the left, I've white-balanced for daylight, and Gary's shirt looks yellowish. On the right, I've white-balanced for incandescent, and the daylight looks blue. Obviously, if I had to go one way or the other, I'd go with the setting on the right, but sometimes you end up with no good setting. I have an example on my website at **bit.ly/mix_temps** where the subject's shirt is white on one side and blue on the other because he used mixed lighting.

I use compact fluorescent or LED lighting for all my shoots nowadays, including webcasts. The lighting mixes well with daylight, although time of day matters since daylight is most yellowish at noon or thereabouts, while bluish in the AM and afternoons. To be totally safe, you should shut the lights and the curtains and light the event yourself.

Use Flat Lighting for Live Events

There are two basic lighting styles: the dramatic, shadow-filled, three-point lighting on the left in Figure 13-11, and the flat, no-shadows flat lighting on the right. Which to use for your live event?

Figure 13-11. Shadowed three-point lighting on the left, flat on the right..

I recommend flat lighting, which is the lighting style used by all news and similar broadcasts on TV and the web; check this out next time you watch the evening news or SportsCenter. Flat lighting is much easier to produce and much less error-prone, and typically is a better match for the matter-of-fact information presented in a meeting or webcast.

In a stage or in a meeting room, you produce flat lighting by pointing multiple lights from many different directions at all speakers; the more lights, the better. I am not an expert here, so if you're in charge of producing a big event and need help with the lighting, you might find a professional in your area who can help.

For a personal webcast, you should download the flat lighting tutorial available at **bit.ly/webcastresources**. The basic message is to use two lights of equal power at about 45 degrees from the speaker's face a bit above eye level. You can see my typical webcast setup in Figure 13-12, including the compact fluorescent lights (CFL) I use, which I picked up for around $150 or so on Amazon (**bit.ly/cheap_cfl**).

Figure 13-12. Producing flat lighting with two CFL softboxes.

While the average rating of the product is around three stars, most of the negative comments focus on construction and assembly. I agree on both points; the plastic parts feel about one strong twist away from breaking and it takes way too long to put together the light boxes. However, I keep the lights assembled and am careful when I adjust them, and they work just fine. While I wouldn't buy these if the predominant use was on the road, they're great for using in and around my office, and you absolutely can't beat the price.

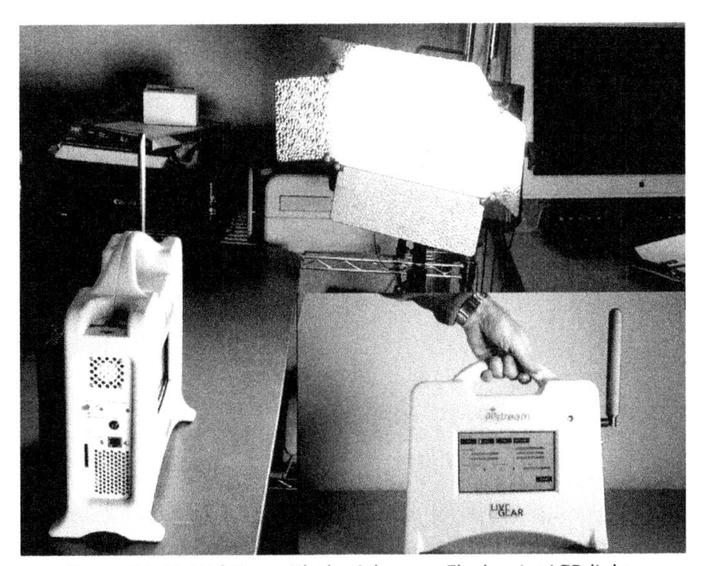

Figure 13-13. Lighting with the Adorama Flashpoint LED lights.

In the meantime, keep your eye on LED lighting, because sometime soon it will replace the bulky compact fluorescent lights shown in Figure 13-12. I recently reviewed two sets of these lights in a review of two Adorama LED lights you can read at **bit.ly/LED_lights**. The lights are smaller and easier to pack up and set up, and the barn doors let you focus the light when needed.

The only real negative about the Adorama light kits was that they were cheaply made, although certainly still usable. In Figure 13-13, I'm lighting a product video that I produced with the LED light on the right and one of the CFL softboxes on the left (and out of the picture), with the actual video shown on the bottom right. The color temperatures of the two lights match, so you can easily integrate LED lights with any existing CFL lights. LED light kits do cost about twice as much as CFLs, but if I were buying lights now and could find a light that seemed sufficiently sturdy, I would definitely buy the LEDs.

Camera-Related Options

This section and the next are more targeted toward small, personal webcasts than large staged events and focus on basic equipment decisions like webcam or video camera, and the best microphone alternatives.

Let's start with camera-related tests. Embedded webcams in notebooks are exceptionally easy to use, but how does the quality compare with consumer and prosumer camcorders? To assess this, I ran two tests. The first compared the quality produced by the webcam on my MacBook Pro to a Canon ZR500, a $500 consumer camcorder, and the Canon XH A1, a $3,500 prosumer camcorder. For this test, I placed the camera about 3 feet away and captured a 640x480 SD video at 1 Mbps, with all three devices using Telestream Wirecast, connecting to the two camcorders via FireWire.

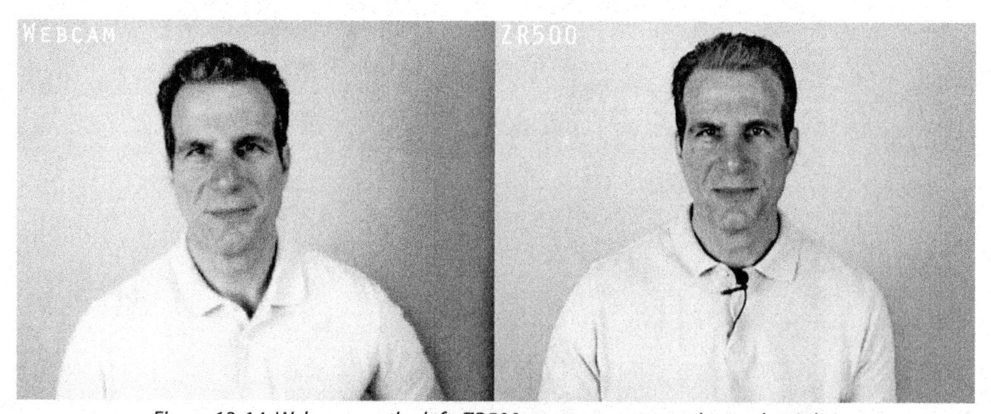

Figure 13-14. Webcam on the left, ZR500 consumer camcorder on the right.

If you're reading from the book, you'll definitely want to download the presentation at **bit.ly/webcastresources** to see the difference here, which is very noticeable in digital form but much less so on paper. Specifically, the camcorder image is noticeably sharper and the colors more distinct. The bottom line is that if you have a simple option to use a camcorder instead of your webcam, you should do it.

How did the ZR500 consumer camcorder compare with the XH A1 that costs six times as much? Have a look at Figure 13-15. Even at 640x480 resolution, the XH A1 produces more clarity, and the color accuracy is significantly better. That's because the XH A1 lets me set the white balance manually, while the ZR500 doesn't. So the background looks true blue, its real color, rather than gray, which is what the ZR500 thought it was. The image is also brighter because I could set exposure by hand, which I couldn't readily do with the Z500.

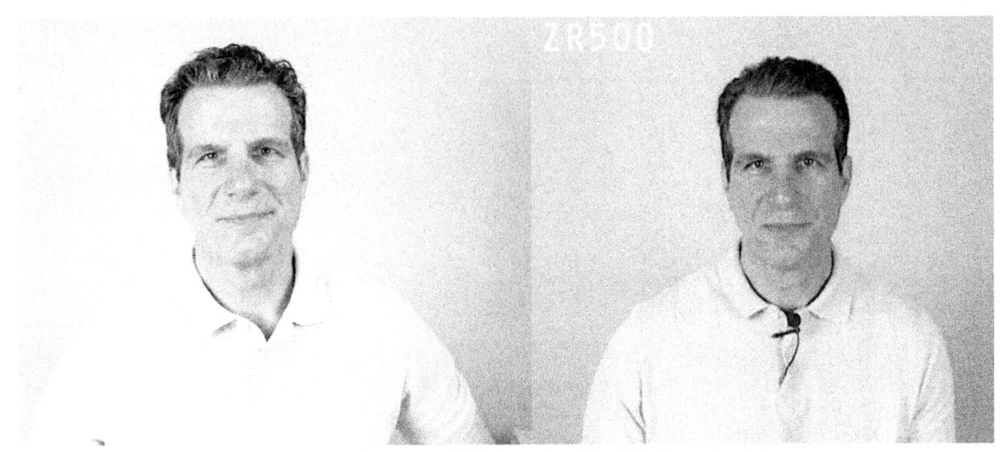

Figure 13-15. Prosumer XH A1 on the left, ZR500 on the right.

The bottom line is that better cameras produce better results, not only because they use bigger, higher-quality imaging components, but because they afford greater control. Webcams are undoubtedly easier, but if you're all about the impression you make, use a camera. If you're going to use a camera, better cameras do make a difference. Note that in an article you can read at **bit.ly/Awesome_wc**, I compared the quality of a stand-alone webcam with that of an embedded webcam, and I reached the same conclusion: The stand-alone webcam is bigger, which enables better optics and produces a superior image.

The second test compared three camcorders about 10 feet away from the subject and capturing at 848x480 @ 1.5 Mbps, again using Wirecast but connecting via a Blackmagic Design DeckLink 2 card. The three cameras here were Panasonic AG-HMC150, an early AVCHD camcorder; my Canon Vixia HF S10 that I use on vacation; and the XH A1. The comparison frames are shown in Figure 13-16.

In terms of clarity, the Vixia more than held its own, but in terms of color accuracy, the two prosumer cameras were clearly superior—although you'd have to be looking at a color image to see that distinction. I'm guessing the Vixia has manual white balance mode that would have addressed this, but I shot in full auto assuming that most newbies would take that approach.

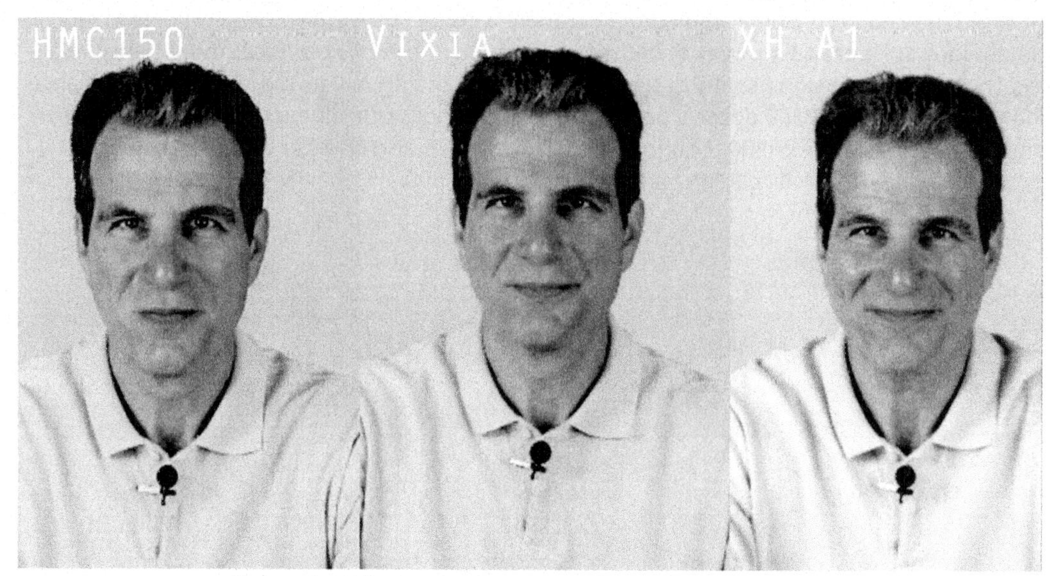

Figure 13-16. Three cameras compared from about 10 feet away.

The bottom line is that better cameras deliver better results, even after all the compression applied for live streaming. I prefer higher-end gear because the cameras are easier to configure.

> **Tip:** *I'm a big believer in setting everything up and getting it running in the office, then tearing down the gear and making sure it's all safely packed and brought to the event. While it's useful to think through what you need to bring, actually making it work is the best way I've found to ensure I do bring all needed gear.*

A couple of additional points, and then we can move on to audio.

Progressive or Interlaced?

First, on the settings that I use for shooting, I always shoot at 30 fps, usually in progressive mode, although there are a couple of exceptions. First is when you're working with a mixer that prefers progressive. For example, when I reviewed the Livestream Studio HD500 (**bit.ly/LS_HD500**), Livestream sent one of its producers down with some extra cameras and the mixer to make sure the experience was a positive one. This guy had a pretty impressive résumé, having done projects with the likes of Justin Bieber, John Travolta and many others.

He recommended shooting in interlaced mode because that was the signal that most professional producers used. I, of course, went with his suggestion.

I've also run into issues working with real-time encoding boxes that only accepted an SD signal. The problem with shooting SD progressive is that technically, 480p is not SD, which is set up for a maximum of 240 lines for NTSC. For example, when I tested the ViewCast Niagara 2120 encoder (**bit.ly/Ozer_2120**), the maximum signal it would accept was 480i, so progressive didn't work.

Otherwise, I always shoot in progressive mode. For an overview as to why, check out my article "Shooting for Streaming—Progressive or Interlaced?" at **bit.ly/prog_inter**.

White Balance

It's best practice to white balance for each shoot. Procedures vary by camera, but typically involve focusing on a totally white object and pressing a white balance button on the camera (after disabling auto-white balance, of course). Check your camera manual for precise procedures.

Configuring Exposure

Moving on to other configuration parameters, I almost always set exposure manually, setting the shutter speed at 60, gain to 0 and the aperture to whichever value gets my zebra stripes to their happy place. In Figure 13-7, that was a setting of 3.4.

I will shoot in automatic exposure mode when I'm controlling two cameras in a dynamic lighting environment like a ballet or some concerts. I frame one camera on the entire stage using automatic exposure settings, or often spotlight mode. Then I follow the action with the other camera, changing exposure as needed according to my zebras. When shooting an event with stable lighting, I would use manual settings on all cameras.

Note that dynamic lighting environments aren't limited to stage productions. Often when you shoot outside the lighting can change dramatically—say, when a cloud obscures the sun. I've seen some outdoor videos, obviously shot with manual exposure, which became unusable in these instances. So be careful setting exposure manually when you're running multiple cameras.

Why the bias against auto-exposure? Because when you use auto-exposure, you can't really control how the critical elements of the scene are exposed. For example, if there's a light in the background, the auto-exposure control may darken foreground objects, like your subject's face, because it doesn't know what portion of the frame is important.

In addition, in auto-exposure mode, camera will adjust exposure even when you don't want it to. For example, if the subject leans forward, perhaps making the frame a bit darker, the camera will respond by brightening the entire frame. If your subject holds up a piece of white paper, brightening the entire frame, the camera will make the entire frame darker. The only way to get exposure where you want it and keep it there is via manual exposure control.

Configuring Focus

When shooting an interview, seminar or other video where the subjects don't move closer to or further from the camera, I would typically set focus manually. In these instances, auto-focus can often lead to disasters as the subjects move laterally, because the camera may decide to focus on the background rather than the subject.

In a dynamic environment like most stage productions, I typically use auto-focus, which can respond much faster than I can to motion on the set. My views here are hardly universal, however, so if your personal preference is different, go your own way.

It sounds corny, but before every shoot, I sit down with each camera and a checklist of items listed below and configure all the controls as I want them to be. This is absolutely critical given I may be switching from an outdoor interview (manual focus, auto-exposure) to an indoor stage event (auto-focus, manual exposure). I have an actual checklist because typically I'm a one-man production team, and if I don't discretely focus on each discipline, something will fall through the cracks. I've included the checklist as Figure 13-17, and I recommend you use it before every shoot.

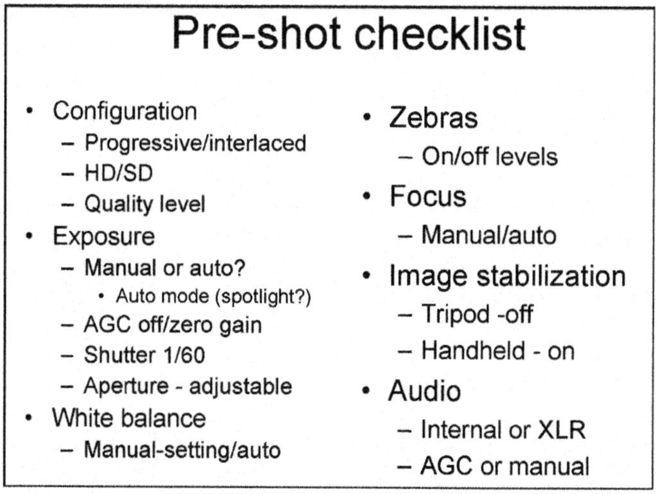

Pre-shot checklist

- Configuration
  - Progressive/interlaced
  - HD/SD
  - Quality level
- Exposure
  - Manual or auto?
    - Auto mode (spotlight?)
  - AGC off/zero gain
  - Shutter 1/60
  - Aperture - adjustable
- White balance
  - Manual-setting/auto
- Zebras
  - On/off levels
- Focus
  - Manual/auto
- Image stabilization
  - Tripod -off
  - Handheld - on
- Audio
  - Internal or XLR
  - AGC or manual

Figure 13-17. Here are the items you should set or check before every shoot.

Taking Care of Audio

While most users expect some visible artifacts with their video, they've come to expect pristine, trouble-free audio. For this reason, if you don't think audio is at least as important as video, you and your clients will be disappointed with your results. Let's start with a quick primer on audio; then we'll take a look at a few common scenarios and get some recommendations about which type of microphone to use and how to get it connected.

Audio Primer

There are multiple aspects you need to know about microphones before you can understand which is best for the job at hand. First is pickup pattern.

Pickup Pattern

Pickup pattern describes where the microphone picks up the bulk of its audio, and it's important to use the right pickup pattern for the particular job at hand. Here are the pickup patterns that you'll be working with most. Note that Audio-Technica has a great reference on these patterns that you can read at bit.ly/AT_pickup.

- ***Omni-directional.*** Captures sound equally from around the microphone. Typically, the microphones used on notebook computers and camcorders are omni-directional, so they pick up lots of ambient noise. You would want to use an omni-directional microphone when several speakers or singers are sharing a microphone, but the ambient sound gets distracting when recording a single speaker.

- ***Cardioid.*** The cardioid pattern captures sound in front of the microphone and is the most common pattern for handheld and lavalier microphones. You can see the pickup pattern in Figure 13-18.

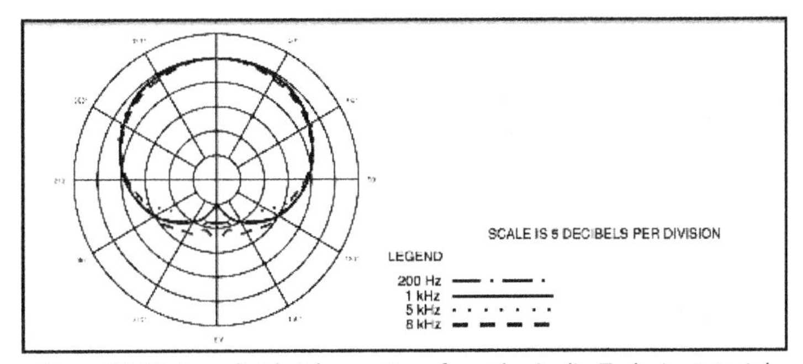

Figure 13-18. A cardioid pickup pattern from the Audio-Technica tutorial.

- ***Hyper-cardioid.*** This pattern is a bit more front-focused, but also picks up some sound from the rear.

- ***Line and gradient.*** This pattern is used on shotgun microphones to pick up sound from directly in front of the microphone. Note that shotgun microphones don't really amplify, or reach out and grab sound from long distances—they just do a better job than other mics at canceling out ambient noise. For this reason, even shotgun microphones work best within 4 feet of the subject, and start capturing poor quality after about 8 to 10 feet.

Condenser and Dynamic Microphones

Microphones fall into two categories: dynamic and condenser. Dynamic microphones are driven by vibration and require no external power. They're typically more rugged than condenser microphones and superior for field use—and they're cheaper. The most significant downside is that they typically don't produce as strong a signal as a condenser microphone, which means more amplification and more noise.

Condenser microphones are driven by internal condensers that require 48 volts of phantom power, which can be supplied by a battery or AC power or the camera. Condenser microphones are more fragile and more expensive than dynamic microphones, but they typically produce higher-quality sound in a controlled environment.

Figure 13-19. Microphone power switches from the Panasonic HMC150 camcorder.

Beyond these basics, choosing a microphone is very subjective. For example, our local radio station recently transitioned back to dynamic microphones because condensers were too fragile, and the quality difference was negligible. In contrast, for narration, I prefer condensers because of the additional volume. There is no right or wrong answer here—only preferences.

The key point is that when you're using a condenser microphone, you have to supply power. Most cameras and audio/video mixers can supply 48 volts of power (called phantom power) to a microphone, although you typically have to flip a switch or configure an option in the menu to do so. Figure 13-19 shows the switches from my Panasonic AG-HMC150 camcorder.

Wired or Wireless

This category refers to how the microphone connects to the camcorder or audio or video mixer. Wired microphones connect directly to the camcorder, which is typically simpler but only works when the subject is close to the camcorder and isn't moving around that much. Wireless connections work best over longer distances or when the speaker is moving around.

All wireless systems transmit to a base unit that connects to your camcorder. There are two technologies used for this: UHF and VHF. UHF systems provide more channels, decreasing the risk of interference with other wireless devices, but work at higher frequencies and require more expensive components. If you're shooting a dual-mic wedding out in the country, VHF should probably perform as well as UHF. If you're shooting a 10-microphone performance in mid-town Manhattan, UHF is the way to go.

If you're looking for a truly portable system, you should get a system with a battery-powered receiver that sits on your camera's cold shoe mount, like the dual channel Azden 330ULT shown in Figure 13-20. The other alternative is an AC-powered system that tethers your shoots to locations with power and limits your camera movement.

Figure 13-20. The dual-channel Azden 330ULT mounts on my camcorder for ultimate portability.

Microphone Connection

Most microphones connect via one of two connectors: the 1/8-inch stereo connector (or 3.5mm) found on the left in Figure 13-21, or the XLRM (for XLR male connector) shown on the right. Most professional gear uses XLR because it produces a balanced signal that can travel further. Note that all 3-pin XLR connectors are mono. For this reason, most prosumer and above camcorders have XLR inputs with phantom power for condenser microphones.

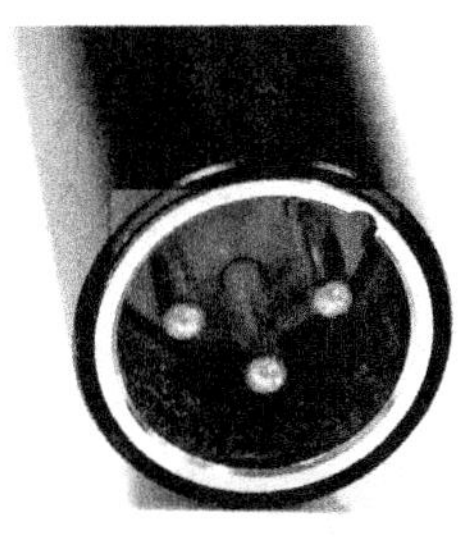

Figure 13-21. A 1/8-inch stereo connector on the left, male XLR on the right.

The key here is to find a microphone with an output that matches the input on your recording device. If you don't, you'll need an adapter of some sort, which is just another piece of gear to bring (or forget to bring), and power. I'll discuss a few adapter options later in the chapter.

Choosing the Right Microphone

Briefly, whenever you produce an event, you will have one or more of the following alternatives.

- ***Internal microphone on your notebook.*** If you're capturing a webcast from your notebook, you can use your notebook's internal microphone. While this is the simplest approach for these types of webcasts, the audio has a noticeably tinny quality and the omni-directional microphone picks up lots of background noise.

- ***Internal microphone on your camcorder.*** If you're capturing a webcast with a camcorder, you can use the internal microphone on the camcorder. This approach is also fairly easy, but has several negatives. First, camera microphones are omni-directional, so will also pick up lots of background noise, which typically isn't what you want when shooting a live event or interviewing someone. In addition, camera microphones sound tinny, and can pick up noise if you adjust controls on the camera body.

Figure 13-22. Shotgun microphones pick up sound in the direction to which it's pointed.

- ***Shotgun microphone on your camcorder.*** If you're capturing a webcast with your camcorder, you can connect a shotgun microphone to the camcorder. Briefly, shotgun microphones are uni-directional and designed to pick up sounds from the direction to which it's pointed. Like the Azden SGM-1X shown in Figure 13-22, they typically ship with a shock mount, so they don't pick up noises from the camera body. Shotguns are convenient since you don't have to attach a microphone to the speaker, but obviously lose clarity and volume as you move away from the speaker.

- ***Lavalier microphone input into your camcorder.*** If you're capturing a webcast with your camcorder, you can connect a lavalier microphone to the speaker like the microphone shown in all frames in Figure 13-16. This can be a bit of a hassle, since you have to hook it up for each speaker, but you end up with the microphone extremely close to the source of the audio, which produces the best result.

- ***Audio capture directly from the soundboard.*** If you're producing an event where the sound is captured via a soundboard or video mixer, often you can take a direct feed from the soundboard or mixer into your camcorder. While this can be a hassle logistically and connection-wise, it guarantees the best overall result.

To a great degree, the quality produced by some of these techniques is affected by distance from the sound source, which is summarized in Table 13-1. If the microphone is within 4 feet of the subject, you can get away with using any of the approaches, though audio from your notebook and internal microphone won't sound as good as the other approaches. At 10 feet, the internal microphones from these devices become unworkable, while the shotgun microphone is sketchy.

	4 Feet or Less	Up to 10 Feet	Beyond 10 Feet
Internal Microphone-Notebook	Suboptimal	No	No
Internal Microphone-Camera	OK	No	No
Shotgun	Very Good	Passable	Suboptimal
Lavalier	Excellent	Excellent	Excellent
Sound Board	Excellent	Excellent	Excellent

Table 13-1. Quality produced by these sources at various distances from the subject.

Beyond 10 feet, the only viable approaches are a lavalier microphone or audio from the soundboard, which is presumably produced by a lavalier or other microphone proximate to the subject. Note that I wrote an article for Onlinevideo.net titled "Microphone Performance Tests: Your Built-in Mic Isn't Good Enough," that you can read at **bit.ly/ozermics**. You can also listen to the sound produced by the different options I tested, which included the internal microphone of my camcorder vs. the shotgun, and two lavalier systems. If you want to put sounds to the words on this page, go and have a listen.

Other considerations include the type of event. If it's a concert or high-quality ballet or opera, anything other than audio from the soundboard simply won't do, even if you're 3 feet from the stage. In particular, most musicians value the audio experience far more than the visual. If you don't absolutely nail the audio, they'll consider the event a bust. If it's a football game and the value of the experience is almost all contained in the visual portion, you can get away with lower-quality sound.

Many years ago, I captured my first concert using the microphone on a Canon XL2 camcorder. The performer was Taimo Toomast, an Estonian opera singer, who was fabulous. It was my first two-camera shoot, and when I showed him the DVD I produced three weeks later, he oohed and ahhed at the video—and appropriately so, because the video quality was marvelous. Fishing for a compliment, I asked him, "So, how's the audio quality?" Taimo looked over and said, "Oh, it's horrible," like he was saying, "Your shirt is blue."

Which should have come as no surprise to me. Objectively speaking, it's virtually impossible to record a concert from the back of the room and hope to get even passable quality, at least from the perspective of a true artist. You're just too far from the source, and end up recording echoes from the walls and ceiling, noise from the HVAC system, crowd noises, and the inevitable camera-handling clatter.

So if you're producing any live event where there's a soundboard, ask to speak with the sound guy about getting a feed from the board. I've never been refused, but keep in mind that there are logistical problems—particularly when the soundboard is off to the side and you're shooting from the back—and/or connection issues, particularly if you don't have a camera with XLR connectors (more below). So chat with the sound guy (they're always guys) well in advance; ask where he'll set up and what connection he can send you.

> **Tip:** *Whenever possible, I prefer to shoot using batteries rather than AC power. This preference dates back to a horror story you can read about at* **bit.ly/audiogonebad**. *The* CliffsNotes *version is that while shooting a concert, interference from a radio station somehow wormed its way into my audio file, potentially when a microphone cable crossed over the power cord to my camcorder. As one experienced audio pro explained to me, all long cords can function as an antenna, and antennas can often pick up the strangest signals. Ever since then, I've worked with batteries whenever I can, especially when using wired microphones.*

Other Webcast Options

If you're producing a webcast from or around your desk, there are a couple of other options you may want to consider. First, even if you're using the webcam on your notebook, you can typically buy an external microphone for under $20 that will produce better results than the internal microphone. If you visit Amazon and search for "microphone for PC," you'll find plenty of options.

The other approach I frequently use is to place a shotgun microphone on a stand and move it close to where I'll be sitting, which you can see in Figure 13-23. This produces about the same sound quality as a lavalier, but is slightly easier to set up, and there's no lavalier in the picture. This could also work if you have two speakers chatting, but only if they're close to each other.

Figure 13-23. Attach the shotgun to a microphone stand and move it close to the subject.

Getting Audio Connected

There are two aspects of getting your audio connected. First you have to connect your microphones and other inputs into the camcorder while recording. Then you have to transfer the sound out to your encoding device. We'll cover both aspects.

Connecting a Microphone—XLR

Connecting a microphone to a camera can be surprisingly challenging, and it's an exercise performed well in advance of the actual live event. At a high level, you have to accomplish multiple steps to make it work, which is why it's so complicated. Adding to the confusion is that the procedures are different for each camera, although the basic steps are all there. For example, sometimes you switch from the camcorder's internal microphone via the menu and sometimes the switch is on the camera body. However, no matter which camcorder you're working with, you will have to tell the camcorder to switch from internal to external sources.

1. ***Physically connect the microphone to the camera.*** If you're connecting two XLR inputs into the camera, this is simple: Insert the left cable into Channel 1, the right in Channel 2. If you're connecting a single XLR connector, typically you want the audio to flow to both tracks. Even if you're encoding in mono, this prevents errors like audio only sounding on side. In this case, insert the cable into Channel 1 and use the switch shown in Figure 13-24 to send the audio to both channels.

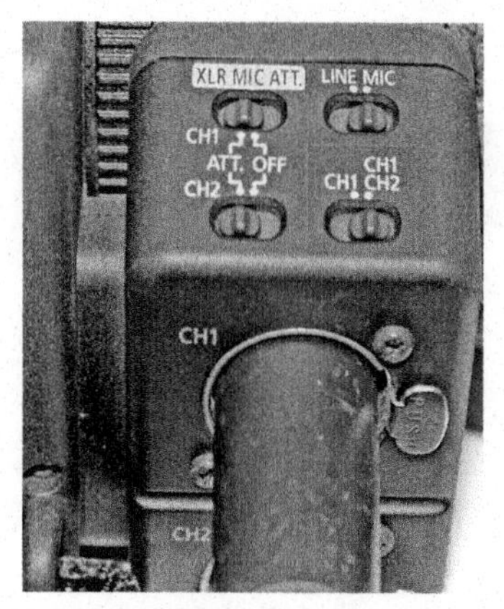

Figure 13-24. Telling the XH A1 to send the audio received from Channel 1 to both channels

2. ***Turn on phantom power if the microphone needs power.*** The XH A1's controls are in front of the XLR jacks as shown. I'm working with the battery-powered Azden shotgun microphone, so I don't need power.

Figure 13-25. Here's where you turn on phantom power when needed.

3. ***Turn the microphone on (if necessary).*** Don't forget this stage or you'll waste several minutes figuring out why you're not hearing anything in your headphones.

4. ***Switch the camcorder from the internal microphone to the XLR connection.*** This is very camcorder-specific—sometimes performed via switches on the camera body, sometimes via menu options. Note that if the audio is too faint, I can boost XLR gain by 12 dB using the XLR Gain Up control.

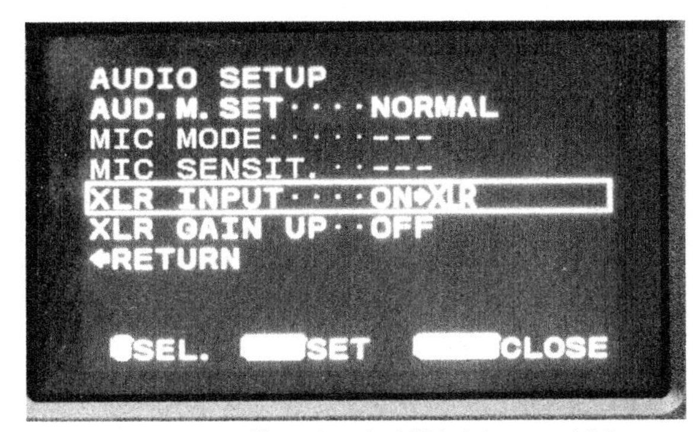

Figure 13-26. Changing the XH A1's input to XLR.

5. Select line or microphone input. Microphone input is much less powerful than the input from an audio mixer or other similar device. To handle both, cameras have microphone/line switches that let you choose the input. In this case, since I'm connecting to a microphone, I'll choose Mic. Note that if you're connecting to a microphone and don't hear anything in the headphones, it may be because you're got this switch configured to Line. If you're connecting to a powered device like a soundboard, choose Line input.

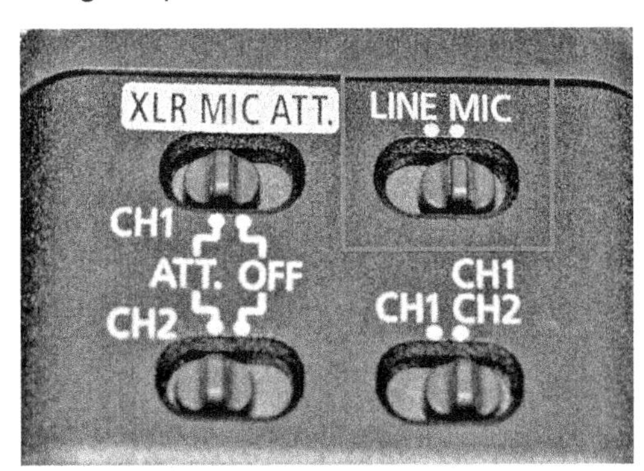

Figure 13-27. Choosing between Line and Mic inputs.

6. Enable attenuation, if needed. Sometimes in either Line or Mic mode, the signal is too powerful for the camcorder to handle without distortion or too much ambient sound. This is the case with the Azden shotgun microphone, so I've enabled attenuation, which reduces the incoming audio levels by 20 dB. How do you know when the signal is too hot? After you enable manual gain control and set the volume controls at mid-level, if the volume is bumping against the top of the volume meter, you should try attenuation.

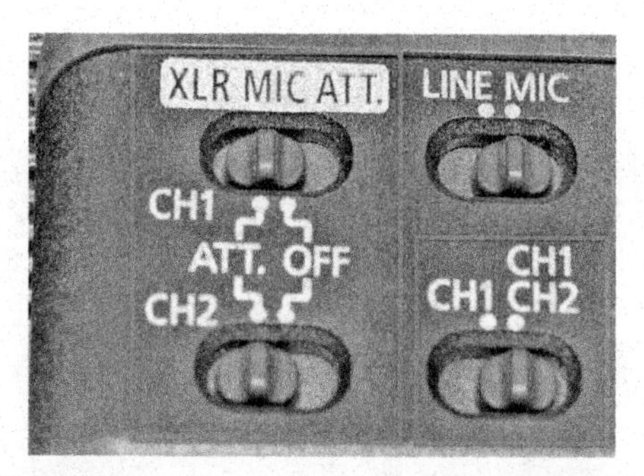

Figure 13-28. Attenuating the signal from the Line and Mic inputs.

7. ***Choose your channel strategy.*** Often you'll only have one microphone, and typically in these instances, you'll want to record the mono signal to both channels. With the XH A1, you have to connect the microphone to Channel 1 and select CH1/CH2 as shown on the lower right of figure 13-28. Every camcorder is different, so check your documentation for the required settings.

8. ***Choose your gain control strategy: manual or auto.*** The switch at the bottom of Figure 13-29 controls whether I use automatic gain control, where the camera controls the volume, or manual, where I use the two dials above the switch to control volume. Typically, if I'm only driving one camera in a fairly static setting, like a seminar or speech, I'll use manual gain control. On the other hand, in dynamic setting like a concert where I have continually try to follow the action and adjust exposure, I'll usually go auto.

Figure 13-29. Choosing between auto and manual volume controls.

Note that the old complaint about auto gain control was that the camera would boost gain during silent periods, creating audible noise when it's supposed to be quiet. However, this isn't a behavior I've noticed with any of my prosumer camcorders. If anything, the camcorders are much faster to adjust to changing levels than I am, and are always paying close attention.

9. ***Check levels with headphones and volume meters.*** The final step is to plug in some headphones and make sure the audio sounds good. The volume meters on most prosumer camcorders will also tell you when you're properly connected and when audio is only flowing into a single channel. You want the bulk of your recordings to appear in the zone shown in Figure 13-30, but always want to err on the side of being too low, rather than too high. You can always boost audio volume in post (or at the encoding station) but if the volume controls continually push against the top of the volume meters, you're clipping your audio signal, which causes distortion.

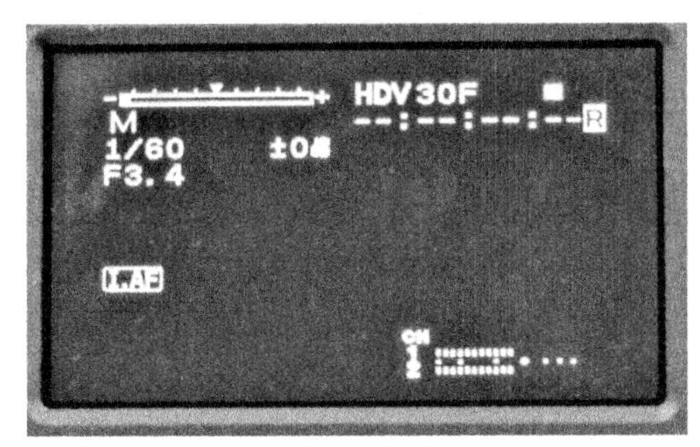

Figure 13-30. The volume meters on the bottom look just about right.

That's the basic procedure. While it will vary from camcorder to camcorder, the overall structure should be fairly similar. Now let's focus on a couple of special situations.

XLR Input Into a 1/8-inch Microphone Port

What do you do if you're working with XLR equipment and have a 1/8-inch microphone input jack? The name to remember is BeachTek, which sells a highly regarded line of XLR to stereo adapters like the DXA-HDV shown in Figure 13-31. The unit fits beneath your camcorder on the tripod, and supplies the phantom power and gain/attenuation controls mentioned above. You plug the XLR cables into the input ports shown and connect the audio out to the stereo microphone input on your camcorder. I still have the DXA-10 I used in the late '90s before I owned camcorders with XLR inputs.

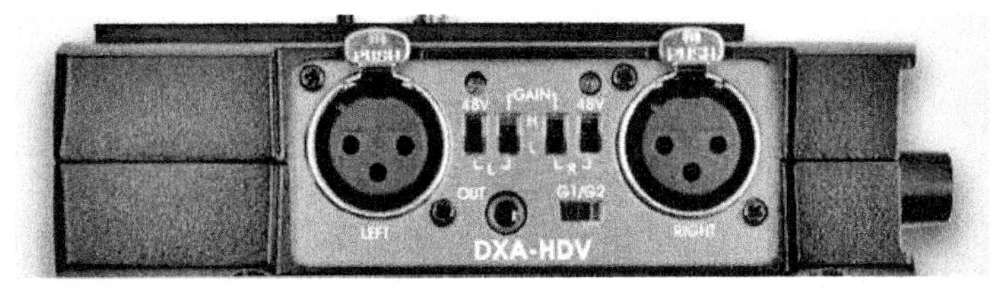

Figure 13-31. This BeachTek XLR to 1/8-inch converter gives you phantom power.

The unit costs $299 at Amazon, although BeachTek does offer cheaper units that may work for you if you don't need phantom power. Of course, if you don't need phantom power, you can by the Hosa XVM-110M cable shown in Figure 13-31 from Amazon for $6.98. This cable takes the incoming mono signal from the XLR and sends it to both channels of the 1/8-inch jack so you won't have a dead channel. Note that the Hosa CYX-403M has two XLR inputs so it can output a true stereo signal into your camcorder.

I've never tested one of these cables, but the Amazon reviews on both are very positive.

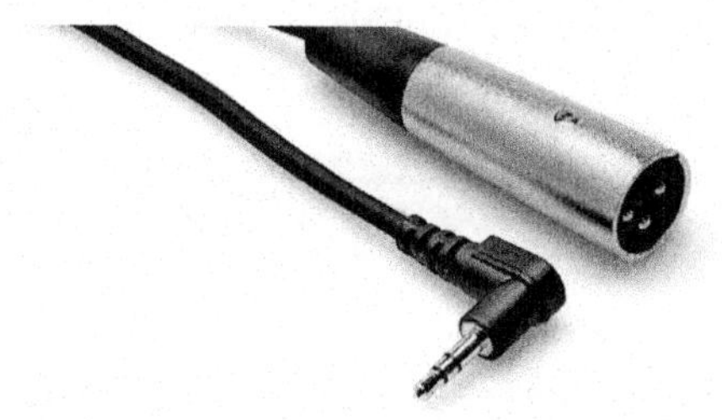

Figure 13-32. If you don't need phantom power, this type of adapter should work just fine.

Connecting to Sound Systems

The basic procedure for connecting to a sound system is the same as the procedure defined above, except that you use Line input rather than Mic. Easy peasy. Typically, if problems arise, they come from one of four directions.

Often the soundboard isn't set up close to the camera position, which can be a real problem. If you ask about this early enough, typically the sound guys or event manager can run a cable over to your position, but advanced notice is required because the cable will have to be taped over or buried. If you show up an hour before the show and start asking questions, it may be too late to get the input.

Another source of problems is when there simply isn't an output that the soundboard can spare. If you get in touch with the sound guy early enough, you can work around this, often by getting some kind of splitter cable that takes the outgoing signal and splits it into two. Figure 13-33 shows the GLS Audio 6 Inch Patch Y Cable Cord that costs $9.99 at Amazon; buy two of these and you can split the left and right XLR outputs and create another output.

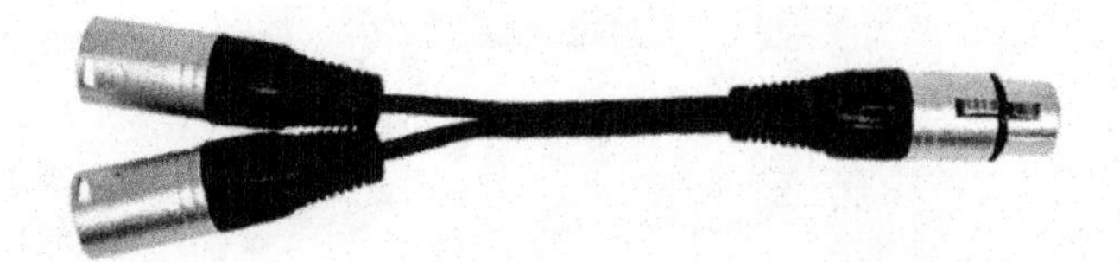

Figure 13-33. An XLR splitter can provide access to the sound board.

The third source of problems typically relates to physical connections. Most soundboards use 1/4-inch connectors out, which can be stereo or mono. If you have mono outputs, you can buy many different adapters to convert from 1/4-inch to XLR, although stereo is tougher. Basically, every situation can be different and sometimes you'll need to buy some kind of adapter to get the job done. Figure 13-34 shows some of the adapters I've acquired over the years, but I still call a few days before the show, ask about the connections available and try to test the connection a day or two before the event.

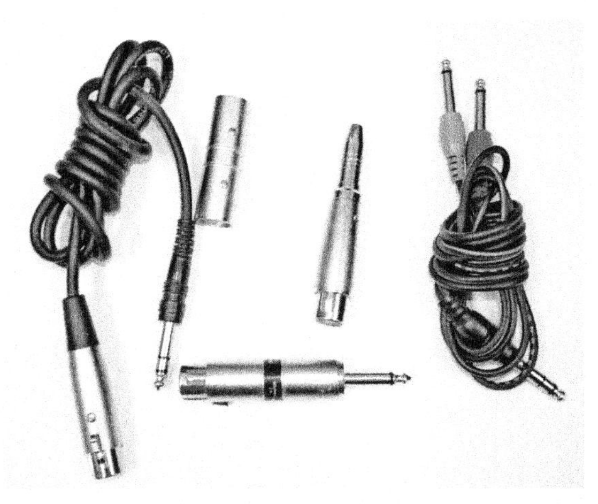

Figure 13-34. Some of the detritus bought to connect to multiple sound boards.

The fourth source of problems, which I've only witnessed from afar, is when the signal coming from the soundboard is too hot for the camera or other capture device to handle. Here you need a signal attenuator like the Shure A15AS that has switchable values for a 15, 20 or 25 dB reduction in signal strength. To make this work, connect it between your output and your input, which means that you suddenly need two cables.

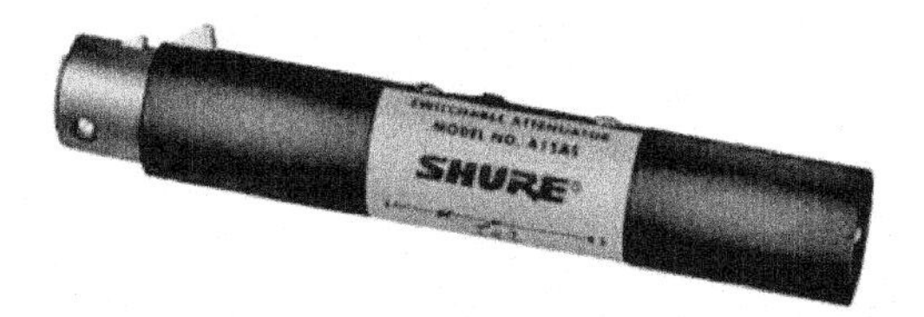

Figure 13-35. Attenuators reduce the levels coming from the sound board.

I saw an attenuator like this save the day at the J Street convention I helped produce in March 2012. Fortunately for me, it wasn't the feed coming into my camera; it was in a different ballroom. Without the attenuator, the signal was too powerful for the camcorder it was plugged into, producing significant noise. They fixed this about 11:00 the night before the convention, and if one of the sound guys didn't have the adapter on hand, it would have gotten ugly. In short, when you're streaming a live event and don't control the sound, contact the sound guy as early as possible, and try to test the connection at least a day or two before the event.

OK, we've beat getting audio into the camera to death; let's spend time getting the audio out.

Audio Outputs

Audio outputs depend on the signal you're using for video. For example, if you're connecting to your capture device via DV, HD-SDI or HDMI, the audio signal is passed with the video. If you're using composite, S-Video or component, audio is separate.

When your audio sources are connected to your camera, the audio output port you'll use will vary by camcorder. Some older camcorders use three RCA outputs, with a composite video output (yellow) and left (white) and right (red) audio connectors. You can tell when the jack is carrying three signals when there are three connection grooves as shown on the bottom left in Figure 13-36. If this port is available, use this cable to connect to your audio—although S-Video, if available, is always preferable for video.

Other camcorders use a 1/8-inch stereo output, which requires the kind of splitter shown in the middle of Figure 13-36, which outputs to red and white RCA connectors. Some capture devices accept RCA input, in which case you're in good shape. Others use BNC connectors, for which you'll need the small adapter shown on the upper right, third to the right. The other two are BNC-to-RCA adapters that you may also need from time to time.

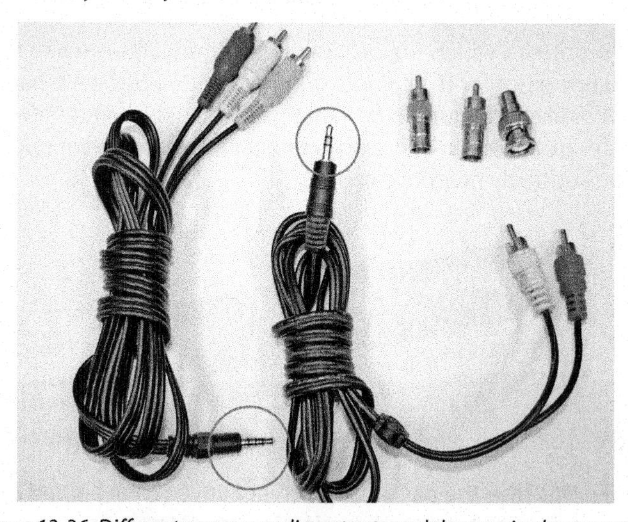

Figure 13-36. Different camera audio outputs and the required connectors.

Issues like these are why I recommend setting up and testing the entire audio capture/shooting/encoding workflow before packing out of the event. Even the most detail-minded person can't keep all these items straight in their heads.

Monitoring Audio During the Shoot

As promised, let's spend a bit of time talking about monitoring audio at the shoot, including volume control. As you'll see, understanding how loud is loud enough is a seemingly simple problem that becomes complex once you start asking a few questions.

To set this up, in March 2012, I spent two days helping to broadcast the J Street convention in Washington, DC. It was a fun event, and I learned a lot because I got to work with some real professionals who had more experience than I did. I learned a lot and want to share that with you.

We'll start with the high-level, audio-related goals.

Goal 1: Monitor for stream breaks and other problems.

During the course of a one-hour webcast, or a series of webcasts throughout the day, your mind is going to wander. You'll check email, check stock prices, check out particularly attractive members of the audience.

If the video stops streaming when you're not staring at the screen, there's only a very small chance that you'll notice it visually. However, if you're wearing headphones and the audio stops, you'll notice it immediately. So get in the habit of wearing your headphones throughout all sessions. It's bad enough when the streaming stops unexpectedly; it's even worse when you don't notice it for a few minutes.

Goal 2: Ensure consistent levels to your audience.

Your web viewers can and will adjust volume at their end to ensure optimal levels. If they have to do this once, it's not a big deal. However, if they have to do this repeatedly throughout the webcast, it will be perceived as a problem. Since the client undoubtedly has folks watching the webcast online, it's a problem that reflects poorly on your work. So one key goal is to produce consistent levels throughout the webcast.

Recognize that audio volumes change every time a different speaker approaches the microphone, whether it's on the podium or on the speaker's table. Different speakers talk at different volumes at varying distances from the microphone. So whenever there's a speaker change, you have to be ready to make a volume adjustment.

Complicating this is that there may be audio technician onsite making similar adjustments, and your shooter may be doing the same thing. Once a speaker changes, you might adjust your levels, but then the audio tech or shooter makes their adjustments, so you have to adjust again. The bottom line is that anytime there's a change, you've got to be riding the volume sliders.

All this is simple enough; let's distill it to a few rules.

1. Wear headphones at all times to monitor any breaks, and also audio volume.

2. Be ready to adjust volume at every speaker change.

3. Be ready to adjust volume at every microphone change.

4. Monitor the adjustments the shooter and audio technician are making and adjust for these as well.

How Loud is Loud Enough?

All this is mechanics; the big question is, How loud is loud enough? More specifically, while you're manning the sliders, what's the proper volume level to target?

Figure 13-37 shows the Livestream Procaster audio mixer. The top slider on the left controls the incoming volume from the capture device I was using (the Canopus AVDC-300) while the master slider on the right controls all audio sources. The volume meter next to master provides the most useful guidance regarding outbound audio volumes. It ranges from -infinity, which is obviously too low, to +12 with a peaking indicator (OVR).

The mixer throws some pretty scary clues your way. First, the region from 0 dB to +6 dB is marked in yellow, while the area from +6 to +12 is marked in red. To me, that means caution and stop. Obviously, the OVR indicator is to be avoided at all costs.

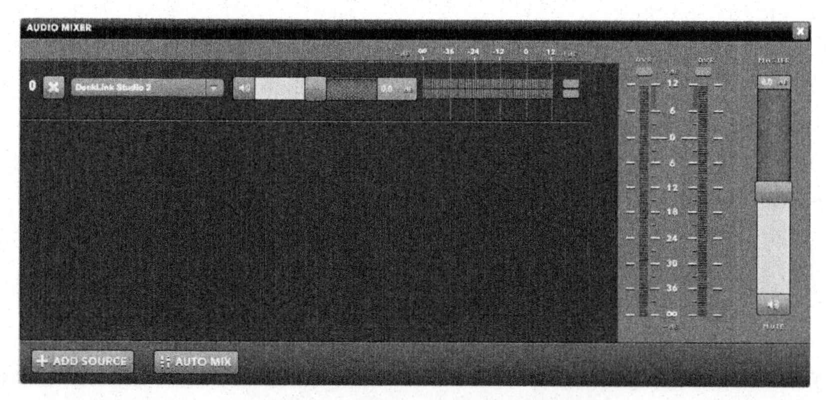

Figure 13-37. How loud is loud enough?

But what is the desired level? We all know that when working with an audio editor like Audition, or video editor like Adobe Premiere, the peak is 0 dB, and that levels beyond this are clipped and distorted. Does distortion start when levels exceed 0 dB in Procaster? Many TV and music producers target around -12 dB to leave headroom for later editing; is that a good target?

For the uninitiated, the more you think about this, the more confused you get. Fortunately, understanding three simple facts simplify the analysis. These are:

1. Analog volume meters are different from digital volume meters. Figure 13-38 shows a waveform in Adobe Audition. You can see on the right that it shares the same bottom end as the Procaster meter, -infinity, but that the loudest end is 0 dB, which is just at the top and the bottom of both channels in the figure.

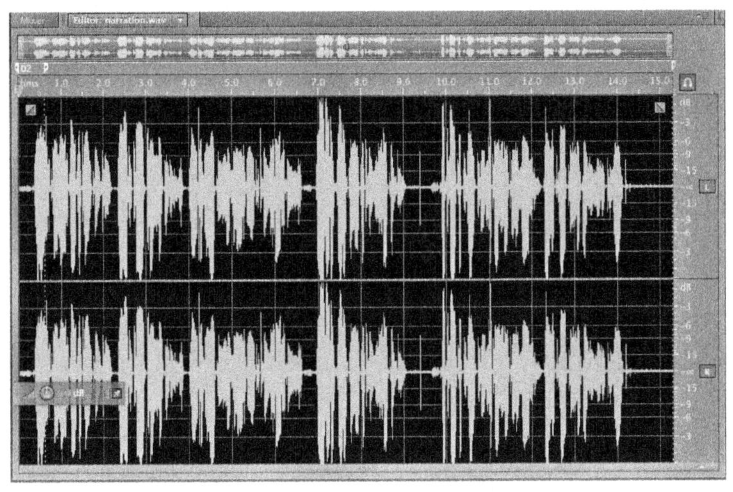

Figure 13-38. With digital volume meters (all the way on the right), 0 dB is as loud as it gets.

In fact, as you may know, if you go beyond 0 dB, you'll produce clipping, which is the slightly flat area in Figure 13-38 on the bottom of both channels right at the 7 second mark. Excessive clipping causes distortion and is to be avoided whenever possible. So, when webcasting, should you limit your webcast audio to 0dB?

The simple answer is no because analog meters are different than digital meters. Here's a bit from the fabulous Sound on Sound website explaining why (**bit.ly/SOS_metering**):

> The majority of digital recorders, mixers and converters use true peak-reading meters whose displays are derived from the digital data stream. As these don't rely on analogue level-sensing electronics they can be extremely accurate.
>
> Analogue meters all have a nominal alignment point—the zero reference—with a notional headroom above. The idea is that signal peaks are routinely allowed into the first 8 dB or so of this headroom, though peaks of +12 dBu will usually start to cause distortion which becomes more and more noticeable with increasing level until clipping occurs, usually at between 18 dBu and 22 dBu.
>
> Digital systems, however, have no headroom above the maximum quantisation level, and therefore a notional headroom must be created by choosing a 'zero' point well below this. Digital meters are scaled such that the maximum quantisation level is denoted as 0 dBFS (full scale), so the alignment level is always a negative value below this point.

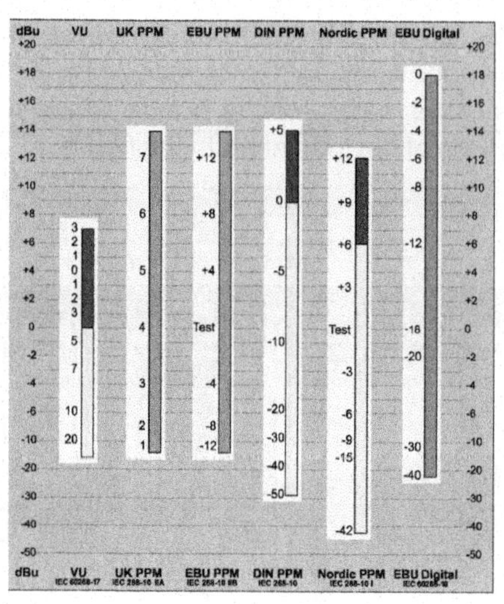

Figure 13-39. Different styles of analog meters.

To simplify, while levels beyond 0 dB cause problems with digital metering systems, with analog style meters, you can go to +12 dB or higher without causing distortion. How high? Well, that depends upon the style of meter, leading us to Fact 2.

2. ***There are multiple styles of analog meters.*** I won't dwell on this, but note that there are multiple styles of analog volume meters, as shown in Figure 13-39, which is a chart from the same Sound on Sound article referenced above. This leads to the third fact.

3. ***You have to know which style your webcasting tool uses.*** If you compare Figure 13-37 with Figure 13-39, it's apparent that Livestream uses the Nordic meter, which I confirmed in an email. I also asked about the recommended approach to setting levels. Here's the reply from my Livestream contact, who runs many of the company's high-profile webcasts:

> If you let the audio just hit 0 in Procaster, it will be too low. You need to get it to between +6 and +12 to get a nice normalized sound that sounds good in laptop speakers. Unless you sustain clip the OVR, you shouldn't clip your audio.
>
> This is specific to Procaster and is somewhat counter-intuitive if you've been working with digital-style metering, but trust me—we ride the meter so that it hits between those all the time and it sounds great. If the input signal is not hot enough, I definitely recommend boosting it in Procaster.

Synthesis

OK, let's take what we learned and synthesize some general-purpose lessons. First, regarding the Livestream mixer, you should target volume so that it stays within the 6-to-12 region, but seldom triggers the OVR light.

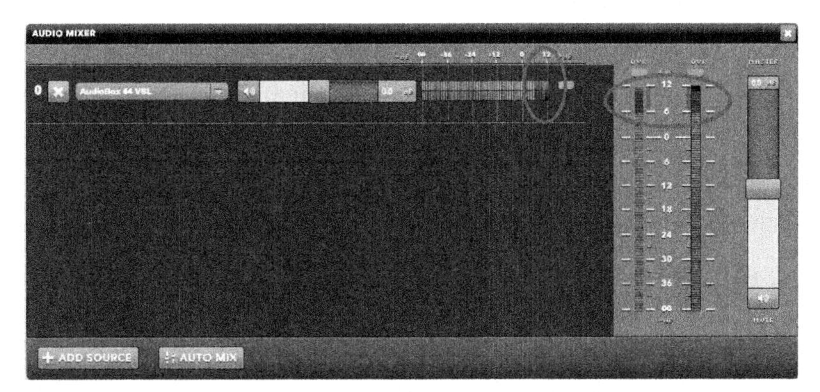

Figure 13-40. Optimal levels for the Livestream mixer.

As with video, you get the cleanest signal when you can increase volume without boosting gain, which is electronic amplification that can introduce noise into the signal. You can see that gain is set to 0 dB on both on the top slider on the left, and the master slider. If you can boost volume at the source without boosting gain there, you should do so. In other words, if the soundboard or the shooter can boost volume, generally that's preferred. What about if you have to boost gain in your mixer?

I asked the Livestream folks whether boosting gain with their tool would introduce audible noise into the audio stream. They responded that it almost never does, except when volumes are boosted so high that it produces clipping.

For working with other mixers, keep these rules in mind:

1. ***Most audio mixers use analog style meters with beyond 0 dB.*** Check the documentation that comes with the mixer (if any) for details, but in the absence of a manual, if there's a clipping indicator, boost your volume so that it's only occasionally triggered. If there is no clipping indicator, make sure you don't flatline the audio volume at the top of the meter for extended periods.

   It's always a good idea to capture some video, input it into an audio editor like Audition and study the waveform. If it looks like Figure 13-38, you're in good shape. If it's very clustered around the center line, the volume is too low; if there are multiple flat areas against the top, it's too loud.

2. ***When you need to boost volume, try to boost from the source if you can do so without boosting gain there.*** Don't sweat if you have to boost gain in your mixing tool, but listen to make sure you're not introducing noise into the signal.

At the Shoot—Framing Your Subjects

At the J Street event, I was just the compressionist, but I got to work alongside a world-class shooter named Manrico Zottig. In this section, I'll detail what I learned by watching Zottig shoot the sessions we worked on together. If you're about to shoot a panel discussion or similar event, I think you'll find it useful. You'll also find it useful if you're unfamiliar with the Rule of Thirds or will soon face any stage production involving multiple participants.

Framing a Panel

Figure 13-41 shows the panel that Zottig was in charge of shooting. It's a typical panel; a podium on the left where each speaker gave introductory thoughts, and a table for discussion. What struck me when I was watching the production was how precisely Zottig framed his shots, never cutting off critical edges and minimizing problem areas in the scene.

For example, there's a metal pole peeking out on the right, and the drape doesn't extend to the right as far you'd like. As you'll see, Zottig adjusted his framing to minimize this and other problems.

Figure 13-41. The basic setup for our conference.

Basic Rule of Thirds

Figure 13-42 shows the shot of the speaker at the podium (from a different session) with Rule of Thirds guides. According to Wikipedia, the Rule of Thirds "proposes that an image should be imagined as divided into nine equal parts by two equally spaced horizontal lines and two equally spaced vertical lines, and that important compositional elements should be placed along these lines or their intersections." The intersections of the lines in the tic-tac-toe-like board are called "saddle points."

Figure 13-42. Classic Rule of Thirds positioning for a speaker facing the camera. Well, perhaps a little high.

Typically, when shooting people, the eyes are the most important compositional elements. When the person is facing the camera, as all shots are here, you place the subject in the middle of the camera with eye level at the top or bottom third. This is your classic anchorman video. If the person were facing your left, you'd place their eyes on one of the right saddle points, leaving "look room" on the left. If facing your right, you'd place their eyes on one of the left saddle points.

Like all video-related rules, it's better to know the rule and break it with intent than to not know it and break it accidentally. For example, in a recent video spoof I shot, I placed my subject on the wrong saddle point to create a slight feeling of tension and to create diversity in my shots. However, I'd never use a shot like this in a seminar or business video.

Figure 13-43. Violating the rule of thirds to create a slight feeling of tension.

Back to Figure 13-42, you'll note that the speaker's eyes are slightly above the top line, and that the J Street logo was shown fully with the bottom of the podium showing. However, the current videos on YouTube, from which I grabbed these streams, are not the videos that actually streamed live at the event. Rather, these were produced from the archived full DV file that we provided to the client after the event.

During the event itself, whether from the capture device or streaming service, the live video was cropped slightly on the top and bottom, so the Rule of Thirds positioning was closer and the bottom edge of the podium was cut off, leaving the logo shown in full. I was positioned right next to Zottig during the shoot, and he glanced frequently at the streaming feed to make sure that positioning was good and that nothing critical was cut off. However, Zottig seldom, if ever, made slight adjustments to adjust his positioning, particularly to correct for Rule of Thirds.

That's because while few viewers will notice that Rule of Thirds is off, all viewers will notice if you make minute adjustments to achieve perfect positioning. For this reason, it's best to get close to optimal positioning as quickly and smoothly as possible and then leave it alone. However, while I would not make minor adjustments for Rule of Thirds, I would if critical branding or other noticeable edges were cut off.

Let's start codifying these into rules:

1. ***Mind classic Rule of Thirds positioning, with the exceptions noted below.***
2. ***Don't cut of logos, branding or other noticeable edges whenever possible.***
3. ***Get close to optimal positioning as quickly and smoothly as possible, and then leave it alone unless critical branding or other noticeable edges are cut off.***
4. ***Observe both what's on your camera and what's in the streaming feed, since sometimes they'll be different.***

Managing the Speaker Transition

Once a new speaker stepped to the podium, Zottig would quickly frame the speaker on his viewfinder and then check what was appearing in my streaming feed. Then he would typically sit down and watch the speaker, waiting for cues that he or she was about to finish. Once the speaker finished, Zottig would follow the speaker back to his or her chair and follow the new speaker up to the podium. This gave the remote viewer the same experience as the viewer in the room.

Figure 13-44. Managing the transition.

Tough Shots at the Table

Once the speakers finished at the podium and the discussion started at the table, the rules changed and became much more fluid. If the speakers were far enough apart, Zottig would frame them normally. You can see this in Figure 13-45, where the speaker to the subject's right is visible, but not distractingly so.

Figure 13-45. Normal Rule of Thirds positioning when the speakers are sitting far apart.

What happened when the speakers sat closer together? Zottig would violate classic positioning to avoid noticeably cutting off the adjacent subject. You can see this in Figure 13-46, where author Gershom Gorenberg is speaking. Rather than place him in the center of the frame, which would very noticeably cut off Dr. Klein to his left, he framed them both, almost like both of them were speaking.

Figure 13-46. Avoid Rule of Thirds when it would noticeably cut off adjacent speakers.

At the other side of the podium, Zottig had the same problem plus another. That is, if he placed Daniel Levy in the middle of the frame, he would cut Dr Barghouti in half and expose the edge of the curtain on the right. To avoid both issues, he framed the shot to show both, even though Levy was doing the talking.

Figure 13-47. And avoid Rule of Thirds when it would noticeably show the edge of the stage.

I've long followed this "don't cut off" rule in my concert and other productions. For example, during a banjo solo, I'd show a two-shot of the banjo player and guitar player if focusing on the banjo player would cut the guitar player in half. Basically, your job is to create the most attractive picture in the frame at all times. While the Rule of Thirds is important, other aesthetic rules can trump it and the Rule of Thirds needs to be applied judiciously.

To add to our rules above:

5. ***Follow speakers to and from the podium.***

6. ***Ignore the Rule of Thirds when it would noticeably cut off an adjacent subject or expose some other unattractive component of the setup.*** Remember: Only pros know what the Rule of Thirds is. All viewers will find it it's awkward if you cut adjacent speakers in half.

You're certainly free to disagree with how Zottig framed his shots. What's most important is to understand the issues that are at stake, make decisions about how to address them and then apply these decisions uniformly.

Conclusion

OK, that's it on live event production. It really is a topic worthy of further exploration, but I hope the details provided in this and the previous three chapters have been of value.

Chapter 14: Introduction to Closed Captions

Although relatively few websites are required to provide closed captions for their videos, any website with significant video content should consider captioning. Not only does it provide access for deaf and hard-of-hearing viewers, but captions and the associated metadata can dramatically improve video search engine optimization. In this introduction to closed captions, you'll learn about who needs to caption and who doesn't (and why you may want to anyway), the available workflows for captioning live events and on-demand files, and a bit about web caption formats and how to marry them to your streaming files.

Let's jump in.

What Are Closed Captions?

Closed captions enable deaf and hard-of-hearing individuals to access the audio components of the audio/video experience. Closed captions incorporate all elements of the audio experience, including identifying background sounds, the identity of the speaker, the existence of background music, descriptions of how the speaker is talking, and similar information. They are also closed, so they can be disabled for viewers with no hearing disabilities. In contrast, open captions are burned into the video stream and can't be disabled.

Subtitles are typically implemented to allow viewers to watch videos produced in different languages. Technically, background sounds and other non-vocal audio don't have to be incorporated into the text description, since subtitles are not designed for the deaf and hard-of-hearing, but these elements are often included. There are many closed captions standards, and several are discussed here. While each has a unique format and structure, the content of all closed caption files is similar, primarily consisting of the textual data and time code information that dictates when it's displayed.

Who Has to Provide Closed Captions?

Two classes of websites caption: those required by law and those who caption voluntarily. Let's start with those legally required to caption.

Section 508 of the Rehabilitation Act

Four laws create the obligation to caption. Starting with federal agencies, Section 508 1194.24(c) of the Rehabilitation Act (29 U.S.C. 794d) states, "All training and informational video and multimedia productions which support the agency's mission, regardless of format, that contain speech or other audio information necessary for the comprehension of the content, shall be open or closed captioned." Beyond these federal requirements, note that states that receive federal funds under the Assistive Technology Act of 1998 must also comply with Section 508 to some degree.

Interestingly, several agencies meet this requirement via YouTube's captioning. For example, the surgeon general videos available from the Department of Health and Human Services website use YouTube (Figure 14-1), as do videos from IRS.gov and my home state of Virginia. More on how to use YouTube follows.

Figure 14-1. Videos from the Department of Health and Human Services website use YouTube's captioning system to meet the Section 508 requirements.

Twenty-First Century Communications and Video Accessibility Act of 2010

The next class of producers who must add closed captions to their streaming videos are broadcasters, but only with regard to content that has been previously played on TV with closed captions. Specifically, under powers flowing from the Twenty-First Century Communications and Video Accessibility Act of 2010, the Federal Communications Commission issued regulations called the IP Closed Captioning rules in August 2012, which went into effect for some classes of video soon thereafter on September 30, 2012.

That is, prerecorded programming that was not edited for Internet distribution must be captioned for the web if it was shown on television on or after September 30, 2012. If content was edited for Internet distribution, the deadline is pushed back a year until September 30, 2013. There are several other classes of content covered, including live content published with captions on TV, which must be captioned on the Internet by March 30, 2013, and older content that predates the act.

The FCC regulations and interpretations thereof make several points clear. If the content is never shown on television with captions, there is no requirement to caption for streaming. The rules also only relate to full-length programming, not clips or highlights of this programming. This last point explained why ESPN's full-length shows such as *Mike & Mike in the Morning* (Figure 14-2) do have captioning, while none of the highlights that I watched do.

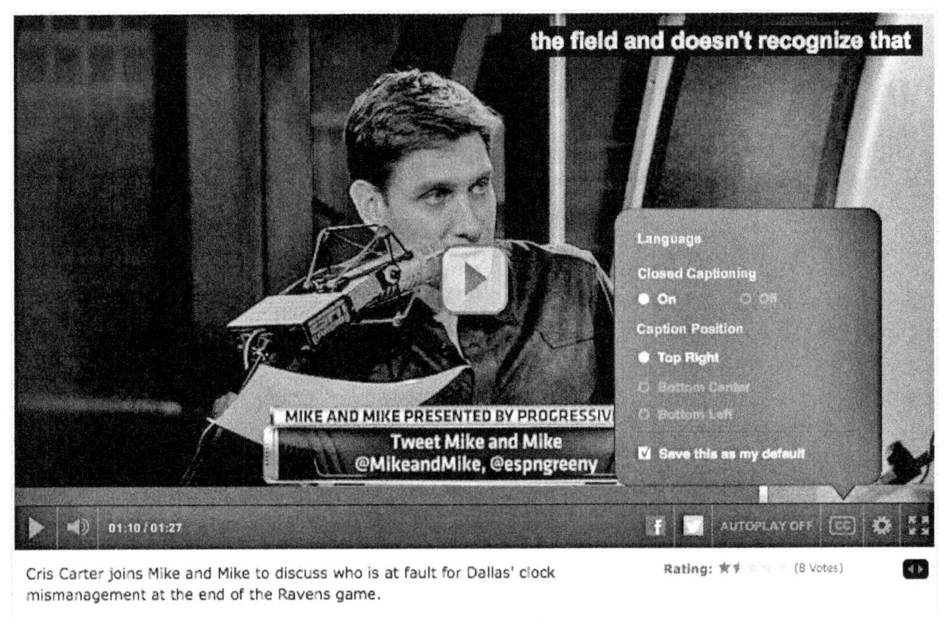

Figure 14-2. ESPN captions programs that displayed in their entirety on TV.

For more information on the regulations, check out "FCC's New Closed Captioning Rules Kick Into Gear" on the FCC Law Blog at **bit.ly/CC_law**.

Other Closed Captioning Provisions

Several additional federal laws may impose captioning requirements on varying classes of publisher. For example, some authorities opine that Title II of the Americans with Disabilities Act ("Title II") and Section 504 of the Rehabilitation Act of 1973 ("Section 504") imposed captioning requirements on public schools, public universities, as well as towns and other municipal bodies. There were conflicting opinions here, though, and I bring it up not to take a position either way but to advise that the obligation may exist. Don't contact me; contact your attorney.

Large web-only broadcasters should also be concerned about a recent ruling in the National Association of the Deaf's (NAD) lawsuit against Netflix, in which NAD asserted that the Americans with Disabilities Act imposed an obligation for Netflix to caption its "watch instantly" site. In rejecting Netflix's motion to dismiss, the court ruled that the Internet is a "place of public accommodation" and therefore is subject to the act. Netflix later settled the lawsuit, agreeing to caption all of its content over a three-year schedule and paying around $800,000 for NAD's legal fees and other costs. In the blog post referenced previously, the attorney stated: "Providing captioning earlier, rather than later, should be a goal of any video programming owner since it will likely be a delivery requirement of most distributors and, in any event, may serve to avoid potential ADA claims." This sounds like good advice for any significant publisher of web-based video.

Voluntary Captioners

Beyond those who must caption, there is a growing group of organizations that caption for the benefit of their hearing-impaired viewers, to help monetize their video, or both. One company big into caption-related monetization is Boston-based RAMP. I spoke briefly with Nate Treloar, the company's VP of strategic partnerships.

In a nutshell, Treloar related that captions provide metadata that can feed publishing processes that enhance search engine optimization (SEO), such as topic pages and microsites, which are impossible without captions. RAMP was originally created to produce such metadata for its clients, and it only branched into captioning when it became clear that many of its TV clients—which include CNBC, FOX Business, Comcast SportsNet, and the Golf Channel—would soon have to have captioning on the web.

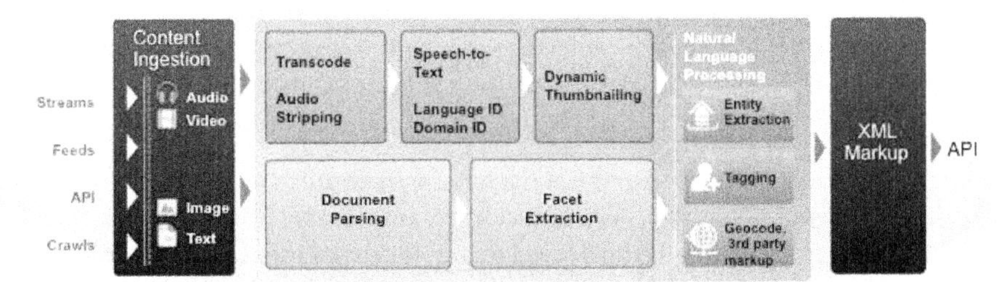

Figure 14-3. RAMP's cloud-based captioning and metadata creation workflow.

As shown in Figure 14-3, RAMP's technology can input closed captions from text or convert them from audio, with patented algorithms that prioritize accuracy for the nouns that drive most text searches.

This content is processed and converted into time-coded transcripts and tags for use in SEO with dynamic thumbnailing for applications like video search. The transcripts can then be edited for use in web captioning.

RAMP's prices depend upon volume commitment and the precise services performed, but Treloar stated that video processing ranges from pennies to a several dollars per minute. With customers such as FOX News reporting a 129% growth in SEO traffic, the investment seems worth it for sites where video comprises a substantial bulk of their overall content.

Now that we know who's captioning for the web, let's explore how it's done.

Creating Closed Captions for Streaming Video

There are two elements to creating captions: The first involves creating the text itself; the second involves matching the text to the video. Before we get into the specifics of both, let's review how captions work for broadcast TV.

In the US, TV captioning became required under the Television Decoder Circuitry Act of 1990, which prompted the Electronics Industry Association to create EIA-608, which is now called CEA-608 for the Consumer Electronics Association that succeeded the EIA. Briefly, Section 608 captions are limited to 60 characters per second, with one Latin-based character set that can only be used for a limited set of languages. These tracks are embedded into the line 21 data area of the analog broadcast (also called the vertical blanking interval), so they are retrieved and decoded along with the audio/video content.

Where CEA-608 is for analog broadcasts, CEA-708 (formerly EIA-708) is for digital broadcasts. The CEA-708 specification is much more flexible and can support Unicode characters with multiple fonts, sizes, colors, and styles. CEA-708 captions are embedded as a text track into the transport stream carrying the video, which is typically MPEG-2 or MPEG-4.

A couple of key points here. First, if you're converting analog or digital TV broadcasts, it's likely that the text track is already contained therein, so the caption creation task is done. Most enterprise encoding tools can pass through that text track when transcoding video for distribution to OTT or other devices that can recognize and play embedded text tracks.

Unfortunately, although QuickTime and iOS players can recognize and play these embedded text tracks, other formats—such as Flash, Smooth Streaming, and HTML5—can't. So to create captioning for these other formats, you'll need to extract the caption file from the broadcast feed and format it in a number of caption-based formats that are discussed in more detail later. Not all enterprise encoders can do this today, although it's a critical feature that most products should support in the near future.

Captioning Your Live Event

If you're not starting with broadcast input, you'll have to create the captions yourself. For live events, Step 1 is to find a professional captioning company such as CaptionMax, which has provided captioning services for live and on-demand presentations since 1993. I spoke with COO Gerald Freda, who described this live event workflow:

> With CaptionMax, you contract for a live stenographer (aka real-time captioner) who is typically off-site and who receives an audio feed via telephone or streaming connection. The steno machine has 22 keys representing phonetic parts of words and phrases, rather than 60-plus keys on a typical computer keyboard. The input feeds through software, which converts it to text. This text is typically sent to an IP address in real time, where it's formatted as necessary for the broadcast and transmitted in real time. The text is linked programmatically with the video player and presented either in a sidecar window or, preferably, atop the streaming video just like TV.
>
> Unlike captions for on-demand videos, there's no attempt to synchronize the text with the spoken word—the text is presented as soon as available. If you've ever watched captioning of a live broadcast, you'll notice that this is how it's done on television, and there's usually a lag of 2 to 4 seconds between the spoken word and the caption.

According to Phil McLaughlin, president of EEG Enterprises, seemingly small variations in how streaming text is presented could impact whether the captioning meets the requirements of the various statutes that dictate its use. By way of background, EEG was one of the original manufacturers of "encoders" that multiplex analog and digital broadcasts with the closed caption text; it currently manufactures a range of web and broadcast-related caption products. The company also has a web-based service for captioning live events.

Here are some of the potential issues McLaughlin was referring to. By way of background, note that language in the hearings related to the FCC regulations that mandated captioning for TV broadcasters discussed providing a "captioning experience ... equivalent to ... [the] television captioning experience." McLaughlin says he doubts that presenting the captions in a sidecar meets this equivalency requirement because the viewer has to shift between the sidecar and video, which is much harder than watching the text over the video. At NAB 2012, McLaughlin's company demonstrated how to deliver live captions over the Flash Player using industry standard components, so sidecars may be history by the time broadcasters have to caption live events in March 2013.

Interestingly, McLaughlin also questions whether the use of the SMPTE-TT (Society of Motion Picture and Television Engineers Timed Text format) allowed in the FCC regulations provides the equivalency the statute is seeking for live captioned content. Specifically, McLaughlin noted that SMPTE-TT lacked the ability to smooth scroll during live playback, as TV captions do. Instead, the captions jump up from line to line, which is harder to read. You can avoid this problem by tunneling 608 data with the SMPTE-TT spec, but not all web producers are using this technique.

McLaughlin feels that using the embedded captioning available in MPEG-4 and MPEG-2 transport streams, like iOS devices can, is the simplest approach and provides the best captioning experience. Note that neither the sidecar or smooth scrolling issues present problems for on-demand broadcasts. With on-demand files, the captions are synchronized with the spoken word and predominantly presented via pop-up captions over the video, which are much easier to follow when they match what's happening on screen.

While this won't meet the FCC requirements, another option for private organizations seeking to provide a feed for deaf and hard-of-hearing viewers is the New York City Mayor Bloomberg approach of supplying a real-time American Sign Language interpretation of the live feed. This was the approach originally used by Lockheed Martin Corp. for its live events. Ultimately, the company found using real-time captioning to be more effective and less expensive.

Captioning On-Demand Files

As you would suspect, captioning on-demand files is simpler and cheaper than captioning live events. There are many service providers such as CaptionMax, where you can upload your video files (or low-resolution proxies) and download caption files in any and all required formats. You can also buy software such as MacCaption from CPC Computer Prompting & Captioning Co. to create and synchronize your captions (Figure 14-4).

Figure 14-4. Using MacCaption to create captions for this short video clip.

For low-volume producers, there are several freebie guerilla approaches that you can use to create and format captions. For a 64-second test clip, I used the speech-to-text feature in Adobe Creative Suite to create a rough transcript, which I then cleaned up in a text editor. Next, I uploaded the source video file to YouTube, and then uploaded the transcript. Using proprietary technology, YouTube synchronized the actual speech with the text, which you can edit online as shown in Figure 14-5.

From there, you can download an .SBV file from YouTube, which you can convert to the necessary captioning format using one of several online or open source tools. I used the online Captions Format Convert from 3Play Media for my tests. Note that YouTube has multiple suggestions for creating its transcripts, many summarized in a recent ReelSEO.com article. If you're going to use YouTube for captioning, you should check this out.

Which approach is best for your captioning requirements? Remember that there are multiple qualitative aspects to captioning, and messing up any of them is a very visible faux pas for your deaf and hard-of-hearing viewers. For example, to duplicate the actual audio experience, you have to add descriptive comments about other audio in the file (applause, rock music). With multiple speakers, you may need to position the text in different parts of the video frame so it's obvious who's talking, or add names or titles to the text. There are also more basic rules about how to chunk the text for online viewing.

Basically, if you don't know what you're doing and need the captioning to reflect well on your organization, you should hire someone to do it for you—at least until you learn the ropes. For infrequent use, transcription and caption formatting is very affordable, though few services publish their prices online.

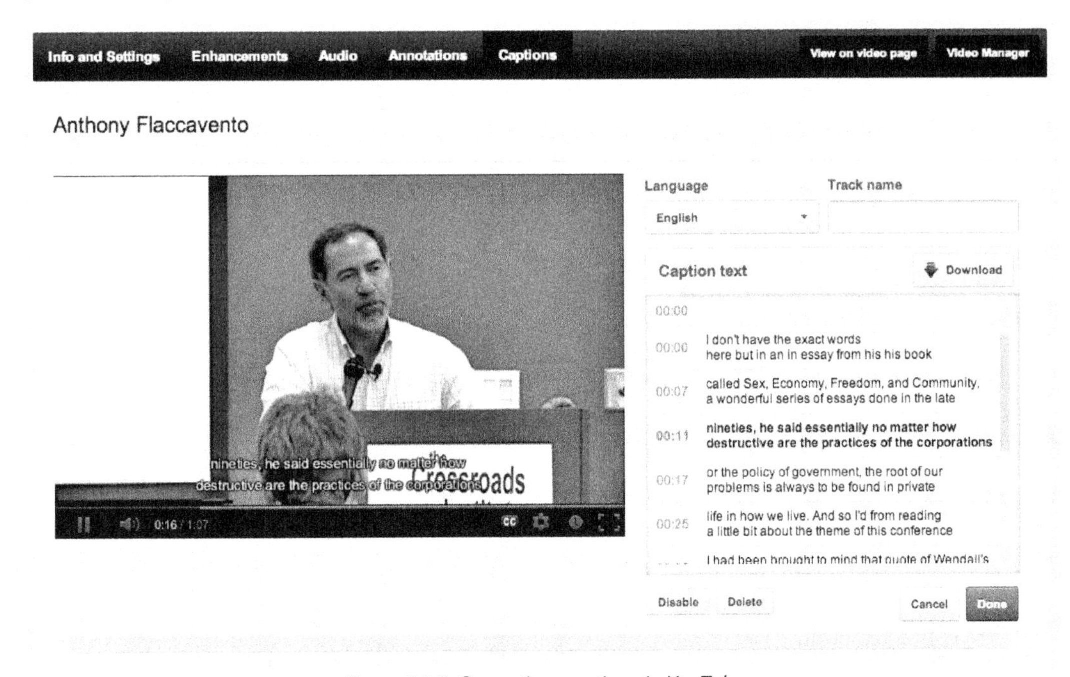

Figure 14-5. Correcting captions in YouTube.

The lowest pricing I found was around $3 per minute, but this will vary greatly with turnaround requirements and volume commitments. Remember, again, that there is a both an accuracy component and a qualitative component to captioning, so the least expensive provider is not always the best.

Captioning Your Streaming Video

Once you have the caption file, matching it to the streaming video file is done programmatically when creating the player, and all you need is the captioned file in the proper format. For example, Flash uses the World Wide Web Consortium (W3C) Timed Text XML file format (TTML—formerly known as DFXP), which you can add via the FLVPLaybackCaptioning component. Brightcove, one of the OVPs used by streaming media, can accept either DFXP or the aforementioned SMPTE-TT format, for presenting captions (Figure 14-6). Most other high-end OVPs, as well as open source players such as LongTail Video's JW Player and Zencoder's Video.js, also support captions with extensive documentation on their respective sites.

In a way that seems almost uniquely native to the streaming media market, the various players have evolved away from a unified standard, adding confusion and complexity to a function that broadcast markets neatly standardized years ago. Examples abound. Windows Media captions need to be in the SAMI format, for Synchronized Accessible Media Interchange, while Smooth Streaming uses TTML.

As mentioned, iOS devices can decode embedded captions in the transport stream, eliminating the need for a separate caption file. With HTTP Live Streaming, however, Apple is moving toward the WebVTT spec proposed by the Web Hypertext Application Technology Working Group as the standard for HTML5 video closed captioning. Speaking of HTML5, it has two competing standards—TTML and WebVTT—although browser adaption for either standard is nascent. This lack of captioning is yet another reason that large three-letter networks, which are forced to caption, can't use HTML5 as their primary player on the desktop.

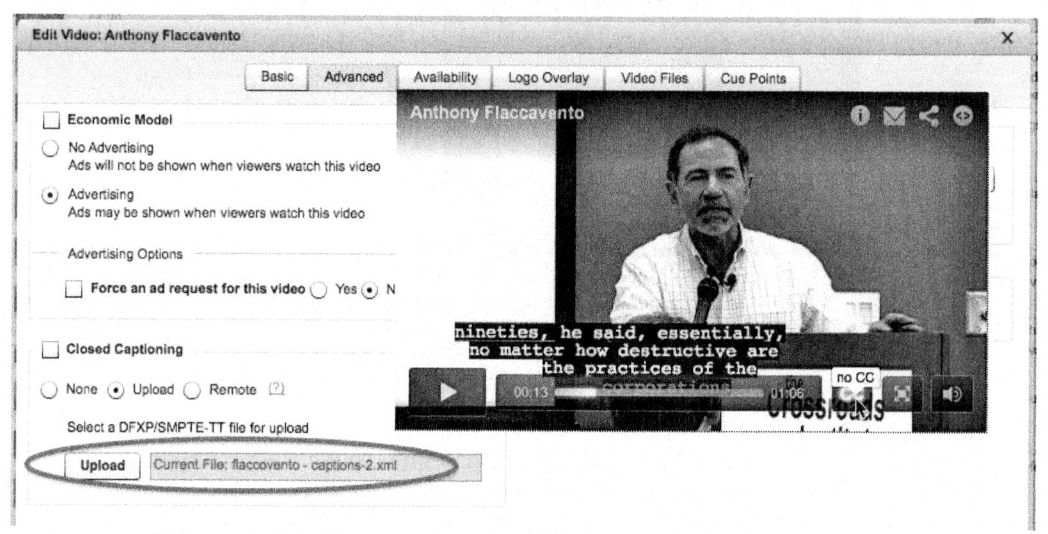

Figure 14-6. Captioning in Brightcove: The player is superimposed over the video properties control where you upload the caption file.

For a very useful description of the origin of most caption-related formats and standards, check out "The Zencoder Guide to Closed Captioning for Web, Mobile, and Connected TV" at **bit.ly/Zencoder_CC**.

Trans-Platform Captioning

What about producers who publish video on multiple platforms—say, Flash for desktops and HLS for mobile? We're just starting to see transmuxing support in streaming server products that can input captions for one platform, such as HLS, and convert the captions for use in another platform, such as Flash.

For example, I spoke with Jeff Dicker from Denver-based RealEyes Media, a digital agency specializing in Flash-based rich Internet applications. He reported that Adobe Media Server

5.0.1 can input captioned streams targeted for either HLS or RTMP and transmux them for deployment for any target supported by the server. For HLS, this means captions embedded in the MPEG-2 transport stream. For delivery to Flash, this means breaking out and delivering a separate caption file with the audio/video chunks.

In late 2012, Wowza announced Wowza Media Server 3.5, which has the ability to accept "caption data from a variety of in-stream and file-based sources before converting captions into the appropriate formats for live and on-demand video streaming using the Apple HLS, Adobe HDS and RTMP protocols." So once you have your captions created and formatted for one platform, these and similar products will automatically convert them as needed for the other target platforms supported by the servers.

I spoke with Scott Kellicker, chief architect of the Wowza Media Server, about the closed caption functionality. He shared that caption-related functionality would be different for live and video on-demand (VOD) files. For live, the server could accept captions embedded in the input stream, either CEA-608 captions or onTextData events; or could inject onTextData captions into the live stream using a Wowza API. The onTextData events would be transmuxed for live output to Apple HLS (via embedded CEA-608 captions) or the onTextData for RTMP and Adobe HDS delivery. Silverlight is not supported for live streams at this time.

For VOD, the server can accept 3GPP data tracks embedded in MP4 files or separate TTML caption files. From this data, the server can output CEA-608 data for HLS output, or onTextData events for either RTMP or HTTP-based Dynamic Streaming. The server can pass through embedded CEA-608 caption data to iOS devices, but can't transmux it for delivery to other formats. Silverlight is not yet supported for VOD files either.

Consultant Robert Reinhardt of VideoRx.com beta-tested the caption-related features of Wowza Media Server 3.5 and can attest to how caption transmuxing simplifies the caption deployment process. During 2012, before the release of the 3.5 version, Reinhardt helped his client the Government of British Columbia create a microsite with adaptive streaming videos that supported 11 different languages with closed captions. In order to serve caption-enabled video to iOS devices, separate VOD content was encoded and delivered outside of the Wowza infrastructure. Now, with Wowza Media Server 3.5, the organization can simplify its encoding and delivery processes.

Although Wowza Media Server 3.0 supported audio and video transmuxing, it didn't transmux captions, so Reinhardt had to create separate groups of adaptive files for Flash and mobile playback. In addition, because Wowza Media Server didn't natively support captions during delivery to iOS devices, he had to create 11 additional files, one for each language, that were sent via progressive download when an iOS client clicked the closed caption button, ending the adaptive streaming experience. With version 3.5, Reinhardt could upload a single version of the adaptive group, which WMS 3.5 will transmux as necessary for Flash and iOS delivery.

Conclusion

Captioning will increase in importance for all streaming producers over the next few years. Even if you're not required to caption, the search engine optimization benefits alone may make captioning worthwhile for many organizations. Now is the time to start assessing how you can integrate captioning into your streaming media.

Chapter 15: Essential Tools

For the most part, streaming players provide a pitiful amount of usable data, which makes programs that provide insight into the content of these files invaluable to compressionists. In this chapter, I'll introduce you to the tools that I use daily, starting with the tool that's installed on all of my computers, Jerome Martinez's MediaInfo. You can see a video describing two the tools, MediaInfo and Bitrate Viewer, at **bit.ly/twovidanalysistools**.

MediaInfo

MediaInfo is a cross-platform tool that offers an extensive and often unique range of data, as well as the ability to export file-based information for printing or further analysis. It also supports pretty much every codec that I've ever tried to load, including Ogg Theora and WebM, plus the normal H.264, VP6, WMV, MOV, MPEG-1 and MPEG-2. And MediaInfo details the bits per pixel for the file, which is the single most important compression-related metric (see "Understanding Bits per Pixel" in Chapter 5 for more information).

The Windows version is available in 23 different languages, including simple and traditional Chinese, and supplies more file-related data, can open multiple instances and offers more data views than the Mac. You can download both versions at **mediainfo.sourceforge.net/en**. Both are free, although donations are gladly appreciated.

Both versions load files using drag-and-drop or via traditional menu or button controls. In the Windows version, you have six different views, including text, HTML, and the Mac-style tree view. You can set which view opens by default by clicking Options > Preferences (I like the Tree view), and click the Explorer Extension to make MediaInfo appear in the right-click menu when you click a file in Windows Explorer (Figure 15-1). Very handy.

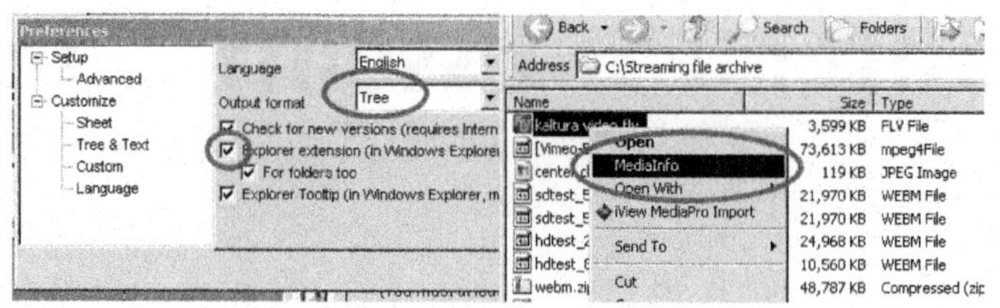

Figure 15-1. Selecting the Tree view and making MediaInfo accessible from the right-click menu.

Click Debug > Advanced Mode in the Windows version, and the program shows about three times the data, although most of the critical data is available in the Basic view.

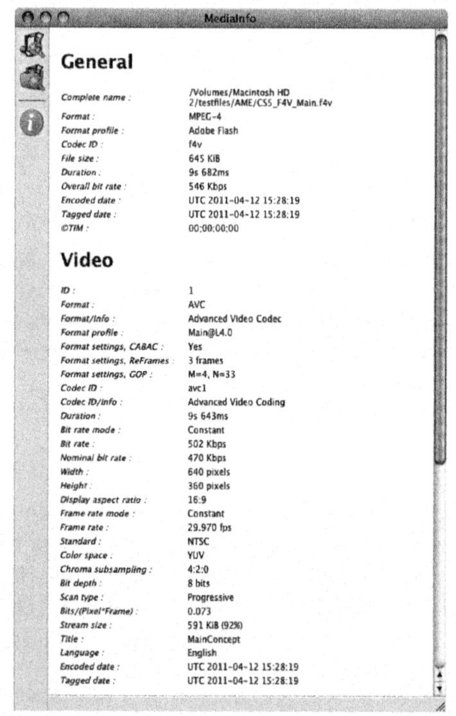

*Figure 15-2. MediaInfo provides info like the entropy encoding (CABAC), B-frame interval (M=4), I-frame interval (N=33), and Bits/(Pixel*Frame), which I call bits per pixel.*

The Mac version of MediaInfo can only export a simple text file, while the Windows version can output CSV, HMTL, text, and custom formats. The Windows version can also analyze multiple files simultaneously, either displaying all results in a single instance of the program, or exporting a consolidated report.

For all files (Figure 15-2), both MediaInfo versions show the resolution, data rate and other basic stats relating to both the audio and video components, including whether they were encoded with constant- or variable-bit-rate encoding. For H.264 files, you can see the profile used; whether the file was encoded with CABAC or CAVLC entropy encoding; the number of reference frames; B-frame interval (M=4, which means a B-frame interval of 3); I-frame interval (N=33); and Bits/(Pixel*Frame), which I call bits per pixel.

Although the Mac version is less full-featured than the Windows version, it's the only option I found for analyzing a broad spectrum of files on the Mac, making it a natural for most producers. The tool also reveals enough unique file characteristics on Windows—like VBR/CBR for Windows Media files, and profile and CABAC/CAVLC for H.264—to make it invaluable for most Windows producers. Let's put it this way: It's the only tool I have on all my computers, from netbook to high-end workstations, Mac and Windows.

Bitrate Viewer

Bitrate Viewer is a Windows-only tool available at **winhoros.de/docs/bitrate-viewer** though I'm not sure how much longer the tool will be free. You can see the tool in all its glory in Figure 15-3.

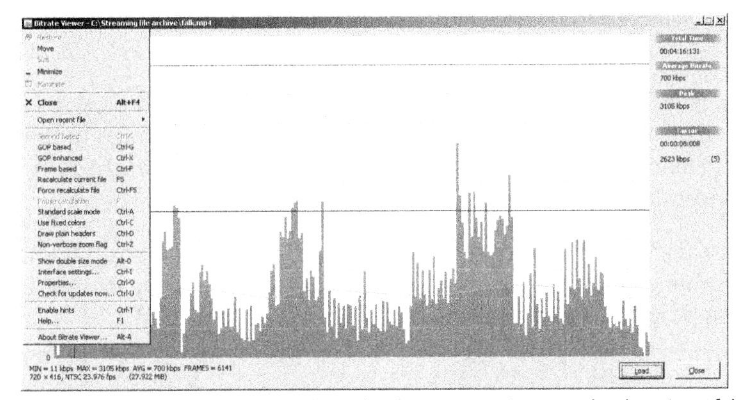

Figure 15-3. Bitrate Viewer shows you how the data rate varies over the duration of the file.

The individual spikes represent the data rate of the file at each 1-second interval, while the wavy faint blue line is the average data rate of the file. On the upper right, you can see the average bit rate, which is always handy, plus the peak bit rate in the file. On the lower left, you can see resolution and frame rate statistics.

I like Bitrate Viewer because it instantly identifies issues that may cause playback problems. For example, in the file shown in Figure 15-3, about 80% of the way through the file, you see a large area that extends above the average bit rate line. This is a data spike that could interrupt playback. You'd have to guess that this file was encoded with variable-bit-rate encoding without a constraint, or with a constraint that was too high. If users reported interruptions during playback, you would know exactly why and exactly how to fix it (either encode using CBR or tightly constrained VBR).

One negative about Bitrate Viewer is that in its default mode, the window is too small to be useful. You can double the size by clicking the icon in the upper left and choosing Show Double Size Mode, but the larger window displays an annoying message about how that feature will only be available in an upcoming for-fee version (Figure 15-4).

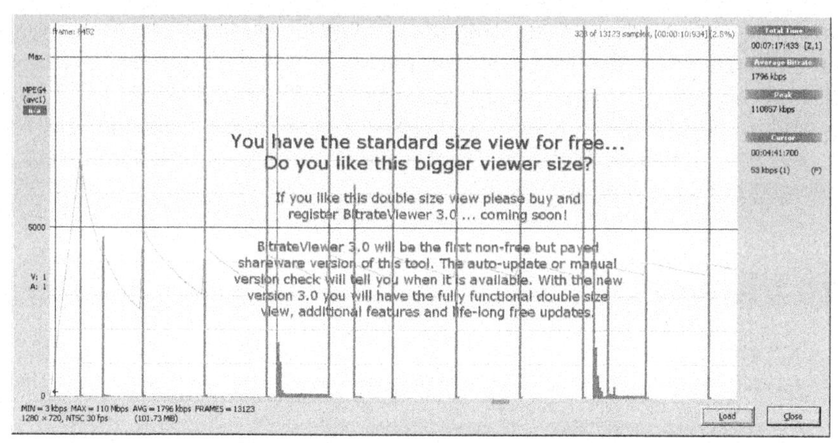

Figure 15-4. Bitrate Viewer displays this annoying message when you scale to double size.

The workaround I found quite by accident was to open multiple instances of the program, double the size of several of them (making the text appear) then returning to the menu and choosing Show in Standard Size. Usually (but not always), this makes the text disappear in the remaining instances that are still double-sized.

I'm not sure this will still work that way when you download the program, but until then, it's a great little utility that provides information you can't get anywhere else for free. Even at $10 to $20 per copy, this program would be easily worth it.

FLV Player

FLV Player is the first utility I download to any new Windows machine because it plays .FLV files of any format (SV3, VP6, H.264). Even better, when you click the info button on the lower left, an info window identifies the audio and video codecs, file resolution and frame rate, and total reported (rather than calculated actual) audio and total data rates. You can download the utility at **mdvisser.nl/flvplayer**.

Be careful when you install the program, because sometimes it comes with shovelware like the Yahoo Toolbar. There is no malicious spyware, but to make a few well-deserved bucks, apparently the developer, Martijn, has hooked up with several companies seeking broader distribution. Pay attention during the install; just say no at the right time, and you'll be fine.

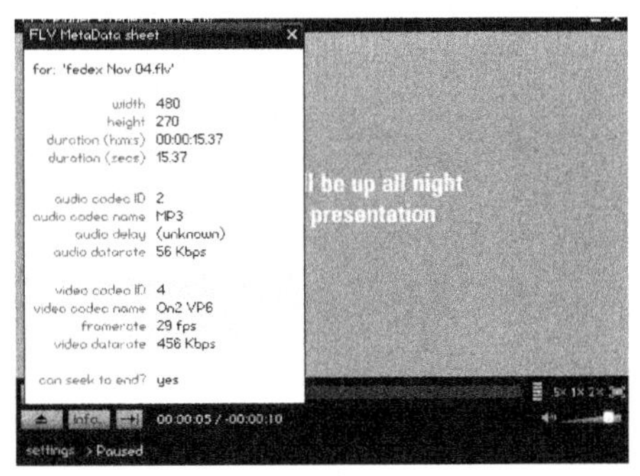

Figure 15-5. Martijn de Visser's free FLV Player gives you lots of good information right off the bat.

QuickTime Player 7

QuickTime Player 7 can open multiple instances on Mac and Windows, which makes it an essential playback tool for many producers. QuickTime comes in two versions, Player and Pro, with Pro offering more encoding and diagnostic features. On the Mac, there's also QuickTime X, which came with Snow Leopard, although it has much fewer diagnostic features than Version 7.

Figure 15-6. Information available in all versions of QuickTime, Mac and Windows.

With the Player versions of QuickTime 7 (Windows and Mac) and both versions of QuickTime X, you can load a file, choose Window > Show Movie Inspector, and get the information shown in Figure 15-6. Nowhere near as much as MediaInfo, but helpful, nonetheless.

If you upgrade to the Pro version of QuickTime 7 ($29.99), you can access unique data relating to hinted streaming files. Choose Window > Show Movie Properties to see the information shown in Figure 15-7, which identifies the hinted streams and shows their respective data rates. Note that data rate is not one of the default columns; you have to right-click the window and choose it and other desired columns from the context menu.

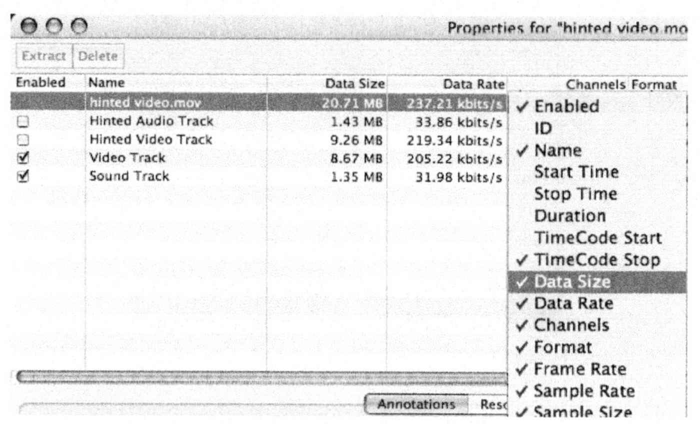

Figure 15-7. QuickTime Pro 7's Movie Properties window.

Other than this unique capability, however, QuickTime Player falls well short of other .MOV analysis tools on the Windows and Mac platforms. Probably the best alternative is MediaInfo, which is similarly available on both platforms.

MPEG Streamclip

As I mentioned back in Chapter 6, MPEG Streamclip is a free encoding tool that's only so-so as an encoder, but offers one critical feature you can't get anywhere else: trimming without re-encoding. The drill goes like this: You spend hours encoding a file and then realize that you left a few black frames at the end, or some garbage frames at the front. Or, someone wants the file with some frames cut off at the beginning, or perhaps even some frames in the middle. MPEG Streamclip is the only tool I'm aware of that can do this without re-encoding.

You can download the tool, which is available for Windows and Macintosh, from **squared5.com**. I've documented how to use the tool in a video you can watch at **bit.ly/trimh264**. The *CliffNotes* version is this:

- ***To trim frames from the beginning and/or end of a clip.*** Move the current-time indicator in the clip to the new starting point and press the letter I to create a new in point. Then move to the new ending point and press the letter O to create a new out point. If you don't set a new start or end point, MPEG Streamclip will use the existing starting or encoding point. Then choose Edit > Trim to trim the frames. Then choose File > Save As, name the file and location, and press OK. This is the critical stage; if you click Export to MPEG-4, the MPEG Streamclip will re-encode the file. You'll notice that it just takes a few moments to save the new clip, which lets you know that the clip isn't being re-encoded.

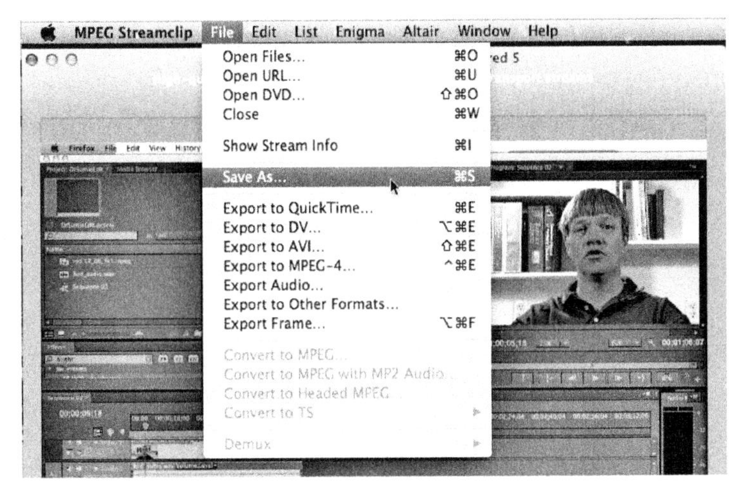

Figure 15-8. Choose File > Save As to save without re-encoding.

- ***To delete frames from the middle of the clip.*** Mark the in and outpoints that you'd like to delete as detailed above. Then choose Edit > Cut to delete the frames, and save the file as detailed above. Again, if you choose Export to MPEG-4 rather than Save As, you'll re-encode the entire video.

Conclusion

That's it. I'm sure there are some tools out there that I've missed, but hopefully you'll find the ones discussed of some benefit.

Next chapter, you'll learn a bit about H.265, and then we are DONE!

Chapter 16: Introduction to HEVC

While I think the impact of H.265/HEVC won't be felt until the next edition of this book is published, I didn't want to publish this one without at least a brief mention of the spec. So, here its.

H.265/HEVC is the successor codec to H.264, which, like H.264, is jointly developed by the ISO/IEC Moving Picture Experts Group (MPEG) and ITU-T Video Coding Experts Group (VCEG). The primary goal of the new codec is 50% better compression efficiency than H.264 and support for resolutions up to 8192x4320.

Technology Background

By way of background, the ITU-T began development of a successor to H.264 in 2004, while ISO/IEC began working in 2007. In January 2010, the groups collaborated on a joint Call for Proposals, which culminated in a meeting of the MPEG and VCEG Joint Collaborative Team on Video Coding (JCT-VC) in April 2010, at which the name High Efficiency Video Coding (HEVC) was adopted for the codec.

In October, 2010, the JCT-VC produced the first working draft specification, with the Draft Standard, based upon the eight working draft specification, approved in July 2012. On January 25, 2013, the ITU announced that HEVC had received first stage approval (consent) in the ITU-T Alternative Approval Process, while MPEG announced that HEVC had been promoted to Final Draft International Standard (FDIS) status in the MPEG standardization process.

In essence, this means that the initial versions of the specification were frozen so that multiple vendors could finalize their first HEVC products. The current implementation includes a "Main" profile supporting 8-bit 4:2:0 video, a "Main 10" profile with 10-bit support, and a "Main Still

Picture" profile for digital pictures that uses the same coding tools as a video "intra" picture. HEVC will continue to advance, with work already starting on extensions for 12-bit video and 4:2:2 and 4:4:4 chroma formats as well as incorporating scalable video coding and 3D-Video into the spec.

How It Works

Like H.264 and MPEG-2, HEVC uses three frame types—I-, B- and P-frames—within a group of pictures, which incorporate elements of both inter-frame and intra-frame compression. HEVC incorporates numerous advances, including:

- ***Coding Tree Blocks.*** Where H.264 used macroblocks with a maximum size of 16x16, HEVC uses coding tree blocks, or CTBs, with a maximum size of 64x64 pixels. Larger block sizes are more efficient when encoding larger frame sizes, like 4K resolution. This is shown in Figure 16-1.

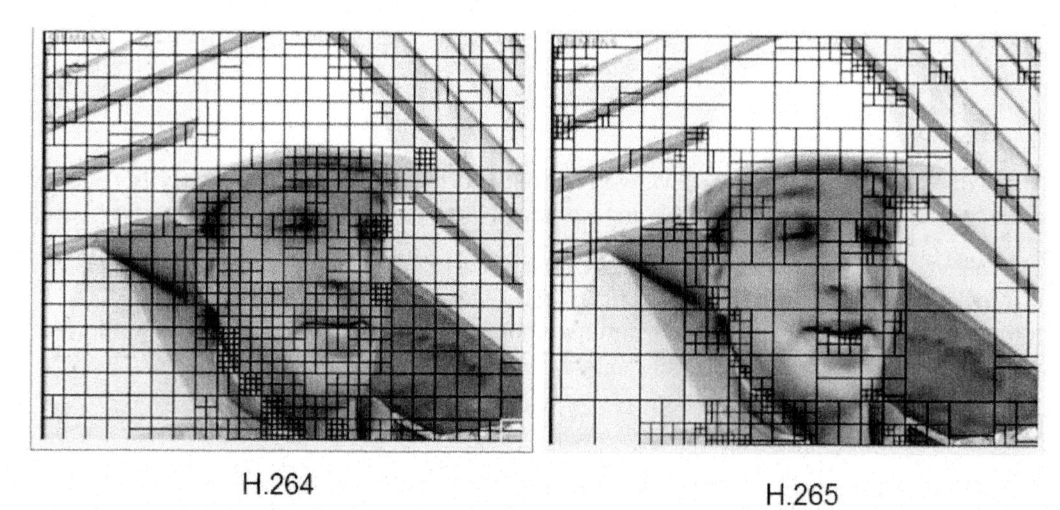

Figure 16-1. Larger blocks sizes enhance encoding efficiency. Images from HEVC webinar by Elemental Technologies, viewable on-demand at **bit.ly/Elemental_HEVC**.

- ***More intra-prediction directions.*** Where H.265 used nine intra prediction directions, HEVC can use more than 35, adding more potential reference pixel blocks that fuel more efficient intra-frame compression (see Figure 16-2, from an Ateme presentation available at **bit.ly/ATEME_HEVC**). The obvious cost is the additional encoding time required to search in the additional directions.

Other advances include:

- ***Adaptive Motion Vector Prediction***, which allows the codec to find more inter-frame redundancies.

- **Luma:** 35 prediction directions (33 + Planar + DC)

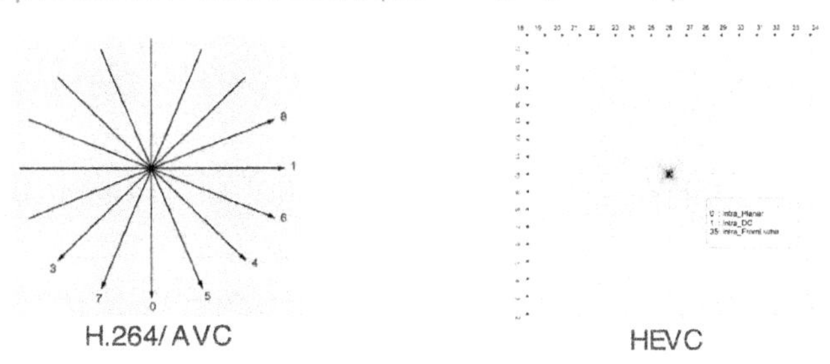

Figure 16-2. Searching is expanded to find more reference pixel blocks.

- ***Superior parallelization tools***, including Wavefront parallel processing, for more efficient encoding in a multi-core environment.

- ***Entropy coding***, which is CABAC-only—no more CAVLC.

- ***Improvements to the deblocking filter*** and the creation of a second filter called ***Sample Adaptive Offset*** that further limits artifacts along block edges.

The Results Please

There are a number of published white papers and presentation that focus on comparing the quality of HEVC vs. H.264 and MPEG-2. One of the most oft-cited sources is an article titled "Comparison of the Coding Efficiency of Video Coding Standards—Including High Efficiency Video Coding (HEVC)," available at **bit.ly/HEVCvH264**, which reports the results of both Peak Signal to Noise Ratio (PSNR) comparisons and subjective evaluations. The report looked at multiple scenarios, including interactive and entertainment applications.

For the entertainment-related comparisons, the study encoded multiple clips ranging in resolution from 832x480 (480p) to 1920x1080 (1080p). For the PSNR-related tests, the study encoded files using four different technologies—HEVC, H.264, MPEG-4 and H.263—until all files had the same PSNR value.

The study then showed viewers multiple files encoded at multiple data rates with H.264 and HEVC and asked them to grade the results. From these tests, the researchers concluded, "The test sequences encoded with HEVC at an average of 53% lower bit rate than the H.264/MPEG-4 AVC HP encodings achieved approximately the same subjective quality."

Encoding	Bit-Rate Savings Relative to			
	H.264/MPEG-4 AVC HP	MPEG-4 ASP	H.263 HLP	MPEG-2/ H.262 MP
HEVC MP	35.4%	63.7%	65.1%	70.8%
H.264/MPEG-4 AVC HP	–	44.5%	46.6%	55.4%
MPEG-4 ASP	–	–	3.9%	19.7%
H.263 HLP	–	–	–	16.2%

Table 16-1. HEVC efficiency compared with H.264, H.263 and MPEG-4 using PSNR values in entertainment applications.

Another article, titled "Subjective quality evaluation of the upcoming HEVC video compression standard," compared H.264 and HEVC using both PSNR and subjective comparisons (**bit.ly/HEVC_quality**). The study concluded:

> The test results clearly exhibited a substantial improvement in compression performance, as compared to AVC. ... For the natural contents considered in this study, a bit rate reduction ranging from 51 to 74% can be achieved based on subjective results while the predicted reduction based on PSNR values was only between 28 and 38%. This difference is mostly due to the fact that PSNR doesn't take into account the saturation effect of the human visual system. PSNR also doesn't capture the full nature of the artifacts: AVC compressed sequences exhibit blockiness while HEVC compression tends to smooth out the content, which is less annoying. For the synthetic content considered in this study, a 75% bit rate reduction can be achieved based on subjective results while the predicted reduction based on PSNR values was 68%.

It's reasonable to be a little skeptical of such results, since the comparisons were produced by technologists contributing to HEVC effort, using encoders that have not been released for sale or (in most cases) even for general-purpose beta testing. Speaking on the condition of anonymity, one CTO of a major encoding vendor estimated that HEVC would enable a 30% reduction in file size at the same quality level at 1080p resolution, with further increases at higher resolutions.

Where will it Play?

Playback statistics are harder to come by. However, multiple companies have demonstrated HEVC playback on a tablet computer, including Qualcomm on an Android tablet powered by a 1.5 GHz Qualcomm Snapdragon S4 dual-core CPU (**bit.ly/HEVC_playback**). Note, however, that the video was only 480p in resolution, which makes sense for tablet display but is far from the 4K video HEVC is designed to enable. At the 2012 PBS Technology Conference, an Ericsson presentation estimated that encoding HEVC could require up to 10x more computational complexity with 2x to 3x the complexity upon decode (**bit.ly/Ericcson_HEVC**).

In a report titled "HEVC Decoding in Consumer Devices," senior analyst Michelle Abraham from the Multimedia Research Group estimated that the number of consumer devices that shipped in 2011 and 2012 that would be capable of HEVC playback with a software upgrade totalled around 1.4 billion, with over a billion more expected to be sold in 2013 (bit.ly/MRG_HEVC).

According to Abraham, in compiling these statistics she assumed that all PCs shipped in each year would be HEVC-capable. The report also includes tables summarizing shipments of HEVC decode-capable tablets, portable media players, streaming media players, video game consoles, Blu-ray players, digital TV sets and set-top boxes.

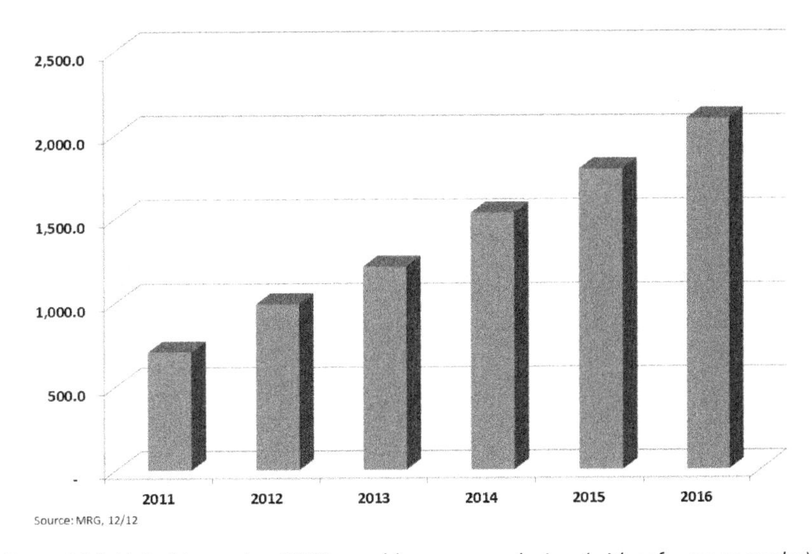

Figure 16-3. Unit shipment so HEVC-capable consumer devices (with software upgrades).

Despite this significant installed base, analyst Frost & Sullivan believes that HEVC adaption is at least five years away for consumer content services. Delaying HEVC adaption includes forces like the recent significant investment in AVC gear made by many Pay TV operators, lack of broad-based support for HEVC in the OTT encoding and deployment ecosystem, and slow HEVC encode and decode chip deployment to preserve the profitability of existing AVC chipsets.

According to Frost & Sullivan VP Dan Rayburn, the initial HEVC rollouts will occur in closed-loop solutions like enterprise video conferencing, Ultra HD services in the far east, and low-bandwidth video on-demand services worldwide, due to the potential cost savings associated with deploying lower-bandwidth HEVC video (bit.ly/Rayburn_HEVC). Rayburn expects Satellite direct to home (DTH) service providers to start rolling out HEVC gear in the 2014-2015 timeframe, with some pilot digital terrestrial television (DTT) channels expected in 2015. Overall, however, he concludes:

> [W]hile certain applications will embrace HEVC much sooner than the norm and HEVC encoding and decoding cores should mature by 2014, we expect it will be around 2017 before a comprehensive ecosystem of first-generation HEVC-enabled products will come to market. Furthermore, we expect AVC to remain in widespread use even in 2018, although it will definitely be considered a commodity technology at that point—much as MPEG-2 is today.

Royalties will Apply

One factor that may slow HEVC adaption is uncertainty surrounding royalties. Like H.264, many of the technologies contributing to HEVC are patented, and patent owners will want compensation for the use of their intellectual property. In June 2012, MPEG LA, a leading packager of patent pools and the organizer of the H.264 patent pool, announced a call for patents essential to HEVC, and a third meeting of the 25 respondents occurred in February 2013 (**bit.ly/MPEGLA_HEVC**).

However, according to MPEG LA officials, there is no set timeframe for the issuance of royalty guidelines or even assurance that a patent group will coalesce, since there are other packagers, or the respective owners could decide to assert their rights individually. Some market segments—most notably chip, encoding and other infrastructure vendors—will likely press on with their HEVC-related efforts in the face of this uncertainty and simply reserve for potential royalties. However, other segments, particularly the distributors of free streaming content chasing the bandwidth savings afforded by HEVC, will almost certainly wait until royalties are known.

Competition In Sight?

These royalties make competitive technologies like Google's VP9 worthy of mention, particularly since Google added VP9 decode to beta versions of Chrome in December, 2012, along with a new decoder for audio streams encoded with the Opus codec (**bit.ly/HEVCinChrome**). According to a Google presentation made at the Internet Engineering Task Force meeting in Atlanta in November 2012, the goal of VP9 was similar quality as HEVC at lower data rates. In a requirements document available on the WebM website, Google stated that VP9 would not be shipped unless it improved quality over VP8 by 50% at a cost of only 40% higher decoding complexity, compared with 200 to 300% for HEVC (**bit.ly/VP9vHEVC**).

As a codec, VP8 compared quite favorably to H.264, producing virtually the same quality at all tested data rates. However, multiple factors doomed VP8 to failure, including the fact that Apple refused to enable VP9 playback on iOS devices or in Safari, that Microsoft refused to include playback in IE9, that H.264 was a joint ITU/MPEG standard, and the fact that it came to market much later than H.264.

On a positive note, one of the issues that plagued VP8—the potential for patent infringement claims from the H.264 camp—won't be an issue for VP9. That's because in March 2013, MPEG LA, who administers the H.264 patents, signed an agreement with Google granting them rights to any technologies that might be essential to either VP8 or VP9 (see **bit.ly/webm_mpegla**).

Going Forward

Of course, as Rayburn from Frost & Sullivan points out, HEVC can't be deployed until the full encode/decode/transport infrastructure is in place. Some encoding vendors, like Elemental Technologies, have announced that all current encoders will be retrofitted for HEVC support via software upgrades sometime in the future. Before purchasing an enterprise-class or even desktop encoder going forward, you should definitely ask about whether the encoder will support HEVC going forward, and how much that support will cost.

Beyond this, precursors to significant availability of HEVC is very market dependent. For example, in video conferencing, it's the availability of real-time HEVC encoders and decoders. In the streaming media space, the playback side is always the driving force, since few producers will encode to a new format until it's clear that it can play reliably for a meaningful group of viewers.

For general-purpose streaming, it's hard to get excited about HEVC without:

- A set royalty policy from MPEG LA.
- Ubiquitous playback via HEVC incorporation into a player like the Flash Player or Silverlight Player. Neither Microsoft nor Adobe responded to my requests for information regarding if or when this might happen.
- The incorporation of HEVC playback into the iOS or Android platforms, either via an app or OS upgrade, and a clear indication about which of the installed base of devices will be HEVC-capable with these upgrades. Given the long-standing policy of both companies not to comment on future technologies, and the black ink-like opaqueness of their general-purpose dealings with the press, I didn't even ask.
- The availability of inexpensive decode silicon that can be incorporated into over-the-top (OTT) set-top boxes (STBs), or the announcement that some current OTT STBs can be retrofitted for HEVC playback via firmware or software upgrades.

Conclusion

Between NAB (April 2013) and IBC (September 2013), expect a flurry of announcements of HEVC-related technologies and products. During that time, the first wave of HEVC-capable encoders and decoders will also come to market, making it possible to gauge the technology's real-world performance, benefits and costs and the interoperability of the encoded HEVC streams.

Until then, as the Frost & Sullivan report suggests, it's most meaningful to consider HEVC adaption in a micro rather than macro sense, since the precursors and economic drivers are very different in each market. While the inevitable general-purpose buzz around HEVC is bound to reach Kaepernick-like proportions, what really matters is how these announcements impact the markets that you serve.

And that's it for the book. It's been a lot of fun; I hope you found it useful (if so, tell all of your friends!). See you next time.

Index

Symbols

A

B

C

D

F

G

H

I

J

K

L

M

N

O

P

Q

R

S

T

U

V

W

CPSIA information can be obtained at www.ICGtesting.com
Printed in the USA
BVOW09s1411041016

463538BV00011B/42/P